Frontiers in Fusion Research

Mitsuru Kikuchi

Frontiers in Fusion Research

Physics and Fusion

Springer

Mitsuru Kikuchi, PhD
Naka Fusion Institute
Fusion Research and Development Directorate
311-0193
Muko-yama 801-1
Naka
Ibaraki
Japan
kikuchi.mitsuru@jaea.go.jp

ISBN 978-1-4471-6055-7 ISBN 978-1-84996-411-1 (eBook)
DOI 10.1007/978-1-84996-411-1
Springer London Dordrecht Heidelberg New York

British Library Cataloguing in Publication Data
A catalogue record for this book is available from the British Library

Cover design: eStudioCalamar, Girona/Berlin

Printed on acid-free paper

Springer is part of Springer Science+Business Media (www.springer.com)

Preface

This book is intended to describe a systematic overview of the latest physical principles of fusion and plasma confinement. Students who learned the basics of quantum and analytical mechanics in university are introduced to the subject and may be interested in it as a future job option. Researchers in other fields may find commonalities and differences in the structure of the scientific principle of magnetic plasma confinement, which has been systematized through the 50 years of fusion research since its inception in 1958 (see Figure 0.1).

Academic fields have been developed in two ways. One is the case where object of purely intellectual curiosity exists and the field is formed to study it. Another is the case where a goal (such as to create a faster airplane, or predict the path of a typhoon) exists and the field is developed to resolve inherent physical issues in it. The relationship between physics and fusion falls into the latter category.

Figure 0.1 Opening ceremony of 22nd IAEA Fusion Energy Conference celebrating 50 years of fusion research on October 13, 2008, at Palais de Nations, Geneva

In any case, triggered by an interesting phenomenon, the physical area (field studies) has formed and traces the history of reduction as fundamental physics applicable to other areas. While new physics can be developed through the development of mathematics such as non-Euclidean geometry in relativistic theory, new mathematics may evolve from new physics such as the theory of distribution, which has developed through the need for the delta function in quantum mechanics.

Physics in the twentieth century has been developed with symmetry as the keyword. Meanwhile, the twenty-first century may be called the era of complexity science, such as life, weather, and social phenomenon, in which governing laws are entangled in complex processes. The ITER plasma, whose structure is formed autonomously through fusion burning, is a new material encapsulated by symmetry and complexity (see Figure 0.2).

The author spent 25 years, the latter half of the 50-year history of fusion, in the world's largest fusion project JT-60. "Plasma physics and fusion" is a "field study" and is systematized through world-fusion research. This "field study" is greatly advanced through interaction with "Fundamental disciplines."

Chapter 1 describes "Fusion" by considering why so much hydrogen fuel was born at the time the universe was created, what is going on in the center of the Sun, and by introducing the beginning of human efforts to realize the "Sun on Earth" and "plasma" as objects of study. In Chapter 2, we introduce the equation governing the "micro world" where fusion occurs. Also, the properties of players in the " Sun on

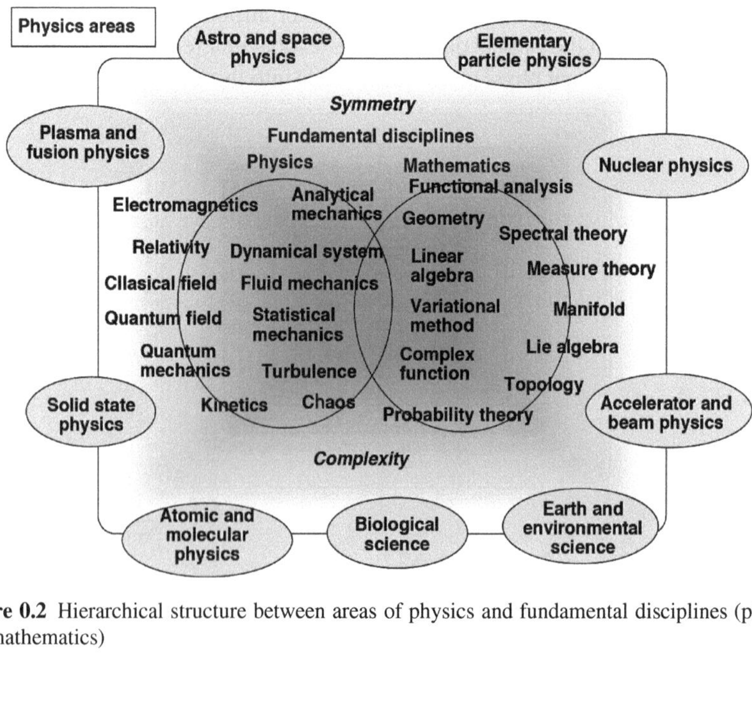

Figure 0.2 Hierarchical structure between areas of physics and fundamental disciplines (physics and mathematics)

Earth" – deuterium, tritium, neutron, and helium are introduced and finally "Fusion cross section" is derived from the theory of nuclear reactions.

Chapters 3 to 9 are devoted to the principle of magnetic plasma confinement. In Chapter 3, the topological requirements of why the magnetic bottle should be a torus are discussed in relation to Poincaré theorem using the Euler index. This is followed by the introduction of "coordinates," "field line dynamics," "magnetic surface," "flux coordinates," and "ergodicity," before describing 2-dimensional and 3-dimensional equilibrium using the machinery of analytical mechanics. Special emphasis is placed on the equivalence of integrability/symmetry in analytical mechanics and the existence of the magnetic surface.

In Chapter 4, particle orbit dynamics on the magnetic surface are discussed using the Lagrange–Hamilton formulation, noting the new variational principle, Littlejohn's variational principle, to satisfy the Liouville theorem in guiding center phase space, as was established in the 1980s; in contrast to the guiding center equation by Alfven, which does not conserve phase space volume. This variational principle is applied to guiding center motion in the flux coordinates and adiabatic invariants are introduced.

In Chapter 5, fundamentals of plasma kinetic theory are described. To clarify the difference between dynamics and kinetics, the conflict faced by L. Boltzmann's Stosszahl Ansatz on the time-reversibility of Newton's equation of motion and time-irreversibility of kinetic equations is discussed in comparison with Poincaré's recurrence theorem. A similar but different problem appears in plasma physics. "Continuous spectrum and phase mixing" is discussed, which produces "Landau damping" from the time-reversible kinetic equation, the Vlasov equation. Coulomb collision in plasma is surveyed as collective phenomena. Finally, modern Gyro Kinetic theory is introduced which is cutting-edge research studying high temperature plasma turbulence.

In Chapter 6, the main subject is "stability." After a general introduction to stability, the energy principle and stability criteria using the Euler–Lagrange equation are given to describe plasma instabilities such as Kink and Ballooning from the linearized magnetohydrodynamics (MHD) equation with an Hermitian linear operator. Dissipation at a rational surface produces "Tearing" – an instability associated with magnetic reconnection. The linearized MHD operator becomes non-Hermitian with equilibrium flow, which is a cutting-edge topic in confined plasma research.

In Chapter 7, the eikonal equation for plasma wave propagation and von Laue form of wave energy are introduced. Then drift waves and Alfven resonance are considered: these are important in practical magnetic confinement research. Furthermore, the Hasegawa-Mima equation is introduced describing the nonlinear interaction among drift waves as a starting point of cutting-edge plasma turbulence research.

In Chapter 8, collisional transport of toroidal plasma in the collisionless regime is discussed. In collisionless plasma, a "self-generated current" or "bootstrap current" is produced by the pressure gradient, which is essential for the efficient steady operation of a tokamak fusion reactor. Neoclassical viscous force is produced by the distortion of velocity distribution function in the collisionless regime. A generalized version of Ohm's law includes bootstrap current expressed in terms of the thermodynamic forces. Cross-field neoclassical transport is also given.

In Chapter 9, the cutting-edge subject of turbulent transport in confined plasma is discussed. While introducing the basic concepts of dynamical systems, a basic picture of plasma confinement is shown. Heat diffusion in low mode confinement can be described by "Self-organized criticality (SOC)," which is an important concept in complexity science, while the Transport Barrier (TB) is a local relaxation of SOC driven by the flow shear to break the turbulent cell.

In Chapter 10, the characteristics of fusion as an energy source and its research status are briefly described.

The author would like to express his appreciations to former director Masafumi Azumi, Dr. Shinji Tokuda, Prof. Kazunobu Nagasaki, and Prof. Zensho Yoshida for their encouragements and inspiring comments. The author is indebted his career to past Prof. Masaaki Sonoda, past Prof. Nobuyuki Inoue, and Prof. Taijiro Uchida. Finally but not least, many thanks to my wife Takako for her patience to allow me to spend time for the book.

Mito City, December 2010 *Mitsuru Kikuchi*

About the Author

The author was born in 1954 in Japan. He took his thesis at the University of Tokyo in 1981. He has worked at JAERI/JAEA since 1983. He was Head of the Plasma Analysis Division, Head of the Large Tokamak Experiment Division I, Head of the Tokamak Program Division, Deputy Director of the Department of Fusion Plasma Research, Director of the Division of Advanced Plasma Research (JT-60), and is Supreme Researcher of the Fusion Research and Development Directorate.

He was Visiting Professor of Kyoto University in 2008, and is Visiting Professor of Kyushu University and Guest Professor of Osaka University. He has been a chairman of the editorial board of the Nuclear Fusion Journal since 2005. The author is the Chairman of the international programme committee of the 50th Anniversary Fusion Energy Conference noted in Figure 0.1.

Contents

Chapter 1
Sun on Earth: Endless Energy from Hydrogen

The Big Bang created the universe 13.7 billion years ago. The universe is brimming with hydrogen, and the beautiful night sky is formed by hydrogen fusion. Here, the fusion is a nuclear reaction creating nuclei such as helium through reactions between light nuclei such as hydrogen.

The Sun is hot dense plasma confined by the gravity of its huge mass, bringing huge energy to the planetary system in the form of solar light produced by massive fusion reactions. All life on our planet depends on solar energy.

A large number of scientists in the latter half of the twentieth century challenged to the dream of creating a Sun on Earth to produce energy from fusion. To do this, we aimed to build ITER creating an energy-producing machine, a Sun on Earth, as well as helping to understand the physics.

A fusion reaction requires a temperature of a few hundred million degrees. Deuterium and tritium fuel take the "plasma state" – the fourth state of matter. Plasma exists in various forms around us. It is necessary to understand and control plasma to develop a usable form of fusion energy.

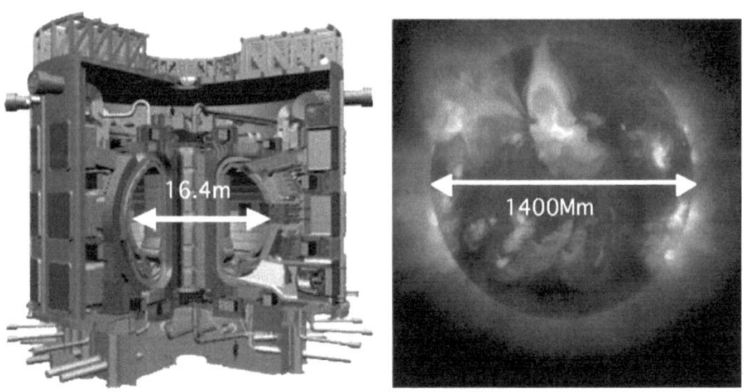

Figure 1.1 The ITER and the Sun. The Earth is blessed from fusion energy of the Sun on orbital path of 150 million km away. ITER, the Sun on the Earth, will realize 10 times the Sun's central temperature using a magnetic force smaller in size by 100 million

1.1 Big Bang: the Mother of Fusion Fuel

A total of 75% of the main elements of the material, except dark matter, in the universe is said to be hydrogen. The stars are formed by the abundant hydrogen. The night sky is full of light produced by fusion reactions in stars. Why are stars made of hydrogen? The structure of the universe we live in is governed by the gravitational field equations of general relativity derived by the German-born physicist Albert Einstein (1879–1955) in 1915 [1–5].

$$R_{\mu\nu} - \frac{1}{2} g_{\eta\nu} R = \frac{8\pi G}{c^4} T_{\mu\nu} \,. \tag{1.1}$$

Here $R_{\mu\nu}$, $g_{\mu\nu}$, R, $T_{\mu\nu}$, G, are the Ricci tensor, metric tensor (infinitesimal distance is given by $\mathrm{d}s^2 = g_{\mu\nu}\mathrm{d}x^\mu \mathrm{d}x^\nu$), Rich scalar, energy and momentum tensor, and the gravitational constant, respectively. Gravity produces curved space where the parallel axiom of Euclidean geometry does not hold. Let a be cosmic scale, then the following Friedman equation is obtained from Einstein's gravitational equation,

$$\frac{1}{2}\left(\frac{\mathrm{d}a}{\mathrm{d}t}\right)^2 - \frac{GM(a)}{a} = -\frac{1}{2}Kc^2 \,. \tag{1.2}$$

Friedmann was a Russian physicist who derived this equation in 1922. Here, G is the gravity constant, $M(a)$ the mass in the cosmic scale, c is the speed of light, and K is a constant. $K > 0$ corresponds to the case of a closed universe with negative total energy. $K = 0$ corresponds to the case of a flat universe, while $K < 0$ if the universe is open and has positive total energy. In any case, Equation 1.1 implies an expanding universe. The beginning of the universe is called the "Big Bang" [6]. The temperature of the initial universe T is uniquely determined as a function of time t if we adopt the Friedmann equation as follows,

$$T\ [\mathrm{k}] = \frac{1.5 \times 10^{10}}{g^{1/4}\sqrt{t\ (\mathrm{s})}} \,. \tag{1.3}$$

Here g is the ultra relativistic correction factor ($g = 1.68$ ($T <$ 6 billion degrees), $g = 5.38$ (6 billion degrees $< T <$ 1 trillion degrees)). Immediately after the Big Bang, the ratio of protons and neutrons obeyed the Boltzmann relation $n_\mathrm{n}/n_\mathrm{p} = \exp(-Q_\mathrm{n}/kT)$ since transition and back transition between proton and neutron could occur through a weak interaction, as pointed out by Japanese physicist Chyushiro Hayashi. Here, Q_n is the mass difference of a neutron and proton, 1.3 MeV (equivalent temperature of 15 billion degrees). The numbers of neutrons and protons are almost same if $kT \gg Q_\mathrm{n}$ and the number of neutrons decreases as the temperature goes down. One second after the Big Bang, the temperature of the universe fell to 10 billion degrees ($\sim$ 0.86 MeV) and the ratio of protons to neutrons fell to $n_\mathrm{n}/n_\mathrm{p} = 0.223$. The ratio of protons to neutrons is fixed ($n_\mathrm{n}/n_\mathrm{p} = 0.157$) if the temperature is lowered to about 8 billion degrees, below which weak interactions do not work and the neutron β decays to a proton with a half-life of 12 min.

If protons and neutrons are in the high temperature state, deuterium can be formed by the reaction, $\mathrm{n} + \mathrm{p} \rightarrow \mathrm{D} + \gamma$, but deuterium is resolved at tempera-

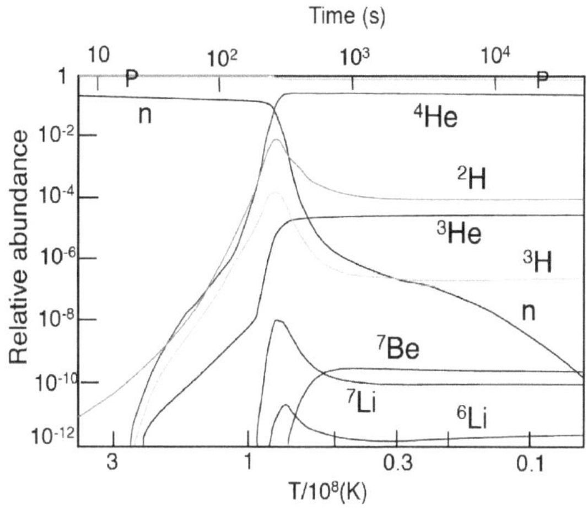

Figure 1.2 Relative abundance of protons, neutrons, ^4He, ^2H(D), ^3He, ^3H, and so forth after the Big Bang versus time (*top-horizontal axis*) and temperature (*bottom-horizontal axis*) [7]

tures higher than the hydrogen binding energy (2.23 MeV: see Section 2.2). This is the temperature corresponding to 26 billion degrees. Below this temperature, the reaction between protons and neutrons can take place to form deuterium, in principle, but high-energy photons filled the early universe destroying deuterium immediately. Protons and neutrons are kept separate until these high-energy photons diminish. Two minutes after the Big Bang, the temperature dropped to 0.1 billion degrees and the number of high-energy photons was reduced. Then the population of deuterium increased as seen in Figure 1.2. Helium nuclei are formed through deuterium reactions, $D + D \rightarrow {}^3He + n$, ${}^3He + D \rightarrow {}^4He + p$. Almost all the remaining neutrons form ^4He and the mass fraction of ^4He is given by $m_{He} = 2n_n/(n_n + n_p) = 0.25$.

Formation of elements in the universe had almost finished a few minutes after the beginning of the universe, the Big Bang, through the production of helium. The observed abundance ratio of helium and hydrogen in the universe almost equals the ratio predicted by Big Bang cosmology. If the binding energy of deuterium were much larger, the dissociation of deuterium may not have happened and the much earlier formation of helium may have led to formation of heavier elements in the initial Big Bang. Stars might not twinkle in the night sky and all life on Earth nurtured by the energy of the Sun might not exist.

Salon: General Relativity and Non-Euclidean Geometry

Einstein (Figure 1.3 (a)), when building his general theory of relativity to include accelerated motion from the theory of special relativity valid only for uniform motion, discovered the need for non-Euclidean geometry where the Euclidean parallel axiom is not viable [8]. In a uniformly rotating disk, contraction of circumference is expected and the ratio of the circumference to radius is no longer considered to be 2π.

The free-fall frame against gravity becomes the inertial frame without gravity. The straight motion of light in the inertial frame (assumption in the theory of special relativity) implies the curved motion of light in the original frame. The same result is obtained for the accelerated frame. Einstein proposed that the straight motion of light is curved by the local gravitational bending of space-time.

Non-Euclidean geometry started from Gauss, Bolyai, and Lobachevskii and led to a general Riemannian geometry developed by the German mathematician Bernhard Riemann (1826–1866; Figure 1.3 (b)) [9]. He gave the distance ds in curved space as $ds^2 = g_{\mu\nu} dx^\mu dx^\nu$ $(\mu, \nu = 0, 1, 2, 3)$. $g_{\mu\nu}$ is termed the metric tensor. In curved space, curvature can be defined by choosing two coordinates x^μ and x^ν (Figure 1.4 (a)). The Riemann curvature tensor $R^i_{j\mu\nu}$ is defined as the difference of a vector V^i circulating the differential area $dx^\mu dx^\nu$ as $\delta V^i = R^i{}_{j\mu\nu} V^j dx^\mu dx^\nu$ [5]. The occurrence of the vector difference after circulating on a curved surface is easily understood considering the vector circulation on a sphere (see Figure 1.4 (b)).

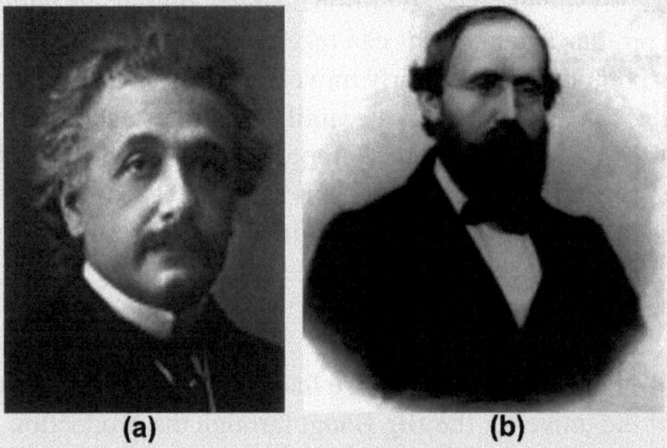

Figure 1.3 (a) Albert Einstein who built a general theory of relativity and (b) Bernhard Riemann who invented Riemann geometry to deal with curved space-time

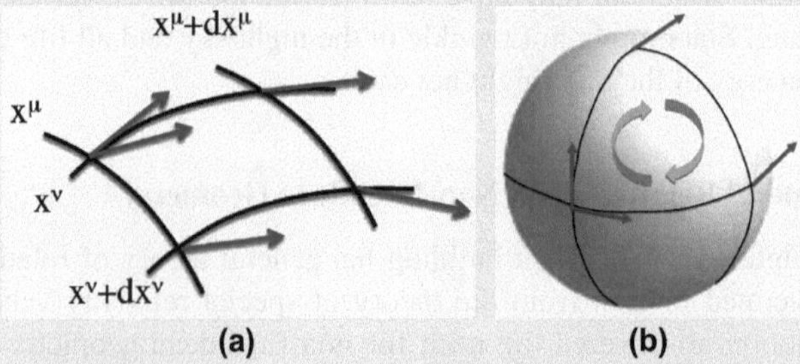

Figure 1.4 (a) Defining coordinates in curved space. (b) Change of vector after circulation on sphere

Note: General Relativity [3–5]

Einstein's equation for general relativity can be expressed in a variational principle using the following action integral (see Section 4.1 for introduction to variational principle).

$$\delta(S_g + S_m) = 0 \tag{1.4}$$
$$S_g = -(c^3/16\pi G)\int R\sqrt{-g}\mathrm{d}^4x\,, \qquad S_m = c^{-1}\int L\sqrt{-g}\mathrm{d}^4x\,.$$

Here, G is a gravitational constant and scalar curvature R is defined as $R = g^{ij}R_{ij}$ where the Ricci tensor R_{ij} is defined as $R_{ij} = R^m_{nmj}$ using Riemann's curvature tensor $R^i_{j\mu\nu}$ introduced in the salon. The value of g, defined as $g = \det[g_{ij}]$ is negative and $\sqrt{-g}$ is the inverse of the Jacobian. L is a Lagrangian of energy and momentum of matter, which is related to the energy and momentum tensor T_{ij} as follows,

$$T_{ij} = \frac{2}{\sqrt{-g}}\left[\frac{\partial(L\sqrt{-g})}{\partial g^{ij}} - \frac{\partial}{\partial x^k}\frac{\partial(L\sqrt{-g})}{(\partial g^{ij}/\partial x^k)}\right]. \tag{1.5}$$

The variation of $S_g(\delta S_g)$ can be expressed using a variation of metric δg^{ij}. Here it is important to note that the gravitational field originates from the curving of space (manifold), which changes the metric. Metric g^{ij} is determined so that action integral becomes extremal. Using the relationship $R = g^{ij}R_{ij}$, the variation for $\delta S'_g = (16\pi G/c^3)\delta S_g$ is given as follows,

$$\begin{aligned}\delta S'_g &= \delta\int g^{ik}R_{ik}\sqrt{-g}\mathrm{d}^4x \\ &= \int\left[R_{ik}\sqrt{-g}\delta g^{ik} + R_{ik}g^{ik}\delta\sqrt{-g} + g^{ik}\sqrt{-g}\delta R_{ik}\right]\mathrm{d}^4x\,.\end{aligned} \tag{1.6}$$

The differential $\delta g = \delta(\det[g_{ij}])$ is a summation of δg_{ij} multiplied by its minor determinant. Since g^{ij} (component of inverse of $[g_{ij}]$) is the minor determinant divided by the determinant g, we have $\delta g = gg^{ij}\delta g_{ij} = -gg_{ij}\delta g^{ij}$. Therefore, Equation 1.6 becomes,

$$\delta S'_g = \int\left[\left(R_{ij} - \frac{1}{2}g_{ij}R\right)\delta g^{ij} + g^{ij}\delta R_{ij}\right]\sqrt{-g}\mathrm{d}^4x\,. \tag{1.7}$$

The second term in the integrant is shown to be zero by converting it to a surface integral [3]. Thus we obtain following variational form for general relativity.

$$\delta(S_m + S_g) = -\frac{c^3}{16\pi G}\int\left[R_{ij} - \frac{1}{2}g_{ij}R - \frac{8\pi G}{c^4}T_{ij}\right]\delta g^{ij}\sqrt{-g}\mathrm{d}^4x\,. \tag{1.8}$$

This gives the Einstein equation (1.1). This is beautiful in a sense that even the structure of the universe can be derived from the variational principle. However, Einstein as the creator of Equation 1.1, reached it through trial and error. He converted the Poisson equation for weak gravitational field $\Delta\phi = 4\pi G\rho$ to $\Delta g_{00} = (8\pi G/c^4)T_{00}$ using $g_{00} = 1 + 2\phi/c^2$ (see note in Section 1.2) and $T_{00} = \rho c^2$. He conjectured that the governing equation of general relativity will be a balance between the 2nd order covariant gravitational tensor X_{ij} and the energy and momentum tensor $(8\pi G/c^4)T_{ij}$ $(X_{ij} = (8\pi G/c^4)T_{ij})$. Since the divergence of T_{ij} is zero, X_{ij} should be divergence free. The Ricci tensor R_{ij} is not divergence free, but he found that $R_{ij} - g_{ij}R/2$ is divergence free.

1.2 Sun: Gravitationally Confined Fusion Reactor

The Sun, bestowing light and energy for our life, is one of 200 billion stars in the Milky Way galaxy. It is 110 times the radius of the Earth ($R_{\mathrm{Sun}} = 70 \times 10^4$ km), and 320,000 times the mass of the Earth ($M_{\mathrm{Sun}} = 2 \times 10^{30}$ kg). It is a giant hydrogen ball (Figure 1.5). The release of energy from the Sun's surface (luminosity) is $L = 3.86 \times 10^{26}$ W. The Sun, formed 4.5 billion years ago (see the salon), released the following gravitational energy from the centralization of mass,

$$E_{\mathrm{g}} = \frac{GM^2}{R_{\mathrm{Sun}}} = 3.8 \times 10^{41}\ \mathrm{J}\,, \tag{1.9}$$

where the gravitation constant is given as $G = 6.67 \times 10^{-11}\ \mathrm{m^3/kg\,s^2}$. This corresponds to an average kinetic energy $\sim$ 1 keV for an ion or an electron (equivalent

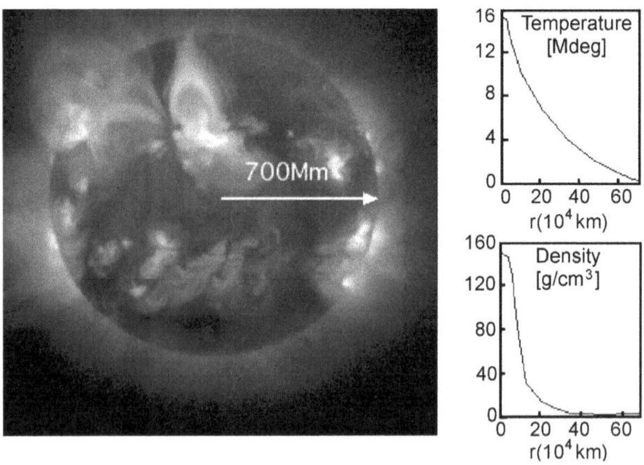

Figure 1.5 The Sun, producing energy from fusion of hydrogen confined by the strong gravitational force and its density and temperature distribution in the center

temperature of 7.7 million degrees). The estimated central temperature of the Sun is 15 million degrees, a similar temperature to that initially achieved from gravitational energy. In such a temperature, atoms are split to nuclei and electrons, and are called "plasma." In other words, the Sun is a giant ball of dense plasma, whose central density is $150\,\text{g/cm}^3$ and half the mass of the Sun is centered within one-fourth of its radius. In 1905, Einstein published the special theory of relativity, and showed that the energy (E) and mass (m) have the same origin and their relationship shows that mass is frozen energy,

$$E = mc^2\,. \tag{1.10}$$

In the Sun and the stars, mass is being converted into energy according to Einstein's equation. In the center of the Sun, the collision of two protons (hydrogen nuclei) do not produce unstable ^2He, but one proton emits a positron through the weak interaction to be converted to a neutron, leading to the formation of deuterium by releasing 0.42 MeV of energy.

$$\text{p} + \text{p} \rightarrow \text{D} + \text{e}^+ + \nu_\text{e} + 0.42\,\text{MeV}\,. \tag{1.11}$$

Here, p, D, e^+, and ν_e are hydrogen, deuterium, a positron and an electron neutrino, respectively. This first proton reaction is similar to those that occurred in the early universe. The time constant of $\sim$ 10 billion years is quite slow with its low temperature for the formation of deuterium. A positron immediately couples with an electron to emit energy of 1.02 MeV. The deuterium fuses with another proton to form ^3He,

$$\text{p} + \text{D} \rightarrow {}^3_2\text{He} + \gamma + 5.49\,\text{MeV}\,. \tag{1.12}$$

Here, γ and ${}^3_2\text{He}$ are gamma rays and helium-3 (isotope of helium), respectively. In addition, helium-4 is generated from two helium-3 atoms,

$${}^3_2\text{He} + {}^3_2\text{He} \rightarrow {}^4_2\text{He} + \text{p} + 12.86\,\text{MeV}\,. \tag{1.13}$$

In the end, helium-4 is born from four hydrogen,

$$4\text{p} + 2\text{e}^- \rightarrow {}^4_2\text{He} + 26.72\,\text{MeV}\,. \tag{1.14}$$

Except for the 0.26 MeV carried by neutrinos, $26.46\,\text{MeV}/4 = 6.55\,\text{MeV}$ is the energy production per hydrogen. The universe has a huge amount of hydrogen produced at the Big Bang and hydrogen fusion makes the night sky twinkle. The combustion rate of hydrogen at the center of the Sun is related to luminosity $L/6.55\,\text{MeV} = 3.7 \times 10^{38}/\text{s}$, that is, 620 million tons of hydrogen per second. Approximately 4.4 million tons of mass is converted into energy per second. This implies that 6% of the total hydrogen is converted to helium in 4.5 billion years. One gram of hydrogen produces 6.4×10^{11} J, equivalent to energy released by 15 tons of oil (75 drums).

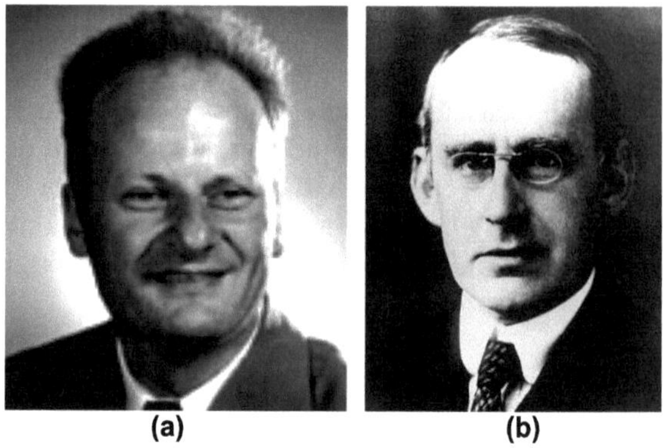

Figure 1.6 (a) H. Bethe and (b) A. S. Eddington who made major contributions to physics of the Sun

For stars with higher temperatures (typically 14–30 Mdegrees for stars larger than the Sun), another fusion mechanism, called the "CNO" cycle, becomes dominant. This was discovered by German-born US physicist H. Bethe (1906–2005) who received the Nobel Prize in physics in 1967 (see Figure 1.6 (a)).

Energy generated at the center of the Sun reaches the surface through convection and radiation. The central part of the Sun is called the "core" where fusion reactions occur. In the middle part ($0.2a$–$0.7a$; a: radius of the Sun) is called the "radiation transport region" where energy is transported through radiation. The surface area ($0.7a$–$1a$) is called the "convective layer" where convective cells (Bénard cells) are the dominant energy transport process. The surface of the Sun is about 5 800 degrees and radiates energy through the black-body spectrum. There are the "chromosphere," "colona," and "prominence" outside of the surface.

Salon: History of the Age and the Energy Source of the Sun

In the nineteenth century, the source of solar energy was not known. To begin with, nobody knew the Sun had burned for several billion years. Knowing the age of the Sun was important in knowing the source of energy. The age of the Solar System was determined by the ratio of radioactive daughter nuclides in meteorites. It was estimated from the accumulation of daughter elements from radioactive decay from dissolved meteorites using the long half-life of potassium 40 $\rightarrow$ argon 40 (half-life of 1.19 billion years). Most measurement of the meteorites shows an age of 4.55 billion years. The Sun is considered to have formed then.

In 1920, the British astronomer Sir A. S. Eddington (1882–1944; Figure 1.2 (b)) suggested that solar energy may be caused by the fusion of hydrogen into helium. In 1929, American astronomer H. N. Russell (1877–1957) examined the spectrum of the Sun to show the atoms in the Sun consisted of 90% hydrogen and 9% by helium. The source of solar energy is nothing but hydrogen fusion [10].

Note: Special Relativity

As you see in later in Section 4.2, the action integral of the variational principle relevant for special relativity must be invariant for the Lorentz transformation. The action form satisfying this condition is an integration of the world interval s ($\mathrm{d}s \equiv (c^2\mathrm{d}t^2 - \mathrm{d}\boldsymbol{x}\cdot\mathrm{d}\boldsymbol{x})^{1/2}$) since s is invariant for the Lorentz transformation. The action integral for free particle motion is then given as follows,

$$S = -m_{a0}c\int_1^2 \mathrm{d}s = -m_{a0}c^2\int_{t_1}^{t_2}\mathrm{d}\tau\ , \tag{1.15}$$

where $\tau = s/c$ is called "proper time" and is a time in a rest frame (time coordinate at $\mathrm{d}\boldsymbol{x}/\mathrm{d}t = 0$) of the object. This action integral has a marked similarity to Fermat's variational principle for light propagation (see Chapter 4). Substituting $\tau = t[1 - (\mathrm{d}\boldsymbol{x}/\mathrm{d}t\cdot\mathrm{d}\boldsymbol{x}/\mathrm{d}t)/c^2]^{1/2}$ into Equation 1.15, we obtain

$$S = \int_{t_1}^{t_2} L\,\mathrm{d}t\ , \tag{1.16}$$

$$L = -m_{a0}c^2\sqrt{1-(\mathrm{d}\boldsymbol{x}/c\,\mathrm{d}t)^2}\ . \tag{1.17}$$

Here m_{a0} is usually called "rest mass." The momentum of the free particle $\boldsymbol{p}$ is defined as $\boldsymbol{p} = \partial L/\partial\dot{\boldsymbol{x}}$ ($\dot{\boldsymbol{x}} = \boldsymbol{v}$).

$$\boldsymbol{p} = m_{a0}\boldsymbol{v}/\sqrt{1-v^2/c^2}\ . \tag{1.18}$$

Energy (or Hamiltonian) is derived from the formula $E = \boldsymbol{p}\cdot\boldsymbol{x} - L$ as follows,

$$E = m_{a0}c^2/\sqrt{1-v^2/c^2}\ . \tag{1.19}$$

Equation 1.19 indicates that the free particle has energy even if $v = 0$ at some inertial frame and is a famous formula by Albert Einstein (Equation 1.10). If a particle with $v = 0$ set in gravitational potential ($\phi(x)$) in x direction, it is no longer in an inertial frame. But it enters an inertial frame in a free-fall frame under ϕ. If we assume the velocity under ϕ is much smaller than c, we can use a non-relativistic formula for L as $L \sim -m_{a0}c^2 + m_{a0}v^2/2 - m_{a0}\phi$. Also, we can assume $t' = t$. Under the gravitational field, $\mathrm{d}x'^2 \sim \mathrm{d}x^2 - 2\phi\mathrm{d}t^2$. In free fall frame (ct, x', y, z), the special theory of relativity applies and $\mathrm{d}s^2 = c^2\mathrm{d}t^2 - \mathrm{d}x'^2 - \mathrm{d}y^2 - \mathrm{d}z^2$, called "Minkowski Space." This can be expressed in initial frame (ct, x, y, z) as

$$\mathrm{d}s^2 = c^2(1 + 2\phi/c^2)\mathrm{d}t^2 - \mathrm{d}x^2 - \mathrm{d}y^2 - \mathrm{d}z^2\ . \tag{1.20}$$

This means the metric g_{00} is no more at unity and $g_{00} = 1 + 2\phi/c^2$ ($\mathrm{d}x^0 = c\,\mathrm{d}t$) under the weak gravitational potential.

1.3 Fusion: Challenge to the Sun on Earth

The question of energy development using fusion reactions was raised in the early twentieth century in response to the development of the theory of relativity and quantum mechanics. German physicist W. Heisenberg (1901–1976; Figure 1.7 (a)) who developed quantum mechanics in the early twentieth century recorded discussions with Danish physicist Niels Bohr (1885–1962; Figure 1.7 (b)) and Lord Rutherford (1871–1937) in his book *Physics and Beyond* (1935–1937, Chapter 13) [11] as follows,

Heisenberg: "May I now put a question to you? Do you think it likely that with more powerful accelerators we may one day be able to use nuclear energy for technical purposes – for instance, for the artificial creation of new chemical elements in appreciable quantities – or to utilize the energy of nuclear bonds much as we exploit the energy of chemical bonds during combustions? –."

Niels: "Of course, things would be quite different if we could raise a piece of matter to so high a temperature that the energy of the individual particles became great enough to overcome the repulsive forces between the atomic nuclei, and if, at the same time, we could keep the density high enough to ensure that collisions did not become too rare. But, this calls for temperatures of something like a thousand million degrees, and long before we reached such temperatures the vessels in which we enclosed our experimental substances would have be evaporated."

Lord Rutherford: "In any case, no one has seriously suggested that energy can be derived from nuclear processes. For though the fusion of a proton or neutron with an atomic nucleus does release energy, a much greater amount is needed to produce the fusion in the first place, for instance, by the acceleration of a very large number of protons, most of which will miss their target. The largest portion of this energy is in any case dissipated in the form of Brownian movement. As far as the liberation of energy is concerned, experiments with atomic nuclei may therefore be called a sheer waste. All who speak of the technical exploitation of nuclear energy are talking moonshine."

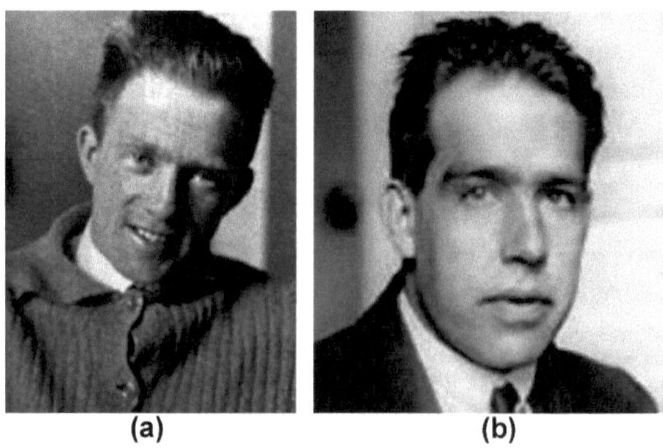

Figure 1.7 (a) W. Heisenberg (In 1960, Institut für Plasmaphysik GmbH (IPP) was founded by a partnership of the Max Planck Society and Prof. Dr. Werner Heisenberg.) and (b) N. Bohr

In 1942, inventor of the fission reactor Italian physicist Enrico Fermi (1901–1954), while having a lunch at the Columbia Faculty Club, suggested to Edward Teller (1908–2003) (inventor of the hydrogen bomb in US) the possibility of burning deuterium to develop a large source of energy. Based on his suggestion, Teller made various calculations and found that fusion between deuterium (D) and tritium (T) is a possibility [12]. Tritium and deuterium react at relatively low energy, creating helium and a neutron. Helium has a higher binding energy, we can generate a huge amount of energy as will be shown in Chapter 2.

$$D + T \rightarrow {}^{4}He\,(3.52\,\mathrm{MeV}) + n\,(14.06\,\mathrm{MeV})\ . \tag{1.21}$$

The most promising magnetic fusion configuration, "tokamak" was initiated in the Soviet Union as recorded in Robin Harman's book [13], *"It was New Year's Eve in 1950. Kurchatov asked his deputy what the theorists thought of Sakharov's new idea. – Golovin said that Sakharov has alerted us to solve a second, no less far-reaching problem of the twentieth century – how to produce inexhaustible energy by burning ocean water! That's a problem man could give his whole life to solve! Kurchatov reportedly remarked. A huge operation will have to be initiated. A problem for peace! Huge! Fascinating! We'll start the new-year not with a weapon but with the Magnetic Thermonuclear Reactor and do a real job on it."* Thus the Soviet Union began tokamak research.

Table 1.1 shows a comparison of parameters of the ITER and the Sun. The confinement of high temperature plasma to produce fusion energy is not an easy task, but then about 46 years later in 1996, the reactor-relevant plasma test device JT-60 in Japan succeeded in producing ultra-hot plasma at 520 million degrees, a temperature that Bohr had predicted would evaporate the vessel containing such a high temperature substance. The JET of the European Union and the TFTR of the United States produced fusion power of 10–16 MW through the D–T reaction. The ITER project is an international project to construct tokamak-type fusion "experimental reactor" proposed at the summit out of the Soviet Communist Party General Secretary Mikhail Gorbachev and US President Reagan in Geneva in November 1985 at the end of the Cold War. The first stage activity of this project, conceptual de-

Table 1.1 Comparison of parameters of the Sun and ITER. There are lots of difference in parameters but both uses fusion of light nuclei

Quantity	ITER	Sun	Ratio
Diameter	16.4 m	140×10^4 km	$\sim 1/10^8$
Central temp	200 Mdeg	15 Mdeg	10
Central density	$\sim 10^{20}/\mathrm{m}^3$	$\sim 10^{32}/\mathrm{m}^3$	10^{12}
Central press.	~ 5 atm	$\sim 10^{12}$ atm	$\sim 10^{11}$
Power density	$\sim 0.6\,\mathrm{MW/m^3}$	$\sim 0.3\,\mathrm{W/m^3}$	$\sim 2 \times 10^6$
Reaction	DT reaction	pp reaction	
Plasma mass	0.35 g	2×10^{30} kg	$1/6 \times 10^{33}$
Burn time const	200 s	10^{10} years	10^{15}

sign activities (CDA), was started on April 3, 1988, and ran for three years among four parties, the US, Europe, the Soviet Union, and Japan. From June 1992, the second stage activity, engineering design activities (EDA), ran for six years and was extended for three years. After the battle to host ITER, Cadarache in France was chosen as the ITER construction site. The dream of Sakharov to realize the Sun on Earth has been postponed to the twenty-first century.

1.4 Plasma: Fourth State of Matter

The term "plasma" used for fusion, in which ions and electrons are moving independently, comes from the Greek language. It was first used as an academic term, "protoplasma" for jelly-like cells (protoplasm), by Czech biologist J. Purkyně (1787–1869). In the medical field, plasma means blood plasma or serum. In 1927, Irving Langmuir (1881–1957; Figure 1.8 (a)) named for nearly neutral ionized gas plasma [14]. Since then, research on this matter is called "plasma physics."

The ionized gas, plasma, has a different nature to the states solid, liquid and gas. Sir W. Crookes (1832–1919) (who studied low-pressure glow discharge) described plasma as the "fourth state of matter" (see Figure 1.8 (b)). Philosophers of ancient Greece, Empedocles (495–435 BC) proposed that the world was made of earth, water, air, and fire, which may correspond to solid, liquid, gas, and weakly ionized plasma. Surprisingly, this idea may catch the essence. The discharge physics of Crookes was extended by Michael Faraday (1791–1867) with the discovery of the glow discharge in 1835, and discovery of electrons by Sir J. J. Thomson (1856–1940) in 1897, followed by the studies of the electrical discharge process by Sir J. Townsend (1868–1957), von Engel, and Penning, known as "gaseous electronics engineering."

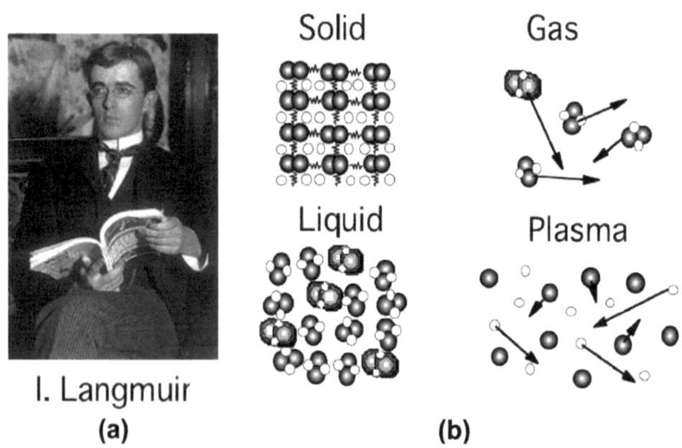

Figure 1.8 (a) I. Langmuir and (b) four states of matter (solid, liquid, gas, plasma). Plasma is electrically neutral as a whole with almost equal numbers of electrons and ions

Meanwhile, in the field of astrophysics, the basis of plasma physics and magnetohydrodynamics were established through the work of Indian physicist M. N. Saha (1893–1956) on the theory of ionization equilibrium concerning the degree of ionization of the solar atmosphere in 1920; British physicist S. Chapman (1888–1970), who clarified the origin of the Aurora and the geomagnetic storm; Pakistani physicist S. Chandrasekhar (1910–1995), who studied black holes and magnetohydrodynamics in astrophysics and won the Nobel Prize in Physics in 1983; US astrophysicist Lyman Spitzer Jr. (1914–1997), who developed the theory of fusion plasma, created the Stellarator concept and proposed space telescopes; and Swedish physicist Hannes Alfven (1908–1995), who developed magnetohydrodynamical concepts such as Alfven waves and was awarded the Nobel Prize in Physics in 1970.

Plasma exists at various temperatures and densities as shown in Figure 1.9. Examples of plasmas on earth include natural lightning, fire, and the beautiful Aurora seen at the poles, while 99.9% of visible matter in the universe is in the plasma state. At the center of the Sun, hydrogen density is $9 \times 10^{25}/cm^3$ (1,000 times solid

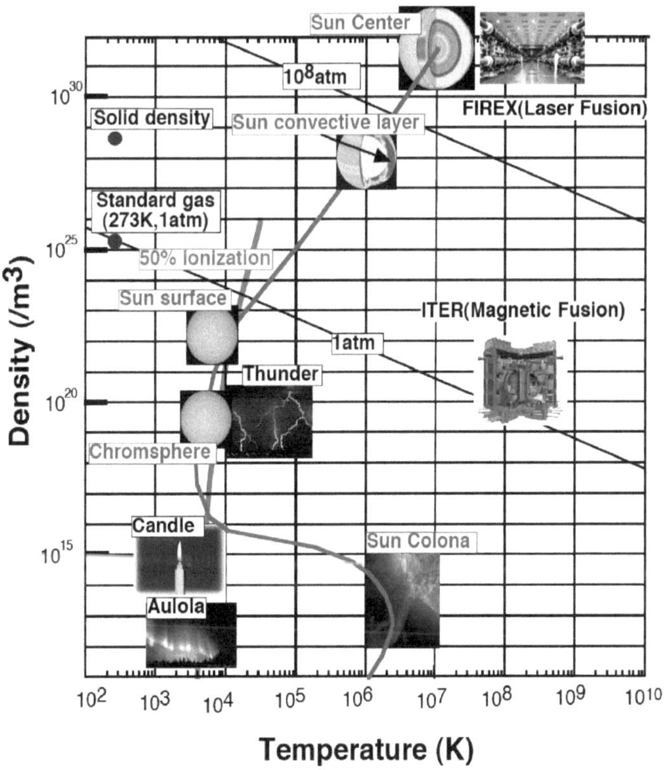

Figure 1.9 Various plasmas at different temperatures and densities. Standard gas (atmospheric pressure, absolute temperature of about 273 K) contains 2.7×10^{25} molecules/m^3. Solid hydrogen contains 4.8×10^{28} atoms/m^3

density, $150\,g/cm^3$), the temperature is 15 million degrees, and the pressure reaches 400 billion atmospheric pressure. In the solar corona, the density is 10^6–$10^9/cm^3$, the temperature is more than 1 million degrees. Typical solar wind parameters are densities of 1–$10/cm^3$ and temperatures of 0.1–0.5 million degrees.

The parameters of the ionosphere 80–500 km above ground are densities of $10^{3-6}/cm^3$ and temperatures of about 1,000 degrees. In this case, only some of the atoms and molecules are ionized forming weakly ionized plasma. The density of weakly ionized plasma in a candle flame is $10^{8-10}/cm^3$ and the temperature is several thousand degrees. From the Saha equation, the degree of ionization $\alpha = n^+/(n^+ + n_0)$ for the ionization equilibrium of a neutral atom and ion ($X \rightleftarrows X^+ + e$) is given below [15]:

$$\alpha = \frac{\rho}{\rho + 1}, \quad \rho = \sqrt{\frac{3 \times 10^{27}}{n_0(m^{-3})}\mathrm{T(eV)}^{3/4} \exp\left(-\frac{V_i}{2kT}\right)}\,. \tag{1.22}$$

For humans to get fusion energy, an engineering scheme has to be developed so that high-temperature plasma can be confined in a limited area under a magnetic field, while plasma is thermally shielded from the vessel, and heat from fusion has to be extracted. Therefore, properties of the plasma must be understood to control the plasma.

References

1. Einstein A (1995) Relativity: The Special and the General Theory. Crown Publ.
2. Aczel AD (2000) God's Equation. Piatkus Books.
3. Landau LD, Lifshitz EM (1994) Classical Theory of Fields. 6th edn. Pergamon Press, Ch. 11.
4. Frankel T (2004) The Geometry of Physics. 2nd edn. Cambridge University Press.
5. Sato K (1980) The Theory of Relativity. Iwanami Books (in Japanese).
6. Weinberg S (1977) The First Three Minutes. Basic Books Inc.
7. Kodama H, Sasaki M (1986) Int. J. Mod. Phys. A1, 265.
8. Parker B (2004) Albert Einstein's Vision. Prometheus Books Inc.
9. Mlodinow L (2001) Euclid's Window. Simon & Schuster Inc.
10. Asimov I (1991) Atom. Truman Talley Books.
11. Heisenberg W (1971) Physics And Beyond. Harper & Row Publishers Inc.
12. Gross RA (1984) Fusion Energy. John Wiley & Sons Inc.
13. Herman R (1990) Fusion: The Search For Endless Energy. Cambridge University Press.
14. Tonks L, Langmuir I (1929) Phys. Rev. 33, 195, 954.
15. Saha MN (1920) Philos. Mag. 40, 472.

Chapter 2
Hydrogen Fusion: Light Nuclei and Theory of Fusion Reactions

Deuterium, tritium, neutron, and helium are the main players in the reaction ($D + T \rightarrow He + n + 17.6\,MeV$; see Figure 2.1) to realize the Sun on Earth. In the microworld of atoms and nuclei, particles exhibit wave characteristics, and nuclear properties and reaction mechanics are governed by the Schrödinger wave equation.

These players have interesting properties. An encounter between deuterium and tritium results in the formation of a compound nucleus through the tunnel effect at a fractional energy of 500 keV Coulomb barrier potential. The compound nucleus has a high reaction probability near 80 keV due to the resonance phenomenon. In this way, nature gives humans a chance to use this reaction.

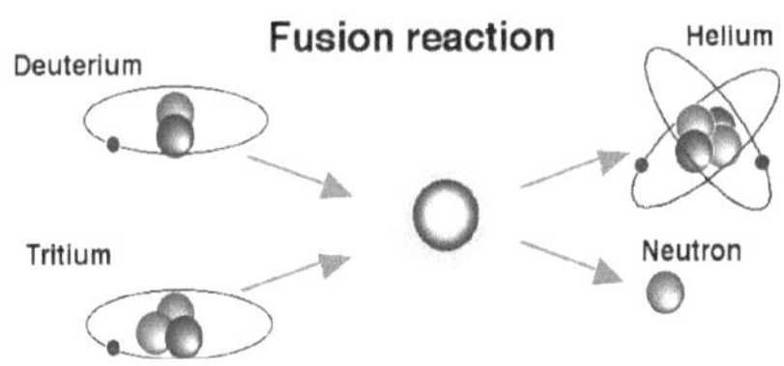

Figure 2.1 The fusion reaction

2.1 Fusion: Fusion of Little Nuts

The fusion that we are currently trying to achieve in ITER is the reaction of deuterium and tritium. This is different from the hydrogen fusion that takes place slowly in the Sun. "Nucleus" is a Latin word meaning "little nuts" [1]. Deuterium and tritium form a "compound nucleus" (a concept originating from N. Bohr), ^{5}He (helium), when they get close enough to each other for the nuclear force to operate beyond the Coulomb barrier when the distance r is less than 3 fermi if we use nu-

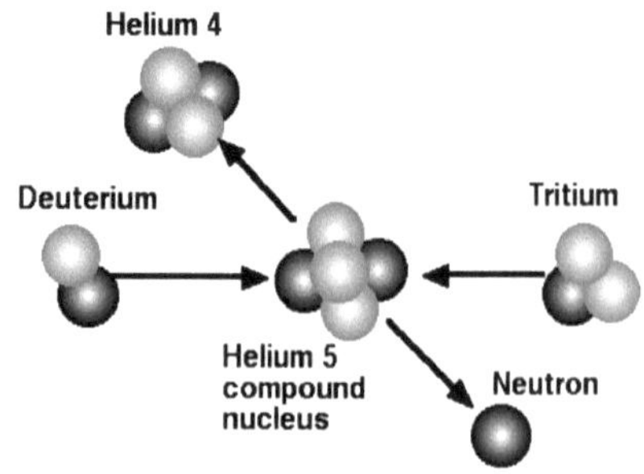

Figure 2.2 Schematic diagram of the DT fusion reaction via a compound nucleus

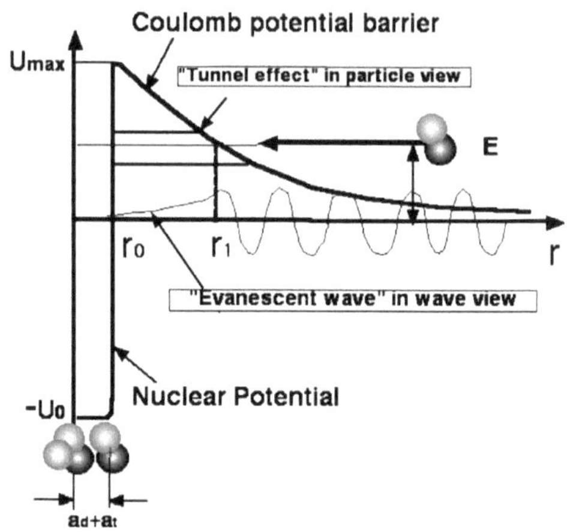

Figure 2.3 Potential and wave function structure in the fusion reaction of deuterium and tritium

clear radius formula $R_c = 1.1\,A^{1/3}$ fermi (1 fermi $= 10^{-15}$ m; A is mass number) as seen in Figure 2.2.

The kinetic energy of the incident nuclei is distributed to nuclei in the compound nucleus. And, the neutron and helium, which have large energies, will, by chance, escape from compound nucleus.

$$\mathrm{D} + \mathrm{T} \rightarrow {}^{5}_{2}\mathrm{H}^{*}_{\mathrm{e}} \rightarrow {}^{4}_{2}\mathrm{He} + \mathrm{n}\;. \tag{2.1}$$

Let $U_{\max}$ be the barrier height for charged particles, $U_{\max} = e^2/(4\pi\varepsilon_0 r) = 0.48$ MeV for $r = 3$ fermi as seen in Figure 2.3. Fusion will occur if the relative energy of deuterium and tritium is larger than 0.48 MeV, but it is difficult to raise the temperature to this level (3.7 billion degrees).

However, thanks to the wave nature of particles, fusion can occur at low energy (several 10 keV) by penetrating the Coulomb barrier. This is called the tunnel effect. Scattering and penetration of the particle beam can be investigated by solving the Schrödinger equation under the Coulomb field (the potential $V = e^2/(4\pi\varepsilon_0 r)$) (see note). This Schrödinger equation was first derived by Austrian physicist Erwin Schrödinger (1887–1961) who was awarded the 1933 Nobel Prize in physics.

In the Sections 2.2–2.5, deuterium, tritium, helium, and neutron will be described as the fuel and products of the fusion reaction.

Note: Schrödinger's Wave Equation [2]

In the world of atoms (radius of 10^{-10} m = 1 Ångstrom (Å)) and nuclei (radius of 10^{-15} m = 1 fermi), the motion of matter exhibits a "particle nature" and "wave nature." In 1905, Einstein, explained the photoelectric effect through the concept of "photons" (light quantization) with energy in proportion to frequency.

$$E = \hbar\omega \ . \tag{2.2}$$

Here, $\hbar$ is Planck's constant, which was introduced by German physicist Max Planck (1858–1947) to describe black-body radiation. The wave nature of the particle ("matter–wave") is given by the French physicist de Broglie (1892–1987) with the following matter–wave relation:

$$\boldsymbol{p} = \hbar\boldsymbol{k} \ . \tag{2.3}$$

From the relation of light quantization and matter–wave, the Austrian physicist E. Schrödinger (Figure 2.4) derived a wave equation to describe the dynamics of the atomic world in 1926. In general, "wave" amplitude ψ is expressed as $\psi = \exp(\mathrm{i}\boldsymbol{k}\cdot\boldsymbol{x} - \mathrm{i}\omega t)$, and relationships $(\partial/\partial\boldsymbol{x})\psi = \mathrm{i}\boldsymbol{k}\psi$ and $(\partial/\partial t)\psi = -\mathrm{i}\omega\psi$ hold. In other words, the wave number $\boldsymbol{k}$ and angular frequency ω correspond to differential operators $(\partial/\partial\boldsymbol{x})$ and $(\partial/\partial t)$, respectively.

$$\mathrm{i}\boldsymbol{k} = \partial/\partial\boldsymbol{x} \tag{2.4}$$

$$-\mathrm{i}\omega = \partial/\partial t \ . \tag{2.5}$$

Let the matter of the atomic world have energy E and momentum $\boldsymbol{p}$. Assuming the energy conservation law of Newtonian mechanics based on macroscopic world $\boldsymbol{p}^2/2m + V(\boldsymbol{x}) = E$ applies similarly to the microscopic world of matter–wave, the microscopic energy conservation law can be obtained by using the Equations 2.2 and 2.3 as follows

$$\hbar^2\boldsymbol{k}^2/2m + V(\boldsymbol{x}) = \hbar\omega \ . \tag{2.6}$$

Substituting wave number and angular frequency from Equations 2.4 and 2.5 into Equation 2.6 results in the following equation of the differential operator consistent with the energy conservation law,

$$-(\hbar^2/2m)\partial^2/\partial\boldsymbol{x}^2 + V(\boldsymbol{x}) = \mathrm{i}\hbar\,\partial/\partial t \ . \tag{2.7}$$

Application of differential operator relation (2.7) to the wave amplitude ψ from left leads to the following Schrödinger's wave equation.

$$\left[-(\hbar^2/2m)\partial^2/\partial\boldsymbol{x}^2 + V(\boldsymbol{x})\right]\psi = \mathrm{i}\hbar(\partial/\partial t)\psi \ . \tag{2.8}$$

Stationary solution of the wave amplitude is obtained by solving the eigen value problem by setting $\psi = u\exp(-\mathrm{i}Et/\hbar)$.

$$\left[-(\hbar^2/2m)\partial^2/\partial\boldsymbol{x}^2 + V(\boldsymbol{x})\right]u = Eu \ . \tag{2.9}$$

In classical mechanics including the theory of relativity, the variables that appear in physics, represents numerical values, and there is a one-to-one correspondence between the variable and the number. Physical quantity is, in principle, measurable and can be regarded as a numerical value. However, this relationship is broken in quantum theory, physical quantity is not the physical number but becomes an "operator", which itself does not correspond to the measurement. Following this idea, momentum $\boldsymbol{p}$ becomes a differential operator $-\mathrm{i}\hbar\, \partial/\partial \boldsymbol{x}$, energy E becomes the differential operator $\mathrm{i}\hbar\, \partial/\partial t$.

Figure 2.4 Erwin Schrödinger who created the Schrödinger equation governing the dynamics of microscopic worlds

2.2 Deuterium: Nucleus Loosely Bound by a Neutron and Proton

The deuterium nucleus consists of one proton and one neutron. Among the combinations of the two nuclei, p-p, n-n and p-n, the only possible bound state is p-n, which is deuterium [3]. Deuterium was discovered in 1932 by American chemist H. C. Urey (1893–1981; Figure 2.5 (a)) [4], who showed that 1 of every 7,000 hydrogen atoms is deuterium. Urey was awarded the 1934 Nobel Prize in Chemistry.

The bound state of a proton and a neutron to form deuterium can be treated as a two-body problem of the nuclear force between a neutron and proton. Then, its state can be solved using the Schrödinger wave equation (2.9) if the potential field $V(r)$ is given. The nuclear force between the neutron and proton is the meson exchange force (see the note) and is represented by a central force given by the potential V as a function of the distance between nucleons r only. This nuclear force is termed a "strong interaction" and has a range of quite a short distance. While long range interaction is explained by the exchange of massless photons, strong interaction is explained by the exchange of mesons with an intermediate mass between that of protons and electrons. Here, the potential is approximated by the graph given in Figure 2.6 for simplicity.

Figure 2.5 (a) H. Urey and (b) M. Oliphant who discovered deuterium and tritium, respectively (Courtesy of the Australian National University)

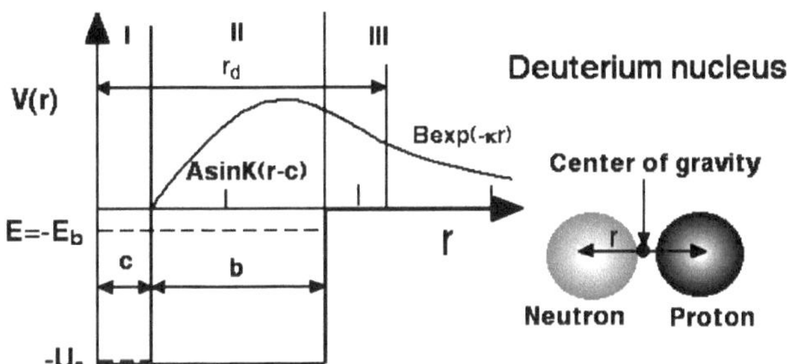

Figure 2.6 Potential model and wave function as a function of distance r from the center of gravity of the neutron and proton in the deuterium nucleus. For deuterium, $c = 0.4$ fermi, $b = 1.337$ fermi, $r_d = 2.1$ fermi, $U_0 = 73$ MeV, $E_b = 2.225$ MeV [3].

Scattering experiments suggest a minimum distance $c \sim 0.4$ fermi between a proton and neutron. Therefore, the wave function is zero at $r < c$ (region I in Figure 2.6) (existing probability is zero). The potential outside (region II) is negative since the attractive nuclear force works. The wave function (ru) in region II is a sine function $A \sin K(r - c)$ (A and K are constants). The wave function in region III (where the nuclear force does not work) is $Be^{-\kappa r}$ (B, κ are constants) considering the boundary condition of $ru = 0$ at $r = \infty$. The width b of region II, and potential U_0 can be determined by the condition of the smooth connection of wave function at the boundary of region II and III then matching the condition to the measured binding energy and nucleus radius r_d of deuterium, leading to $b = 1.337$ fermi, $U_0 = 73$ MeV, respectively. Here, the physical relation between wave function in region III and the binding energy E_b is interesting. The decay rate κ of the wave function is related to the binding energy as $\kappa \sim E_b^{0.5}$, which implies radial decay of the wave function is weak with a probability that two nucleons will stay in region III ($\sim |ru_0|^2 \sim e^{-2\kappa r}$ from Born relation) and becomes larger if E_b is small. In other words, the two nucleons are not strongly bound and are easily separated. Conversely, if the binding energy E_b is large, the wave function decays rapidly in region III, resulting in a low probability that the nucleons will separate . The binding energy of deuterium $E_b = 2.225$ MeV is small and the wave function decays slowly in the r direction, while the wave function in III decays rapidly for helium-4 since $E_b = 28.3$ MeV.

Now, let us consider U caused by the attractive nuclear force. A nucleus confined in a small space (a few fermi) via Yukawa's meson nuclear exchange force, will have a very large wave number $K = (m_r(U_0 - E_b))^{0.5}/\hbar = 1/1.09$ fermi), which implies large kinetic energy of the nucleon ($E = p^2/2m_r \sim K^2\hbar^2/2m_r \sim 50$ MeV: m_r is reduced mass). This means that the nucleon is moving around at breakneck speed in the nuclear potential. To keep this nucleon confined, U_0 must be several tens of MeV.

Heavy water (chemical symbol D_2O, chemically bonded two deuterium and one oxygen) exists at about 158 ppm (ppm: part per million) in seawater, and $\sim$ 140 ppm in the freshwater. There is approximately about 5×10^{13} ton of deuterium on the planet since the volume of seawater is 1.8×10^{18} m^3. Since the amount of combustion of deuterium in a 1 GW level fusion reactor is about 73 kg per year, there exists 70 billion years of resource even if one assumes 10,000 nuclear fusion reactors. Heavy water can be manufactured by electrolysis. Commercially, the GS [GS: Girdler–Spevack] method is used to produce heavy water from fresh water using an isotopic exchange method. Specifically, hydrogen is exchanged in a countercurrent between H_2S (hydrogen sulfide) and H_2O (water). At room temperature, deuterium is exchanged with hydrogen in the water ($H_2O + HDS \rightarrow HDO + H_2S$) and at high temperature (100 °C or more), deuterium is exchanged with hydrogen in the hydrogen sulfide ($HDO + H_2S \rightarrow H_2O + HDS$). Thus heavy water production is mediated by hydrogen sulfide. The energy consumed in the manufacturing process is quite small.

Note: Yukawa's Meson Theory [3,5]

To form a "little nut" by binding neutrons and protons, a new force, the "nuclear force," was needed; this is different from gravity or electromagnetic forces, which were known in the nineteenth century. The first answer to this question was given by Japanese physicist, Hideki Yukawa (1907–1981; Figure 2.7). This force is called a "strong interaction." In the Coulomb interaction, the particle produces a "field" in the space and the force is formed by the interaction of the field with another particle in the space. From a quantum mechanical perspective, it seemed to Yukawa that emission and absorption of massless photons between particles produces this long-range Coulomb force, the so-called "exchange force." To explain the short-range nuclear force, Yukawa considered it to be an exchange force associated with a finite mass particle or "field with mass."

The wave equation of the electric field $[\partial^2/\partial \boldsymbol{x}^2 - \partial^2/(c^2\partial t^2)]U = 0$ can be obtained by applying quantization rules for the momentum $\boldsymbol{p} \rightarrow -i\hbar\,\partial/\partial \boldsymbol{x}$ and energy $E \rightarrow i\hbar\,\partial/\partial t$ to the photon momentum equation $-\boldsymbol{p}^2 + (E/c)^2 = 0$. Then, the "field equation" with rest mass m has to be obtained by applying the quantization rules to the relativistic momentum balance equation of the particle, $-\boldsymbol{p}^2 - m^2c^2 + (E/c)^2 = 0$ as follows,

$$\left[\hbar^2 \frac{\partial^2}{\partial x^2} - m^2c^2 - \hbar^2 \frac{1}{c^2}\frac{\partial^2}{\partial t^2}\right] U = 0\,. \tag{2.10}$$

The static solution of this equation is called the Yukawa potential $U = ge^{-\kappa r}/r$. The fundamental difference to the long range Coulomb field by photon exchange force is that this nuclear force decays quickly with radius at $r > r_0 \sim 1/\kappa = \hbar/mc$ due to finite mass. Yukawa predicted the mass of the exchange particle to be 200 times the electron mass considering a 2 fermi range of interaction of nuclear force. This is π meson. The π meson was discovered by the British physicist C. Powell (1903–1969) in cosmic rays, and was found to be 273 times the mass of the electron. For this achievement, Yukawa and Powell received the 1949 and 1950 Nobel Prize in Physics, respectively.

Figure 2.7 Hideki Yukawa who explained strong force by the exchange force of meson (Courtesy of Yukawa Archival Library, Yukawa Hall, Yukawa Institute for Theoretical Physics, Kyoto University)

2.3 Tritium: Nucleus Emitting an Electron and Neutrino

Hydrogen with mass number 3 is called tritium. The word "tritium" comes from the Greek word meaning "third." The nucleus of tritium consists of one proton and two neutrons. Tritium rarely exists in nature, and is made only in the atmosphere by cosmic rays. This element decays into helium-3 by emitting a high-energy electron beam (see Figure 2.8). This is called beta decay and has a half-life of 12.26 years. Tritium was first produced in the laboratory by Australian physicist M. Oliphant (1901–2000; Figure 2.5 (b)) in 1934 by colliding deuterium [6]

Tritium as fuel of DT fusion is generated by the nuclear reaction of neutron with lithium. The DT fusion produces neutrons and these neutrons can be used to generate tritium by the reaction with lithium (Figure 2.9). Therefore, high temperature plasma is covered with a device called a "blanket" containing lithium to generate tritium.

Lithium has two isotopes (^{6}Li and ^{7}Li) and the abundance of ^{6}Li and ^{7}Li in natural Li is 7.4 and 92.4%, respectively. Tritium is created by neutron absorption. The

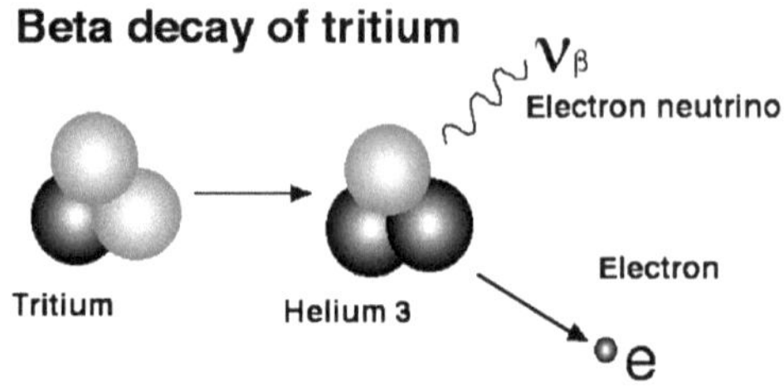

Figure 2.8 Beta decay of tritium into ^{3}He, an electron, and a neutrino

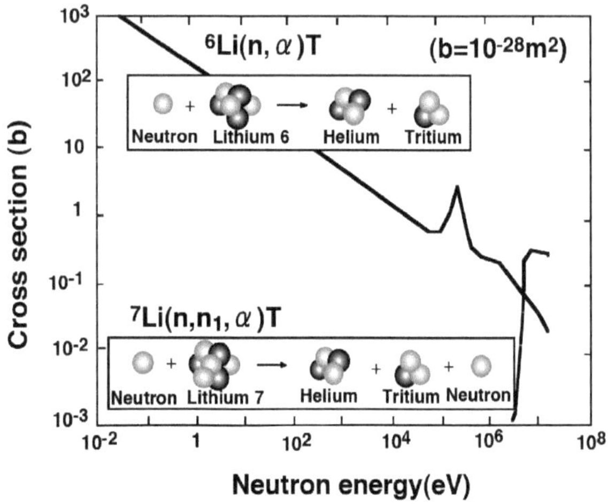

Figure 2.9 Reaction cross-section versus neutron energy for ^{6}Li(n,α)T and ^{7}Li(n,n',α)T reactions

^{6}Li reaction $^6\text{Li} + \text{n} \rightarrow {}^3\text{T} + {}^4\text{He} + 4.8\,\text{MeV}$ is an exothermic reaction, while the ^{7}Li reaction $^7\text{Li} + \text{n} \rightarrow {}^3\text{T} + {}^4\text{He} + \text{n}' - 2.5\,\text{MeV}$ is an endothermic reaction. The cross-section of the ^{6}Li reaction is of the $1/v$ type and it readily reacts at low energy. Meanwhile, the ^{7}Li reaction is called the "threshold reaction" whose cross-section becomes nonzero above a critical energy. The reaction rate for ^{6}Li is much larger than that from ^{7}Li. Therefore, a net energy production occurs in the blanket. There is 233 billion tons of lithium in seawater and the resources can be thought of as infinite if low-cost technologies for the recovery of lithium from seawater are established.

- Lithium resources (reserve base): 9.4 million tons (3.7 million tons of reserves)
- Resources in the sea: 233 billion tons
- Annual production: 21 thousand tons/year (1996) (low demand)

Salon: Beta Decay and Neutrinos

A strange thing happens in the beta decay. While the sum of the mass and kinetic energy is conserved in other nuclear reactions, it seems that the sum of the energy and mass are not conserved for beta decay. In other words, the beta-rays do not have energy equivalent to the mass lost by the beta decay. It is a famous story that Prof. Bohr, one of founders of the quantum mechanics, almost gave up the energy conservation law. This issue troubled physicists for a long time, but in 1931, Swiss physicist W. Pauli (1900–1958) proposed the theory that an electrically neutral particle is emitted by the beta decay. It was named "neutrino" (meaning "little neutron" in Italian) by E. Fermi. Based on this theory, Fermi constructed a theory of beta decay, considering that during beta decay, a neutron is converted to a proton, electron, and neutrino in the nucleus. By assuming the existence of the neutrino, energy, momentum, and angular momentum are conserved. The existence of this neutrino was confirmed, 25 years after Pauli proposed it, by the proton–neutrino reaction in a water tank by F. Reines and C. L. Cowan in 1956. Given the existence of the neutrino, the beta decay of tritium is given as $^3H \rightarrow {}^3He + e^- + \nu + 18.6\,keV$. The mass lost by the beta decay of tritium ΔM_N can be calculated using the fact that mass of a beta-ray $M(\beta)$ is the electron mass $M(e)$

$$\begin{aligned} \Delta M_{\mathrm{N}} &= M_{\mathrm{N}}(^3\mathrm{H}) - (M_{\mathrm{N}}(^3\mathrm{He}) + M(\beta)) \\ &= M_{\mathrm{a}}(^3\mathrm{H}) - M_{\mathrm{a}}(^3\mathrm{He}) = 2 \times 10^{-5} u = 18.6\,\mathrm{keV}(= E_0)\,. \end{aligned} \quad (2.11)$$

Here, M_N and M_a are the masses of the nucleus and atom, respectively. The maximum and average energy of beta-rays emitted from tritium are 18.6 keV and 5.7 keV, respectively. The difference between lost mass (mass defect) and beta-ray energy is the energy of the neutrino. Whether a neutrino has a mass is a long-discussed question in physics. Beta decay of tritium, has been used to determine the maximum value of the neutrino mass due to its low electron kinetic energy. In 1995, A. I. Belesev [7] showed that neutrino mass to be $m_\nu < 4.35\,eV$. In 1998, the existence of neutrino mass was confirmed by the observation of neutrino oscillation between μ neutrino and τ neutrino [8]. Decay phenomena such as beta decay are generally very long compared to the duration of a nuclear reaction, which implies the interaction to be a weak force rather than a strong force binding the protons and neutrons in the nucleus. Recently, studies of neutrinos are expanding as the field Neutrino Astrophysics.

2.4 Neutron: Elementary Particles with No Charge

The neutron was discovered by the British physicist J. Chadwick (1891–1974) in 1932. A particle with a mass similar to a proton and no charge was considered to explain the nuclear spin by Rutherford around 1920, and was named "neutron" by the American chemist W. D. Harkins in 1921. Neutrons have no net charge, but their center has a slightly positive charge distribution, which is cancelled by a slightly negative charge distribution in the periphery. The mass is distributed within a radius of about 0.8 fermi.

A neutron is slightly heavier than a proton and the difference is about twice the electron mass (1.29 MeV). Neutron alone cannot exist stably, and decay to a proton emitting an electron and neutrino with a half-life of about 12 min ($n \rightarrow p + e + \nu$). The mass of the neutron is greater than the sum of the masses of a proton and

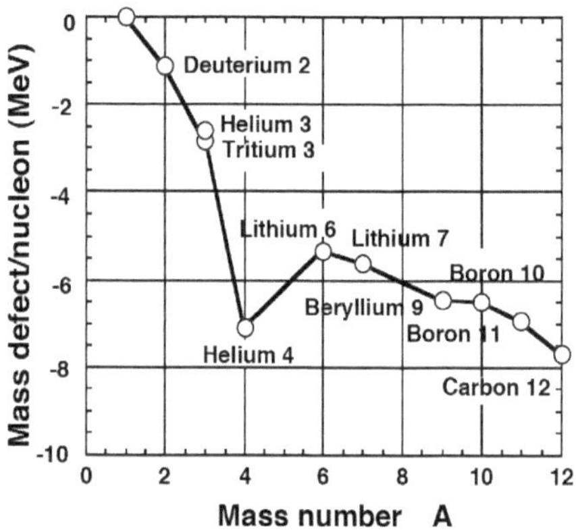

Figure 2.11 Mass defect per nucleon for low mass number elements

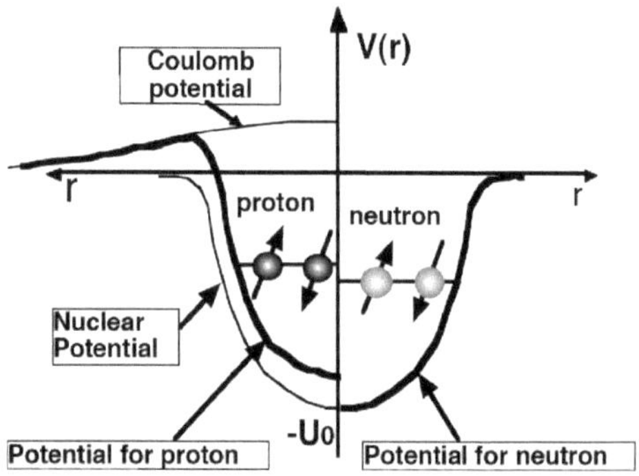

Figure 2.12 Potential structure and energy states of nucleons (neutron and proton) for ^{4}He. A closed-shell structure is formed by the proton and neutron occupying the same two orbits (spin angular momentum of $+1/2$ and $-1/2$) in the ground state S orbit (orbital angular momentum $= 0$)

plained by the nuclear "shell model," where neutrons and protons form energy levels under the average nuclear potential and become stable when a particular energy level is filled up. Helium is composed of two protons and two neutrons, forming spherically symmetric potential. The potential shape for a neutron is different from that for a proton as shown in Figure 2.12 since only nuclear potential operates on neutrons while Coulomb potential is superimposed for protons.

Under these potentials, protons and neutrons can take discrete energy states, independently. This energy level can be obtained by solving the Schrödinger's wave equation. In the case where the potential is proportional to the square of the distance (harmonic oscillator: $V(r) = cr^2$), the solution for such a potential is well-known as $u = R(r)Y(\theta, \phi)$, where $Y(\theta, \phi)$ represents the spherical harmonics and $R(r)$ is the function including Laguerre polynomials. The energy level is expressed by $E_n = -U_0 + (n_1 + (3/2)\hbar\omega)$ $(\hbar\omega = (2U_0\hbar^2/mR_c^2)^{1/2})$, where $n_1 = 2(n-1) + l$ where n is the principal quantum number and l the azimuthal quantum number, respectively. Ground state (lowest energy level) is $E_0 = -U_0 + (3/2)\hbar\omega$ for $n_1 = 0$ $(n = 1, l = 0)$. For the non-zero azimuthal quantum number l, the nucleon has its angular momentum around the center of the nucleus (orbital angular momentum: $L = rmv = l\hbar$). The number of nucleons to the l state is the number of degenerate states of the level $(2l + 1)$ and is 1 for the ground state.

In the nucleus, there is a spin of the nucleon itself in addition to their angular momentum to the center of the nucleus. The value of the spin for neutrons and protons is also quantized and the spin angular momentum associated with rotation of the nucleon itself is 1/2. The states of the spin are clockwise and counterclockwise rotations with respective values of $+1/2$ and $-1/2$.

In helium, two nucleons sit in the ground state (principal quantum number $n = 0$), defined for neutrons and protons independently. In the ground state of ^{4}He, each nucleon takes the orbit with the orbital angular momentum 0 (called the 1S orbit). It is mysterious in some sense that protons and neutrons form an independent energy state. Although protons and neutrons form the nucleus via meson exchange, they behave as if there is no relationship between them. The recent nuclear experiment shows that they are not completely independent.

^{4}He is the first one in the closed-shell state and "2" is the smallest "magic number $(2, 8, 20, \ldots)$". This leads to large binding energy for ^{4}He. Thus, the ^{4}He nucleus is particularly stable and abundant in the universe created by the Big Bang.

Salon: Production and Uses of Helium

Commercially available helium gas is refined from natural gas from wells, while you might get the helium gas from the atmosphere. Natural gas fields in Algeria and the United States contain high concentrations of helium (1–7%). Lunar rock contains a large amount of helium blown in from the Sun as the solar wind and ^{3}He is expected to be a raw material for fusion for the use of bases on the Moon. Helium is a light (1/7 of air or less) and stable (not burning) gas and is used for balloons and airships. However, helium atoms are small, and permeate glass and rubber. This is the reason why balloons inflated with helium gas deflate. In addition, the cryogenic liquid helium is used as a coolant to achieve ultra low temperatures (below −269 °C) for magnetic fusion devices and linear motor vehicles using superconducting magnets. In fusion reactors, such as the Sun on Earth, hydrogen isotopes are fused to produce this stable helium and the lost mass is converted to energy. In future magnetic fusion reactors, energetic helium will be created in the hot plasma and the superconducting magnet of the magnetic fusion reactor will be cooled by cold helium.

2.6 Fusion Cross-section: Tunneling Effect and Resonance

In previous sections, we reached a certain degree of understanding about the nature of deuterium, tritium, neutron, and helium, which are fuel and products for the fusion reactions we are currently investigating. Now, we will talk about the wave dynamics and the cross-section of fusion reactions following Mott and Massey [9]. The classical orbit in Coulomb scattering is given by the hyperbolic curve, $r = r_0/(1 - \alpha \cos\phi)$ (in the center of mass coordinate) shown in Figure 2.13.

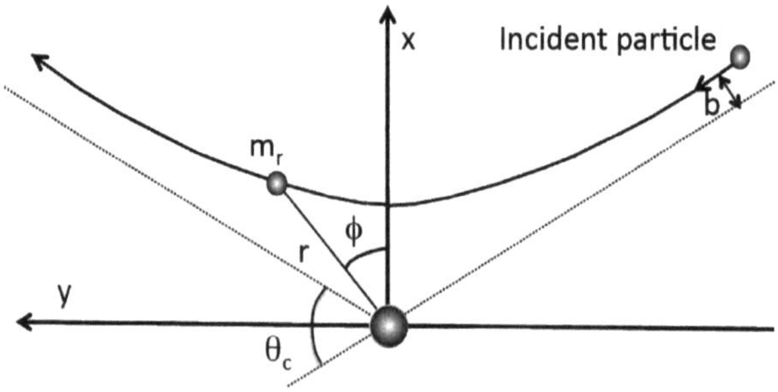

Figure 2.13 Particle orbit in Coulomb potential ($r = r_0/(1-\alpha\cos\phi)$) $r_0 = b^2/b_0$, and $\alpha = (1+(b/b_0)^2)^{1/2}$

Note: Particle Motion in Central Field [10]

If the potential energy between two particles depends only on distance, the force field is called the **central force field** and is given as $\boldsymbol{F} = -(\mathrm{d}U/\mathrm{d}r)\boldsymbol{r}/r$. In the center of the mass system, angular momentum $\boldsymbol{M} = \boldsymbol{r}\times\boldsymbol{p}$ is conserved. Since M is constant, r stays on a plane perpendicular to M. In the polar coordinates (r,ϕ), the Lagrangian is given as,

$$L = m(\dot{r}^2 + r^2\dot{\phi}^2) - U(r)\,. \tag{2.12}$$

Since L is independent of ϕ, canonical momentum conjugates to ϕ, $M = p_\phi = mr^2\dot{\phi}$ is conserved. Substituting this relation into the energy conservation relation $E = m(\dot{r}^2 + r^2\dot{\phi}^2) + U(r)$, we obtain

$$\frac{\mathrm{d}r}{\mathrm{d}t} = \sqrt{2(E-U(r))/m - M^2/(m^2r^2)}\,. \tag{2.13}$$

In the case of repulsive Coulomb potential $U(r) = e_ie_j/(4\pi\varepsilon_0 r)$, the particle orbit is given by following hyperbolic curve.

$$r = \frac{r_0}{1-\alpha\cos\phi} \tag{2.14}$$

where

$$r_0 = \frac{M^2}{m_r(e_ie_j/4\pi\varepsilon_0)}\,,\quad \alpha = \sqrt{1+\frac{2EM^2}{m_r(e_ie_j/4\pi\varepsilon_0)^2}}\,. \tag{2.15}$$

Since $M = bm_r u$ and $E = m_r u^2/2$ (b is impact parameter in Figure 2.13, u is particle speed at infinity), we can convert these parameters as follows,

$$r_0 = \frac{b^2}{b_0}\,,\quad \alpha = \sqrt{1+\frac{b^2}{b_0^2}} \tag{2.16}$$

where,

$$b_0 = \frac{e_i e_j}{4\pi\varepsilon_0 m_r u^2} \,. \tag{2.17}$$

The wave front of the incident wave should be perpendicular to this hyperbolic curve. Let k be the de Broglie wave number, the wave front satisfying this criteria is,

$$z + b_0 \ln k(r - z) = \text{const.} \tag{2.18}$$

where $b_0 = e_i e_j/(4\pi\varepsilon_0 m_r u^2) = 7.2\times 10^{-10} Z_i Z_j/E_r$ (eV) (m), called the Landau parameter. The incident wave is already distorted at infinite distance before encountering the nucleus since the Coulomb field operates to infinity. So, the incident wave is given by $\exp[ik\{z + b_0 \ln k(r - z)\}]$.

For a non-relativistic collision $E_r = p^2/2m_r = \hbar^2 k^2/2m_r$, and the Schrödinger's wave equation (2.9) becomes $[\partial^2/\partial \boldsymbol{x}^2 + (k^2 - \beta/r)]\psi = 0$, where $\beta = m_r e_i e_j/2\pi\varepsilon_0\hbar^2$. Substituting $\psi = \exp(ikz)F(\boldsymbol{x})$, F should have the form $F = F(r-z)$. Then, the wave equation becomes $\zeta \mathrm{d}^2F/\mathrm{d}\zeta^2 + (1-ik\zeta)\mathrm{d}F/\mathrm{d}\zeta - (\beta/2)F = 0$ for $\zeta = r - z$. Taylor expanding F ($F = \sum a_n \zeta^n$ ($a_0 = 1$)) and substituting into the wave equation gives a_n and F is found to be the hyper geometric function F as follows,

$$\psi = \exp(-\pi\alpha/2)\Gamma(1 + \mathrm{i}\alpha)\mathrm{e}^{ikz} F(-\mathrm{i}\alpha, 1; \mathrm{i}k\zeta) \,. \tag{2.19}$$

Here $\alpha = \beta/2k$, $F(a,b;z) \equiv \sum \Gamma(a+n)\Gamma(b)z^n/\Gamma(a)/\Gamma(b)/\Gamma(n+1)$ is a hypergeometric function, Γ is the gamma function. Permeability of the Coulomb barrier P is then given from wave function at the origin as,

$$P(E/E_c) = \frac{\sqrt{E_c/E}}{\exp\sqrt{E_c/E} - 1} \tag{2.20}$$

$$E_c = \frac{m_r e^4}{8\varepsilon_0^2\hbar^2} = 0.98 A_r \,(\text{MeV}) \,. \tag{2.21}$$

Here, m_r is the reduced mass ($m_a m_b/(m_a + m_b)$) between particles a and b, A_r is a mass number of the reduced mass, ε_0 is the vacuum permittivity ($= 8.854 \times 10^{12}$ F/m), and $\hbar$ is Planck's constant. Figure 2.13 shows permeability of the Coulomb barrier P against E/E_c. The critical energy E_c given by Equation 2.21. For the T(d,n)^{4}He reaction, it is 1.18 MeV, while numerically fitting E_c to reproduce the measured fusion cross-section is 1.27 MeV showing good agreement.

A particle with reduced mass jumping into the nuclear potential, will have higher kinetic energy than the original one (de Broglie wavelength shorter than the original de Broglie wavelength). Such big changes in wave number will cause a resonant interaction within the nucleus. For the DT fusion reaction, resonance energy and resonance width are $E_r = 78.65$ keV and $\Gamma = 146$ keV, respectively (Figure 2.15). The probability amplitude of a material wave is large since the compound nucleus ^{5}He has an energy level corresponding to a certain boundary condition.

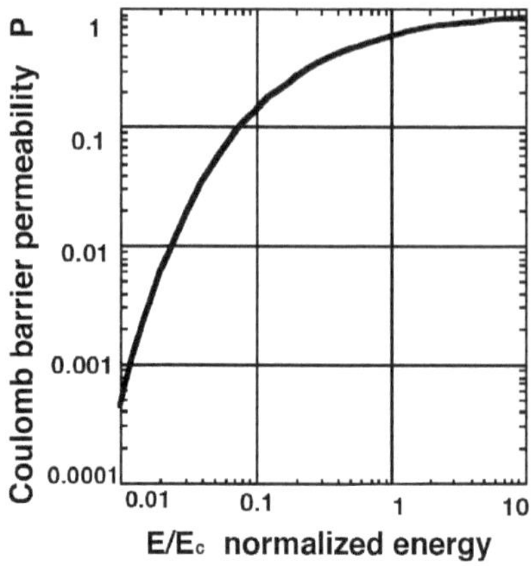

Figure 2.14 Energy dependence of the Coulomb barrier permeability. The Coulomb barrier permeability P is well-behaved ($P \to 1$) in the high energy limit ($E/E_c \gg 1$) while the formula given by Gamov $P(E/E_c) = (E/E_c)^{-1/2} \exp(-(E/E_c)^{-1/2})$ [11] is valid only at low energy

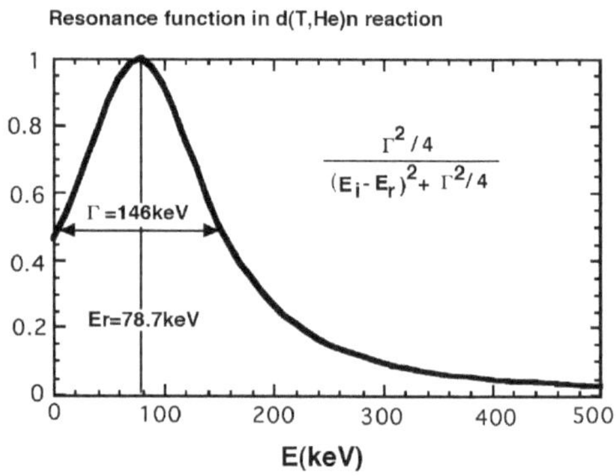

Figure 2.15 Normalized resonance function for DT reaction in laboratory frame

The compound nucleus is unstable and will decay at a decay time constant. The decay time constant τ and the resonance width Γ have the relationship $\Gamma\tau = \hbar$, the decay time is calculated as $\tau = 4.5 \times 10^{-21}$ s since $\Gamma = 146$ keV. This time is much longer (100 times) than the transit time of a nucleon with Fermi energy, $\tau_F = 2R/v_F = 4.4 \times 10^{-23}$ s.

The fusion cross-section, considering the above, is given as,

$$\sigma_r = \pi \lambda\!\!\!^{-2} P(E/E_c) \frac{\Gamma_i \Gamma_f}{(E - E_r)^2 + \Gamma^2/4} . \tag{2.22}$$

Here, $\lambda\!\!\!^{-2} = \hbar^2 (2ME)^{-1}$, and the second factor comes from the Breit–Wigner formula of resonance cross-section [12], here $\Gamma_i \sim k \sim 1/E^{0.5}$. In fact, the measured values of the fusion cross-section are given in a form of Equation 2.22 [13].

$$\sigma_r = \sigma_0 \frac{E_{cl}}{E_l [\exp \sqrt{E_{cl}/E_l} - 1]} \left[\frac{1}{1 + 4(E_l - E_{rl})^2/\Gamma_l^2} + \alpha \right] . \tag{2.23}$$

Here, $\sigma_0 = 23.79$ b, $E_{cl} = 2.11$ MeV, $E_{rl} = 78.65$ keV, $\Gamma_l = 146$ keV, $\alpha = 0.0081$ for the T(d,n)^{4}H reaction [12]. Figure 2.16 shows DT cross-section

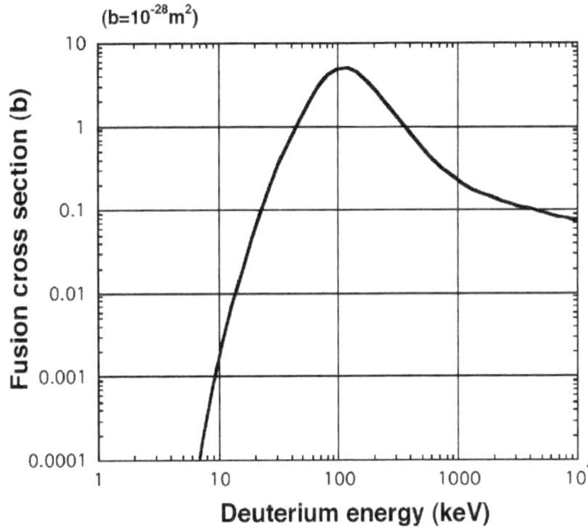

Figure 2.16 DT-fusion cross-sections in laboratory frame (in b (barn $= 10^{-28}$ m^2))

of nuclear fusion reactions (in the laboratory system) illustrating the dependence on deuterium energy. Deuterium energy in the laboratory frame E_l is related to the energy in the center of mass frame E as $E = m_t/(m_d + m_t)E_l$ and thus $E_c = 0.6E_{cl} = 1.27\,\text{MeV}$.

Recently Li et al. [14] gave more clear analysis of fusion cross section. Using the Landau's analysis of resonant scattering of charged particles [12], they derived different form of fusion cross section using the optical potential ($= U_r + iU_i$) for D + T reaction as follows,

$$\sigma_r = \frac{\pi}{k^2\theta^2}\frac{-4w_i}{w_r^2 + (w_i - \theta^{-2})^2} \tag{2.24}$$

where $\theta^2 = (\exp((E_c/E)^{1/2}) - 1)/2\pi$ and $w = w_r + iw_i = \cot(\delta_0)/\theta^2$ where δ_0 is phase shift due to nuclear potential. Using approximate expression $w = C_1 + C_2E_\ell + iC_3$, fusion cross-section in laboratory frame is given as follows,

$$\sigma_r(E_\ell) = \frac{\pi\hbar^2}{2m_d E_\ell\theta^2}\frac{-4C_3}{(C_1 + C_2E_\ell)^2 + (C_4 - \theta^{-2})^2}\,. \tag{2.25}$$

For DT fusion, $C_1 = -0.5405$, $C_2 = 0.005546$, $C_3 = -0.3909$ gives good agreement with measured fusion cross-section.

References

1. Asimov I (1991) Atom. Nightfall Inc.
2. Schiff LI (1974) Quantum Mechanics. 3rd edn. McGraw-Hill Book Company Inc.
3. Enge H (1966) Introduction to Nuclear Physics. Addison-Wesley Pub. Co.
4. Urey HC et al. (1932) Phys. Rev. 39, 164L.

5. Yukawa H (1935) Proc. Phys. Math. Soc., Japan, 17, 48.
6. Oliphant ML et al. (1934) Nature, 133, 413.
7. Belesev AI et al. (1995) Phys. Rev. Lett., B350, 263.
8. Fukuda Y et al. (1998) Phys. Rev. Lett., 81, 1562.
9. Mott NF and Massey HSW (1965) The Theory of Atomic Collisions. 3rd edn. Clarendon Press, p. 57, eq. 17.
10. Landau LD, Lifschitz EM (1969) Mechanics. Pergamon Press.
11. Gamov G, Critchfield CL (1951) The Theory Atomic Nucleus and Nuclear Energy Sources. Clarendon Press, Ch. X, eq. 2.
12. Landau LD, Lifschitz EM (1987) Quantum Mechanics. 3rd edn., Pergamon Press.
13. Duane BH (1972) Fusion Cross-section Theory. Brookhaven National Laboratory Report BNWL-1685, pp. 75–92.
14. Li ZX, Wie MQ, Liu B (2008) Nucl. Fusion 48, 125003.

Chapter 3
Confinement Bottle: Topology of Closed Magnetic Field and Force Equilibrium

In the natural fusion reactor, the Sun, dense hot plasma is confined by a gravitational field. Characteristic of this force is that it is a central force field and acts in the direction of field line. For this reason, the confinement bottle has the topology of a sphere (Figure 3.1 (a)). In the man-made fusion reactor, high temperature plasma is confined by trapping charged particles with the Lorentz force in a magnetic field to sustain reaction in a small dimension of 100 millionth of that of the Sun. Characteristic of this force is that it acts in a direction perpendicular to the field line. For this reason, the confinement bottle has the topology of a torus (Figure 3.1 (b)). In this chapter, force equilibrium is treated to confine high temperature plasma in the topology of a torus. Practically, the magnetic field line dynamics is treated using the methodology of analytical mechanics and symmetry involved in the force equilibrium is discussed.

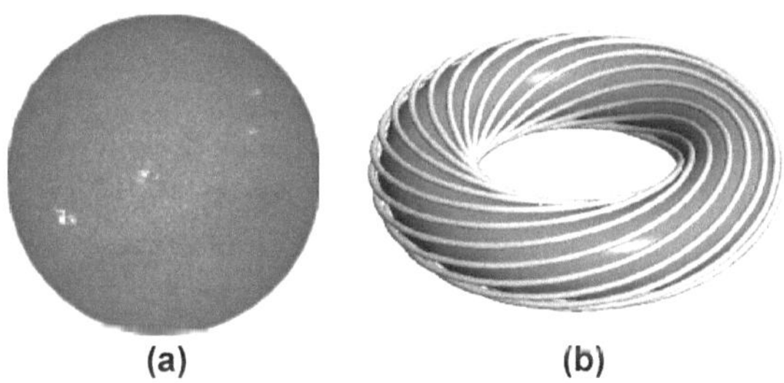

Figure 3.1 (a) Sphere and (b) torus

3.1 Field: Magnetic Field and Closed Magnetic Configuration

Danish physicist H. C. Oersted (1777–1851), during the years 1819–1820, discovered that a compass needle directs to a fixed point by some force when it is placed near a wire carrying a current from a battery invented by Italian physicist A. Volta (1745–1827) (Figure 3.2 (a)). This power is called the magnetic force. British physicist Michael Faraday (1791–1867) gave a novel interpretation of this phenomenon, suggesting that the space around the wire is in a special state by driving the current, instead of the explanation that a remote force works between the current and compass needle. The key here is the way of thinking "the space itself would be different." "Field" as a nature of the space was a concept that attracted Einstein's attention. "Vacuum" as state of nothing originated from Greek philosopher Democritus (460–370 BC). The vacuum can have a different state with a "magnetic field" and this state can have energy.

This magnetic field has a direction, and various phenomena can be explained if we assume that a virtual line, "magnetic line of force," exists along the direction of the magnetic N-pole as shown in Figure 3.2 (a). Magnetic field lines form a circle when circular coils are arranged in a donut shape as shown in Figure 3.2 (b). The donut-shaped space, the torus can be arranged inside the coil. Longitudinal current in the torus produces the magnetic field lines linked to the torus as shown in Figure 3.2 (c). A tokamak-type magnetic configuration which is said to be closest to the fusion reactor, is shown in Figure 3.2 (d), a superposition of the magnetic field lines shown in (b) and (c). The equation describing the magnetic field lines is $\mathrm{d}x/B_x = \mathrm{d}y/B_y = \mathrm{d}z/B_z$ or $\boldsymbol{B} \times \mathrm{d}\boldsymbol{x} = 0$. This equation is equivalent to the condition of the extremum of the path integral called the "action integral" (variational principle) as will be described in Section 3.4. In this case, the action integral is expressed by the path integral of the vector potential $\boldsymbol{A}$ (proof will be given in Section 3.4).

$$\delta \int \boldsymbol{A} \cdot \mathrm{d}\boldsymbol{x} = 0 \,. \tag{3.1}$$

Faraday, in 1821, found that force acts on the current in the magnetic field ($\boldsymbol{F} = \boldsymbol{J} \times \boldsymbol{B}$, where $\boldsymbol{F}$ is force, $\boldsymbol{J}$ the current, $\boldsymbol{B}$ the magnetic field. Orientation is determined by Fleming's left-hand rule). Consider a charged particle with charge q (Coulombs) and velocity$\boldsymbol{v}$ (m/s) moving in a magnetic field $\boldsymbol{B}$ (Tesla), the current is given by $q\boldsymbol{v}$ and the force acting on the charged particle is $\boldsymbol{F} = q\boldsymbol{v} \times \boldsymbol{B}$. This is called the Lorentz force. Ions and electrons in the magnetic field have a circular motion (Larmor motion) with a radius defined by the balance between the centrifugal and Lorentz forces. On the other hand, they move in a straight line in the direction of the magnetic field. A combination of these two motions appears to be helical (Figure 3.2 (d)). It is difficult for the charged particles to escape perpendicular to the magnetic field. The method of confining hot plasma using this principle is called "magnetic confinement." Magnetic confinement becomes efficient if the field line is closed since charged particles move freely along the field.

The study of the structure of closed magnetic field lines to confine the hot plasma is called the theory of plasma equilibrium. Famous examples are the axisymmetric magnetic configuration called the tokamak invented by A. Sakharov (1921–1989) of Kurchatov Institute in the former Soviet Union [1] (Figure 3.2 (d)) and the stellarator (helical in general) invented by Lyman Spitzer Jr. (1914–1997) of Princeton University in the USA [2] (Figure 3.3).

In both configurations, there exists some region where field lines are closed and do not intersect with material walls and the high temperature plasma is confined.

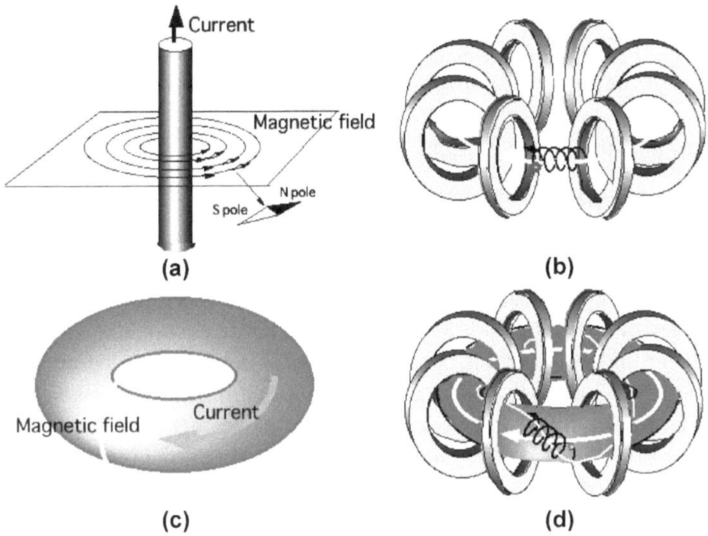

Figure 3.2 (a) When the current flows, a magnetic field is generated around the current. (b) Arranging circular coils around the torus and energizing the coil produces a magnetic field in the toroidal direction. (c) The toroidal plasma current produces a magnetic field linking the torus. (d) A combination of (b) and (c) creates twisted magnetic field lines and is called a tokamak

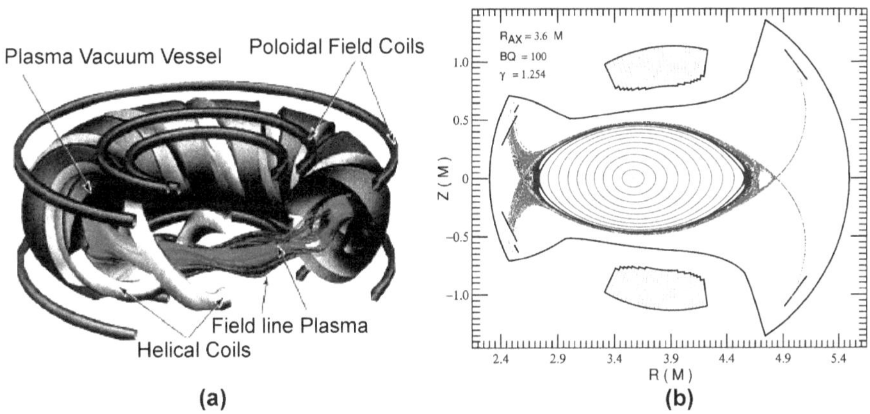

Figure 3.3 (a) Schematic view of helical device LHD showing how twisted magnetic field lines are formed without a toroidal plasma current using helical coils and (b) a typical example of flux surface shape in LHD [3]

Its shape is torus. It is difficult to create a closed magnetic field structure other than the torus. For closed surfaces other than the torus, a null-field point will exist according to the fixed-point theorem of mathematics. Important characteristic of the closed magnetic configuration is that a toroidal magnetic field line densely covers a constant pressure toroidal surface. A simple closed line cannot confine the plasma with a pressure difference. The closed surface should be formed by the magnetic field line and torus-shaped plasma has to be confined in it. Thus the problem of covering the surface with a magnetic field line becomes important. If the magnetic field line trajectory is on the surface, it is called "integrable."

Note: Integrable System [4–6]

"Integrable" is a term in classical mechanics, having its origin in the many-body-problem of celestial mechanics, and is plainly explained in Diacu and Homes [4]. French mathematician J. Liouville (1809–1882; Figure 3.4) gave its mathematical definition. For an equation of motion for particles in dynamical systems to be solvable or integrable, it is key to find some kind of symmetry (in dynamical systems, to find an ignorable coordinate). An N degrees of freedom system of Newtonian mechanics follows the Hamilton equation, and phase space flow is known as an incompressible flow. $N = 1$, *i.e.*, a one degree of freedom Hamilton system (x, v_x) is always integrable since the Hamiltonian of the system is always conserved. The system moves along $H(x, v_x) =$ constant contour.

Motion of $N = 2$, *i.e.*, a two degrees of freedom Hamilton system (x, y, v_x, v_y) is described as dynamics in 4D phase space (position $\boldsymbol{q} = (x, y)$ and momentum $\boldsymbol{p} = (p_x, p_y)$). Its solution is limited on the hyper-surface of $H(\boldsymbol{q}, \boldsymbol{p}) =$ constant $= E$ (3-dimensional manifold with constant energy). If an additional first integral $\Phi(\boldsymbol{q}, \boldsymbol{p}) =$ constant exists, the flow line of phase space motion is on the curved surface (2-dimensional manifold) limited by $H(\boldsymbol{q}, \boldsymbol{p}) =$ constant and $\Phi(\boldsymbol{q}, \boldsymbol{p}) =$ constant. Such a case is called an "integrable system." Known 2-dimensional integrable systems are a 2-dimensional Kepler problem, 2-dimensional harmonic oscillator, and the movement of the mass point in the 2-dimensional central force field. If we cut this surface (2-dimensional manifold) in a plane (for example, the $x = 0$ plane), the flow line becomes a line in the plane. On the other hand, flow lines cut in a plane have a 2-dimensional spread for the non-integrable system.

Figure 3.4 J. Liouville who studied the mathematical nature of integrable systems. He derived the famous theorem, the "Liouville-Arnold theorem" [5]. He is more famous for his theorem, the "Liouville theorem," in phase space dynamics, which is given in Chapter 5

3.2 Topology: Closed Surface Without a Fixed Point

Considering the confinement of hot plasma in a region of 3-dimensional space, the boundary must be a closed surface. Figure 3.5 shows the characteristics of the flow on a torus and a sphere as typical closed surfaces. In the torus, there is no point where flow field vector becomes zero (called a "fixed point") as shown in Figure 3.5 (a) and (b). On the other hand, the flow field on the sphere necessarily has a fixed point, as shown in (c) and (d). Since the hot plasma will leak from the fixed point where the magnetic field is zero, the sphere cannot be used for magnetic confinement.

We can provide a little more familiar example of this nature, "when the wind is blowing on the Earth, there is always somewhere to rest." This is commonly true for the Earth, rugby balls and coffee in a cup.

Mathematically speaking, all surfaces "homeomorphic" to a sphere will have fixed points. This means that the sphere and torus have different topologies. The surface property of a sphere and torus does not change even if they are bent or stretched. A geometrical property, which is not changed by continuous deformations, is called the "topology" of the object. It should not vary during continuous deformations of bending and stretching.

French mathematician Henri Poincaré (1854–1912) proved the theorem: "A closed surface that can be covered with a vector field without a fixed point is restricted to a torus." This is called the "Poincaré theorem" [7–10]. The Poincaré theorem is important for high temperature plasma confinement. Consider the boundary surface of the magnetic confinement, the plasma will leak from the zero point of magnetic field vector. To confine the hot plasma, the surface must be covered by a non-zero magnetic field. This is why we use toroidal geometry for magnetic confinement.

The people of ancient Greece aware that the regular polyhedrons are limited to regular (4) tetrahedron, regular (6) hexahedron, regular (8) octahedron, regular (12) dodecahedron, and regular (20) icosahedron. Let the number of vertices of a polyhedron be p, the number of sides q, and the number of the polygon r, the Swiss mathematician L. Euler (1707–1783) found the relation $p - q + r = 2$ for

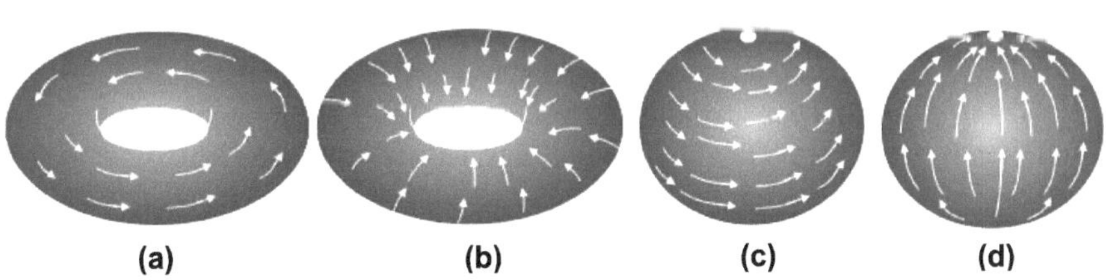

Figure 3.5 Topological properties of torus and sphere. (a) and (b) flow in the torus surface, where flow field without a fixed point can be formed. Flows in (a) and (b) are said to be commutable. (c) and (d) flow fields in a sphere always have a null-vector point (fixed point: ∘)

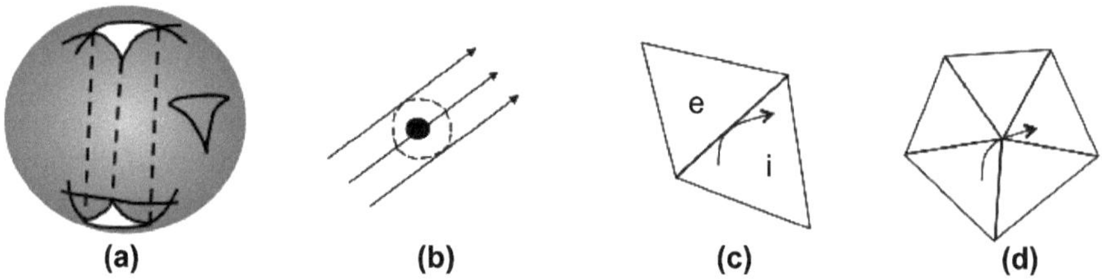

Figure 3.6 (a) Elimination of a triangular prism from a sphere produces a torus and fixed points can be eliminated. The number of vertices is the same as the sphere, the three sides (*the sides of the triangular prism*) increases by for each face. (b) A regular point and its enclosed loop (*dashed circle*). (c) Vector field touches a side from inside the triangle i. (d) Vector field through the vertex of the triangle

any regular polyhedron. For example, in a regular tetrahedron, the number of vertices $p = 4$, of sides $q = 6$, of polygon $r = 4$, gives the index $K = p - q + r = 2$. This relationship holds not only for regular polyhedrons but also for polyhedrons homeomorphic to a sphere, and is called "Euler's polyhedron theorem."

$$K = p - q + r \tag{3.2}$$

is the "Euler index." The relationship always holds irrespective of any division of the sphere by triangles.

Then, what will happen to the Euler index if the torus is covered with triangles? Drill the sphere from top to bottom to eliminate the triangular prism as in Figure 3.6 (a) it becomes homeomorphic to the torus. Then, the following relations hold between p, q, and r of the sphere and p', q', and r' of the drilled-sphere,

$$p' = p\,, \quad q' = q + 3\,, \quad r' = r - 2 + 3\,. \tag{3.3}$$

Thus, $p' + q' + r' = p - q + r - 2 = 0$. In other words, the Euler index of the torus is 0.

Poincaré showed that Euler index is related to the characteristics of the vector fields on the close surface. Poincaré defined the "index" of the vector field. The non-zero point shown in Figure 3.6 (b) is called the "regular point." The index value defined by Poincaré becomes zero for a regular point. Here, the index is defined by $k = (I - E)/2 + 1$, I is the number of vector lines from the inside in contact with a sufficiently small loop around the point (dashed line in Figure 3.6 (b)), E is the number of vector lines from the outside. For regular point, $I = 0$ and $E = 2$. So the index $k = 0$. The small loop in Figure 3.6 (b) can be continuously deformed to an infinitely small triangle. In this case, $I = 0$ and $E = 2$. On the other hand, the index at the singular point takes a non-zero value. The index of the flow surface is defined by the sum of the flow index of all points in the surface. Since the index of the regular point is zero, the index of the surface flow is the sum of the index of the singular points in the plane. The index of a closed surface becomes a sum of the index of each polygon when the surface is divided to several polygons (additivity).

Consider the flow contacting the side of a polygon in Figure 3.6 (c)), this flow has a positive contribution to the index of polygon i, which includes this flow but has a negative contribution to the index of polygon e. In the end, the flow tangent to the side does not contribute to the index of the closed surface (this discussion holds only for closed surfaces). So, the contribution to the index of the flow comes only from vertices of the polygon. Consider the vertex A associated with N sides, we see N polygons share the vertex A. The number of vector fields contacting A from outside is $N-2$ as can be seen in Figure 3.6 (d)). Now, the number of contacts from inside is 0. Thus, total number of contacts of the entire closed surface (external – internal) is the sum of $(N-2)$ for all vertices. By taking the sum for all the vertices, $\sum(E-I) = \sum_{\text{vertices}}(N-2) = (\sum_{\text{vertices}} N) - 2p = 2q - 2p$. Here, we use that sum of N for all vertices are twice the number of sides. Now, the sum of the flow index for the entire closed surface $\sum k = -\sum(E-I)/2 + \sum_{\text{polygon}} 1 = p - q + r$. Here,

$$K = p - q + r \tag{3.4}$$

is the Euler index, and flow index of the closed surface is equal to the Euler index. The Euler index is equal to the sum of the index of the closed surface, we can say that the necessary and sufficient condition that a flow without a fixed point can exist in a closed surface is that the Euler index of the closed surface is 0 and the surface is a torus. It is known that orientable 2-dimensional closed surfaces are limited to the sphere S^2, torus T^2, and n-holed torus $\sum_n$ $(n = 2, 3, \ldots)$. Poincaré's theorem tells us that the torus has a special nature as a 2-dimensional closed surface [11, 12].

Salon: L. Euler and H. Poincaré

Leonhard Euler (Figure 3.7 (a)) was a famous mathematician and physicist born in Swiss. He made a huge contribution to mathematics and physics. He solved the "Königsberg bridge problem" in 1736, starting graph theory as related to topology. He gave the so-called Euler identity $e^{i\pi} + 1 = 0$, which was described as "the most remarkable formula" by R. Feynman.

Henri Poincaré (Figure 3.7 (b)) was a famous French mathematician and physicist. His works appear in this book as "Poincaré theorem on topology" and also "Poincaré recurrence theorem" in Chapter 5. He also created a graphical method to analyze dynamical systems in which he discovered a phenomenon now called "Chaos." He is also famous for the "Poincaré conjecture," recently solved by Russian mathematician G. Perelman [12].

(a) (b)

Figure 3.7 (a) L. Euler and (b) H. Poincaré

3.3 Coordinates: Analytical Geometry of the Torus

Torus topology can be described without using "coordinates" as in Euclidean geometry. However, coordinates should be introduced to understand the physics of the torus, quantitatively. French philosopher René Descartes (1596–1650; Figure 3.8 (a)) published Discourse on Method in 1637, and in its appendix "Geometry," described how to assign numbers, called "Descartes coordinates," to a geometric shape. Descartes made the greatest contribution to the science by appointing numbers to all points in the plane by introducing $x - y$ coordinates (Figure 3.8 (b)).

Efforts to provide the most appropriate coordinates for the torus produced "Hamada coordinates" as a typical example (Section 3.6). Here, we consider general curvilinear coordinates, where space is expressed by 3-dimensional curvilinear coordinates [13, 14]. If we express the most fundamental Cartesian coordinates as (x, y, z) and, the position vector is given by $\boldsymbol{x} = x\boldsymbol{e}_x + y\boldsymbol{e}_y + z\boldsymbol{e}_z$. Let the general curvilinear coordinates (u^1, u^2, u^3) be given by the relations, $u^1 = u^1\ (x, y, z)$, $u^2 = u^2\ (x, y, z)$, $u^3 = u^3\ (x, y, z)$ (see Figure 3.9). The following relation is satisfied between the gradient vector ∇u^i and the tangent vector $\partial \boldsymbol{x}/\partial u^j$, and is called the "orthogonal relation."

Here

$$\nabla u^i \cdot \frac{\partial \boldsymbol{x}}{\partial u^j} = \frac{\partial u^i}{\partial u^j} = \delta_{ij} , \tag{3.5}$$

$$\nabla u^i = \left(\frac{\partial u^i}{\partial x}\right) \boldsymbol{e}_x + \left(\frac{\partial u^i}{\partial y}\right) \boldsymbol{e}_y + \left(\frac{\partial u^i}{\partial z}\right) \boldsymbol{e}_z , \tag{3.6}$$

$$\frac{\partial \boldsymbol{x}}{\partial u^j} = \frac{\partial x}{\partial u^j}\boldsymbol{e}_x + \frac{\partial y}{\partial u^j}\boldsymbol{e}_y + \frac{\partial z}{\partial u^j}\boldsymbol{e}_z . \tag{3.7}$$

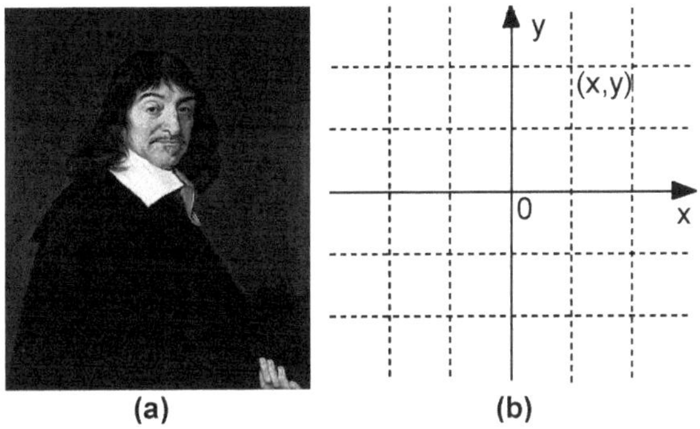

Figure 3.8 (a) René Descartes who introduced (b) "coordinates" in geometry. His most famous philosophical quotation is "cogito, ergo sum"

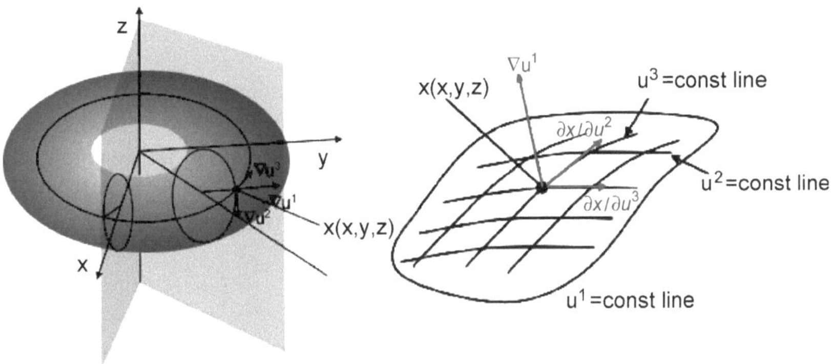

Figure 3.9 General curvilinear coordinate system in a torus. ∇u^1 is a gradient vector normal to u^1 surface. $\partial \boldsymbol{x}/\partial u^2$ and $\partial \boldsymbol{x}/\partial u^3$ are tangent vector on the u^1 surface and are perpendicular to ∇u^1

For example, the tangent vector $\partial \boldsymbol{x}/\partial u^2$ is a differentiation under u^1 and $u^3 =$ constants, so it is tangent to $u^3 =$ constant line on u^1 surface (see Figure 3.9). Naturally, $\partial \boldsymbol{x}/\partial u^2$ is orthogonal to ∇u^1 (and ∇u^3), which is perpendicular to the u^1 (and u^3) plane. A similar relation holds for $\partial \boldsymbol{x}/\partial u^3$. Then, a useful expression $\nabla u^1 = J^{-1}(\partial \boldsymbol{x}/\partial u^2 \times \partial \boldsymbol{x}/\partial u^3)$ is obtained (#1). Here, J is called the Jacobian (#2). Similar relation $\partial \boldsymbol{x}/\partial u^1 = J\nabla u^2 \times \nabla u^3$ are also obtained. Including a similar relationship for u^2 and u^3 yields, the following relations and are called "dual relations." Let $(i, j, k) = (1, 2, 3), (2, 3, 1), (3, 1, 2)$ (#3),

$$\nabla u^i = \frac{1}{J}\left(\frac{\partial \boldsymbol{x}}{\partial u^j} \times \frac{\partial \boldsymbol{x}}{\partial u^k}\right) \tag{3.8}$$

$$\frac{\partial \boldsymbol{x}}{\partial u^i} = J\nabla u^j \times \nabla u^k \tag{3.9}$$

here,

$$J \equiv \frac{\partial \boldsymbol{x}}{\partial u^1} \cdot \left(\frac{\partial \boldsymbol{x}}{\partial u^2} \times \frac{\partial \boldsymbol{x}}{\partial u^3}\right) \quad \text{(Jacobian)} \tag{3.10}$$

Orthogonal and dual relations are fundamental to the geometry of general curvilinear coordinates. Other formulas can be obtained from them. Any vector field (*e.g.*, magnetic field) can be expanded using the gradient and tangent vectors in general curvilinear coordinate system by using orthogonal relation

$$\boldsymbol{B} = \sum_i B^i \frac{\partial \boldsymbol{x}}{\partial u^i} \quad \text{(Contravariant form)} \tag{3.11}$$

$$\boldsymbol{B} = \sum_i B_i \nabla u^i \quad \text{(Covariant form)} \tag{3.12}$$

$$B^i = \boldsymbol{B} \cdot \nabla u^i \tag{3.13}$$

$$B_i = \boldsymbol{B} \cdot \frac{\partial \boldsymbol{x}}{\partial u^i} \ . \tag{3.14}$$

Here, B^i is called contravariant component and B_i is called covariant component.

Consider the trajectory of the magnetic lines of force in a general curvilinear coordinate system. Let s be the position coordinate along the magnetic field lines, the magnetic field orbit is given by $\mathrm{d}\boldsymbol{x}/\mathrm{d}s = \boldsymbol{b}$ ($\boldsymbol{b} = \boldsymbol{B}/|\boldsymbol{B}|$). Since $\mathrm{d}\boldsymbol{x}/\mathrm{d}s = \sum(\partial \boldsymbol{x}/\partial u^j)\mathrm{d}u^j/\mathrm{d}s$, the inner product between $\mathrm{d}\boldsymbol{x}/\mathrm{d}s = \boldsymbol{b}$ and ∇u^i and the orthogonal relation leads to,

$$\mathrm{d}u^i/\mathrm{d}s = \boldsymbol{b} \cdot \nabla u^i \ . \tag{3.15}$$

Since $\{\nabla u^j\}$ is not an orthogonal system in the general curvilinear coordinate system, the inner product of a vector is expressed as $\boldsymbol{A} \cdot \boldsymbol{B} = \sum A_i B^i = \sum A^i B_i$ using the orthogonal relation. Applying vector rotations $\nabla \times \nabla u^i = 0$ and $\nabla B_i = \sum \partial B_i/\partial u^j \nabla u^j$ to Equation 3.12, the following relation for the rotation of a vector is obtained using the dual relation (Equation 3.9) (summation runs for (i, j, k): right-handed),

$$\nabla \times \boldsymbol{B} = \sum_{i=1,3} \sum_{j=1,3} \frac{\partial B_i}{\partial u^j} \nabla u^j \times \nabla u^i = J^{-1} \sum_{k=1,3} \left[\frac{\partial B_j}{\partial u^i} - \frac{\partial B_i}{\partial u^j} \right] \frac{\partial \boldsymbol{x}}{\partial u^k} \ . \tag{3.16}$$

Applying the dual relation (Equation 3.9) to the vector expansion (Equation 3.11), the divergence of the vector is obtained taking $\nabla \cdot (\nabla a \times \nabla b) = 0$ into account,

$$\nabla \cdot \boldsymbol{B} = \nabla \cdot \sum_i B^i \frac{\partial \boldsymbol{x}}{\partial u^i} = J^{-1} \sum_i \frac{\partial J B^i}{\partial u^i} \ . \tag{3.17}$$

Here, $\nabla \cdot \boldsymbol{B} = \nabla \cdot \sum B^i \partial \boldsymbol{x}/\partial u^i = \nabla \cdot \sum J B^i \nabla u^j \times \nabla u^k = \sum (\partial J B^i/\partial u^i)[\nabla u^i \cdot \nabla u^j \times \nabla u^k]$. The relation between covariant component B_i and contravariant component B^j, $B_i = \sum g_{ij} B^j$ is obtained by substituting Equation 3.11 into Equation 3.14, where $g_{ij} = (\partial \boldsymbol{x}/\partial u^i) \cdot (\partial \boldsymbol{x}/\partial u^j)$ (#4). Substitution of Equation 3.13 to Equation 3.12 gives $B^i = \sum g^{ij} B_j$, where $g^{ij} = \nabla u^i \cdot \nabla u^j$. Matrix $[g_{ij}]$ is the inverse matrix of $[g^{jk}]$ as seen from $B_i = \sum g_{ij} B^j = \sum g_{ij} g^{jk} B_k = \sum \delta_{ik} B_k$. The formulas for line, surface and volume integrals are given by,

$$\int \boldsymbol{B} \cdot \mathrm{d}\boldsymbol{x} = \int \boldsymbol{B} \cdot (\partial \boldsymbol{x}/\partial u^i)\mathrm{d}u^i \tag{3.18}$$

$$\int \boldsymbol{B} \cdot \mathrm{d}\boldsymbol{a} = \int \boldsymbol{B} \cdot \nabla u^k J \,\mathrm{d}u^i \mathrm{d}u^j \tag{3.19}$$

$$\int f \,\mathrm{d}V = \int fJ \,\mathrm{d}u^1 \mathrm{d}u^2 \mathrm{d}u^3 \,. \tag{3.20}$$

Here, $\mathrm{d}\boldsymbol{x} = (\partial \boldsymbol{x}/\partial u^i)\mathrm{d}u^i$ for line integral, $\mathrm{d}\boldsymbol{a} = (\partial \boldsymbol{x}/\partial u^i) \times (\partial \boldsymbol{x}/\partial u^j)\mathrm{d}u^i \mathrm{d}u^j = \nabla u^k J \mathrm{d}u^i \mathrm{d}u^j$ for surface integral, $\mathrm{d}V = J \mathrm{d}u^1 \mathrm{d}u^2 \mathrm{d}u^3$ for the volume integral are considered.

#1: Expand ∇u^1 as $\nabla u^1 = a_1(\partial \boldsymbol{x}/\partial u^2) \times (\partial \boldsymbol{x}/\partial u^3) + a_2(\partial \boldsymbol{x}/\partial u^3) \times (\partial \boldsymbol{x}/\partial u^1) + a_3(\partial \boldsymbol{x}/\partial u^1) \times (\partial \boldsymbol{x}/\partial u^2)$ and take the inner product with $\cdot\partial \boldsymbol{x}/\partial u^1, \cdot\partial \boldsymbol{x}/\partial u^2, \cdot\partial \boldsymbol{x}/\partial u^3$.
#2: Jacobian J is originally defined to measure the volume of the coordinate system. The volume between (u^1, u^2, u^3) and $(u^1 + \mathrm{d}u^1, u^2 + \mathrm{d}u^2, u^3 + \mathrm{d}u^3)$ is given by $\mathrm{d}V = [(\partial \boldsymbol{x}/\partial u^1)\mathrm{d}u^1] \times [(\partial \boldsymbol{x}/\partial u^2)\mathrm{d}u^2] \times [\partial \boldsymbol{x}/\partial u^3 \mathrm{d}u^3] = J \mathrm{d}u^1 \mathrm{d}u^2 \mathrm{d}u^3$ consistent with its definition.
#3: $(i, j, k) = (1, 2, 3), (2, 3, 1), (3, 1, 2)$ is called right-handed.
#4: Metric tensor is originally defined to measure the distance of two points in the space. Infinitesimally small distance between two points is given by $\mathrm{d}\boldsymbol{x} = \sum(\partial \boldsymbol{x}/\partial u^i)\mathrm{d}u^i$. So, $(\mathrm{d}\boldsymbol{x})^2 = \sum g_{ij} \mathrm{d}u^i \mathrm{d}u^j$ consistent with the original definition of the metric g_{ij}.

3.4 Field Line Dynamics: Hamilton Dynamics of the Magnetic Field

A magnetic field is a vector field without source and sink, and therefore is incompressible as a flow field ($\nabla \cdot \boldsymbol{B} = 0$). For the incompressible flow, the volume of the fluid element is conserved along with the flow. The dynamic system is similar to incompressible flow in that the phase space flow is incompressible. From this similarity, the theory of magnetic field lines flow can be constructed using the Hamilton form in analytical mechanics (Note 1) [6, 15].

Let ζ be the toroidal angle of the torus and θ the poloidal angle (choice of ζ and θ is arbitrary). In general, the magnetic vector potential $\boldsymbol{A}(\nabla \times \boldsymbol{A} = \boldsymbol{B})$ is given by $\boldsymbol{A} = \phi\nabla\theta - \psi\nabla\zeta + \nabla G$ (G is the gauge transformation part) (#1), then, the magnetic field $\boldsymbol{B}$ can be expressed by following symplectic form.

$$\boldsymbol{B} = \nabla\phi \times \nabla\theta - \nabla\psi \times \nabla\zeta \,. \tag{3.21}$$

It is easy to show this expression satisfies $\nabla \cdot \boldsymbol{B} = 0$. Let us choose our coordinates (ϕ, θ, ζ) and find the orbit of the magnetic field along the toroidal angle ζ. Using Equations 3.15 and 3.21, the following are obtained:

$$\begin{aligned} \frac{\mathrm{d}\theta}{\mathrm{d}\zeta} &= \frac{\boldsymbol{B} \cdot \nabla\theta}{\boldsymbol{B} \cdot \nabla\zeta} = \frac{\partial\psi}{\partial\phi} \,, \\ \frac{\mathrm{d}\phi}{\mathrm{d}\zeta} &= \frac{\boldsymbol{B} \cdot \nabla\phi}{\boldsymbol{B} \cdot \nabla\zeta} = -\frac{\partial\psi}{\partial\theta} \,. \end{aligned} \tag{3.22}$$

This seems to be a Hamilton equation in the dynamical system if we regard ψ as the Hamiltonian, θ as the canonical coordinate, ϕ as the canonical angular momentum, and ζ as time. Thus, the magnetic field line has the same mathematical structure as the Hamilton system. This property is derived from the incompressibility of the magnetic field. In canonical Equation 3.22, ψ is, in general, not only a function of ϕ (*i.e.*, $\psi = \psi(\phi, \theta, \zeta)$), so the magnetic field lines are not necessarily integrable and its structure can be complex. Integrability in a dynamical system has a close relation with the existence of magnetic surfaces in magnetic confinement and chaos is closely related to the disruption of plasma.

In analytical mechanics, the variational principle is formulated using Hamilton's action integral $S = \int [\boldsymbol{p} \cdot \mathrm{d}\boldsymbol{x}/\mathrm{d}t - H]\mathrm{d}t$ leading to the Hamilton equation (Note 2). If we use relations $\boldsymbol{p} \to \phi$, $\mathrm{d}\boldsymbol{x}/\mathrm{d}t \to \mathrm{d}\theta/\mathrm{d}\zeta$, $H \to \psi$ introduced for Equation 3.22, we reach $S = \int [\phi \mathrm{d}\theta/\mathrm{d}\zeta - \psi]\mathrm{d}\zeta = \int [\phi\nabla\theta - \psi\nabla\zeta] \cdot \mathrm{d}\boldsymbol{x} = \int \boldsymbol{A} \cdot \mathrm{d}\boldsymbol{x}$ (gauge part of vector potential ∇G does not contribute to the integral since it becomes the difference in boundary values after integration, that is zero). Thus, the variational principle to give the magnetic field line orbit is,

$$\delta S = \delta \int \boldsymbol{A} \cdot \mathrm{d}\boldsymbol{x} = 0\,. \tag{3.23}$$

Actually, since

$$\delta S(\theta, \phi) = \int \left[\left(\frac{\mathrm{d}\theta}{\mathrm{d}\zeta} - \frac{\partial \psi}{\partial \phi} \right) \delta\phi - \left(\frac{\mathrm{d}\phi}{\mathrm{d}\zeta} + \frac{\partial \psi}{\partial \theta} \right) \delta\theta + \frac{\mathrm{d}(\phi\delta\theta)}{\mathrm{d}\zeta} \right] \mathrm{d}\zeta\,. \tag{3.24}$$

$\delta \int \boldsymbol{A} \cdot \mathrm{d}\boldsymbol{x} = 0$ gives Equation 3.22 (total derivative, 3rd term of right-hand side is zero after integration since boundary value is fixed in the variational principle). This coordinate system (ϕ, θ, ζ) is termed the "magnetic coordinates."

#1: Any vector $\boldsymbol{A}$ can be expressed as $\boldsymbol{A} = A_u \nabla u + A_\theta \nabla\theta + A_\zeta \nabla\zeta$ in the general curvilinear coordinates (u, θ, ζ). If we define a scalar G by $G = \int A_u \,\mathrm{d}u$ ($\partial G/\partial u = A_u$) and consider $\nabla G = \partial G/\partial u \nabla u + \partial G/\partial\theta \nabla\theta + \partial G/\partial\zeta \nabla\zeta$, $\boldsymbol{A}$ can be expressed as $\boldsymbol{A} = \nabla G + (A_\theta - \partial G/\partial\theta)\nabla\theta + (A_\zeta - \partial G/\partial\zeta)\nabla\zeta$. If we define $\phi = A_\theta - \partial G/\partial\theta$ and $\psi = -A_\zeta + \partial G/\partial\zeta$, we reach general expression for the vector potential $\boldsymbol{A} = \phi\nabla\theta - \psi\nabla\zeta + \nabla G$.

Note 1: Hamilton Equations in Dynamical Systems [15]

British physicist Isaac Newton (1642–1727) showed in *Principia* (1687, 1723) that the motion of the object can be described by Newton's equations of motion, $\mathrm{d}p_i/\mathrm{d}t = \partial V/\partial x_i, \mathrm{d}x_i/\mathrm{d}t = p_i/m$. Then about 100 years later, another British physicist, W. Hamilton (1805–1865) in 1835 derived the following equation from Newton's equation, now known as the Hamilton equation.

$$\begin{aligned} \mathrm{d}x_i/\mathrm{d}t &= \partial H/\partial p_i \\ \mathrm{d}p_i/\mathrm{d}t &= -\partial H/\partial x_i \end{aligned} \tag{3.25}$$

Here, H is the Hamiltonian, the sum of the kinetic energy T and the potential energy V ($H = T + V$). p_i and x_i are called the canonical momentum and canonical coordinate, respectively.

Note 2: Variational Principle in Hamilton Form (see Section 4.1)

Lagrangian function L is defined by $L = T - V$. Define generalized momentum p_i by $p_i \equiv \partial L/\partial \dot{q}_i$ and the Hamiltonian by $H(\boldsymbol{q}, \boldsymbol{p}, t) = \sum p_i \dot{q}_i - L(\boldsymbol{q}, \dot{\boldsymbol{q}}, t)$, the variational principle is expanded from position space ($\boldsymbol{q}$) to phase space ($\boldsymbol{q}, \boldsymbol{p}$), where the action integral S is defined under independent variables q_i and p_i,

$$S(\boldsymbol{q}, \boldsymbol{p}) = \int_{t_1}^{t_2} \left[\sum p_i \dot{q}_i - H(\boldsymbol{q}, \boldsymbol{p}, t) \right] \mathrm{d}t \,. \tag{3.26}$$

Taking its variation leads to $\delta S = \int \sum [\delta p_i \{\mathrm{d}q_i/\mathrm{d}t - \partial H/\partial p_i\} - \delta q_i \{\mathrm{d}p_i/\mathrm{d}t + \partial H/\partial q_i\}]\mathrm{d}t$. Since variation δp_i and δq_i are independent, the following Hamiltonian equation is obtained.

$$\frac{\mathrm{d}q_j}{\mathrm{d}t} = \frac{\partial H}{\partial p_j}, \quad \frac{\mathrm{d}p_j}{\mathrm{d}t} = -\frac{\partial H}{\partial q_j}. \tag{3.27}$$

One should be aware that there is other variational principle, where only position coordinate $\boldsymbol{q}$ is an independent variable.

3.5 Magnetic Surface: Integrable Magnetic Field and Hidden Symmetry

In plasma force equilibrium, the plasma's expansion force (∇P) is balanced with the Lorentz force ($\boldsymbol{J} \times \boldsymbol{B}$). Here, $\boldsymbol{J}$ is the current flowing in the plasma, $\boldsymbol{B}$ is the magnetic field, P is the pressure of the plasma. This is the basic principle of the magnetic confinement fusion.

$$\boldsymbol{J} \times \boldsymbol{B} = \nabla P \,. \tag{3.28}$$

From this equation, we obtain,

$$\boldsymbol{B} \cdot \nabla P = 0 \,, \tag{3.29}$$

$$\boldsymbol{J} \cdot \nabla P = 0 \,. \tag{3.30}$$

In other words, the magnetic field lies on a constant pressure surface (P = constant) in force equilibrium. This constant pressure surface is called the "magnetic surface." Similarly, current field also lies on a constant pressure surface. The magnetic surface is a surface formed by independent vectors $\boldsymbol{B}$ and $\boldsymbol{J}$. It is a special state that the field line orbit always lies on a surface. We choose coordinates (u^1, u^2, u^3) such that $u^1 = u$ is the label of magnetic surface, and u^2 ($= \theta$) and u^3 ($= \zeta$) are arbitrary poloidal and toroidal angles, respectively. In the (u, θ, ζ) coordinates, the magnetic field is expressed by the linear combination of tangent vectors, $\partial \boldsymbol{x}/\partial \theta$ and $\partial \boldsymbol{x}/\partial \zeta$. Using the dual relation ($\partial \boldsymbol{x}/\partial u^i = J \nabla u^j \times \nabla u^k$),

$$\boldsymbol{B} = b_2 \nabla \zeta \times \nabla u + b_3 \nabla u \times \nabla \theta \,. \tag{3.31}$$

Substituting this into $\nabla \cdot \boldsymbol{B} = 0$ and using vector formula $\nabla \cdot (\nabla a \times \nabla b) = 0$, we obtain $\partial b_2/\partial\theta + \partial b_3/\partial\zeta = 0$, which leads to the existence of a "stream function" h for the flow $\boldsymbol{B}$ on the magnetic surface.

$$b_2 = -\frac{\partial h}{\partial \zeta}, \quad b_3 = \frac{\partial h}{\partial \theta}. \tag{3.32}$$

On the other hand, b_2 and b_3 are periodic function of (θ, ζ), so the flow function h is given by

$$h(u, \theta, \zeta) = h_2(u)\theta + h_3(u)\zeta + \tilde{h}(u, \theta, \zeta). \tag{3.33}$$

Here, $\tilde{h}(u, \theta, \zeta)$ is a periodic function of θ and ζ. Define $\lambda = \tilde{h}(u, \theta, \zeta)/h_2(u)$ and introduce new variable $\theta_{\mathrm{m}} = \theta + \lambda$, we obtain $h(u, \theta_{\mathrm{m}}, \zeta) = h_2(u)\theta_{\mathrm{m}} + h_3(u)\zeta$. Flow coefficients $h_2(u)$ and $h_3(u)$ are related to the toroidal magnetic flux $2\pi\phi(u) = \int \boldsymbol{B} \cdot \mathrm{d}\boldsymbol{a}_\zeta$, and the poloidal magnetic flux $2\pi\psi(u) = -\int \boldsymbol{B} \cdot \mathrm{d}\boldsymbol{a}_\theta$ as follows:

$$\frac{\mathrm{d}\psi}{\mathrm{d}u} = -h_3(u), \quad \frac{\mathrm{d}\phi}{\mathrm{d}u} = h_2(u). \tag{3.34}$$

Then,

$$\boldsymbol{B} = \nabla\phi \times \nabla\theta_{\mathrm{m}} - \nabla\psi \times \nabla\zeta = \nabla\phi \times \nabla(\theta_{\mathrm{m}} - \zeta/q). \tag{3.35}$$

Here, $q = \mathrm{d}\phi(u)/\mathrm{d}\psi(u)$ is called a safety factor. The first expression of Equation 3.35 coincides with Equation 3.21, but there is an essential difference in that ϕ and ψ are functions only of the magnetic surface label u. $\alpha = \theta_{\mathrm{m}} - \zeta/q$ is called the "surface potential." The expression $\boldsymbol{B} = \nabla\phi \times \nabla\alpha$ is called "Clebsch form." The coordinate u is equivalent to the toroidal magnetic flux ϕ and $(u, \theta_{\mathrm{m}}, \zeta)$ is called the flux coordinates. In the flux coordinates $(u, \theta_{\mathrm{m}}, \zeta)$, the magnetic field lines on the magnetic surface become a straight line in the $(\theta_{\mathrm{m}}\ \zeta)$ coordinates, whose gradient is given by

$$\frac{\mathrm{d}\theta_{\mathrm{m}}}{\mathrm{d}\zeta} = \frac{\boldsymbol{B} \cdot \nabla\theta_{\mathrm{m}}}{\boldsymbol{B} \cdot \nabla\zeta} = \frac{1}{q(\psi)}. \tag{3.36}$$

This gradient can be regarded as the "oscillation frequency" of angle variable θ_{m} when we regard ζ as the "time" variable. In fact, the magnetic field in the force equilibrium is given by Equation 3.35 and the vector potential $\boldsymbol{A}$ is given by $\boldsymbol{A} = \phi\nabla\theta_{\mathrm{m}} - \psi\nabla\zeta$. The action integral S to give the magnetic field line trajectory is given by

$$S = \int \boldsymbol{A} \cdot \mathrm{d}\boldsymbol{x} = \int [\phi\, \mathrm{d}\theta_{\mathrm{m}} - \psi\, \mathrm{d}\zeta]. \tag{3.37}$$

The integrant of this action integral has a form of action-angle variables in classical mechanics $(J\mathrm{d}q - H(J)\mathrm{d}t)$ where ϕ and θ_{m} play the roles of "action" and "angle," respectively. Similar to previous section, ψ and ζ play the roles of the

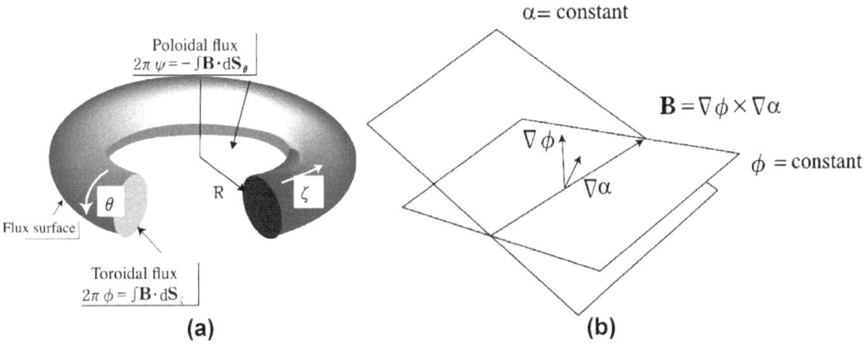

Figure 3.10 (a) Definition of the magnetic surface and fluxes of toroidal plasma and (b) the geometric meaning of the Clebsch expression of the magnetic field

Hamiltonian and time, respectively. In classical mechanics as the time t advances, $\mathrm{d}H/\mathrm{d}J$ gives the "oscillation frequency of the motion" if the system have periodic motion in θ direction. In magnetic field line dynamics, the oscillation frequency of the motion is $\mathrm{d}\psi/\mathrm{d}\phi = 1/q$ [16]. The system Lagrangian L is given in the flux coordinates $(\phi, \theta_\mathrm{m}, \zeta)$ from Equation 3.37 by

$$L = \phi \frac{\mathrm{d}\theta_\mathrm{m}}{\mathrm{d}\zeta} - \psi(\phi)\,. \tag{3.38}$$

The coordinate θ_m becomes a cyclic coordinate. The Hamiltonian ψ and canonical momentum ϕ conjugate to θ_m, and the surface function α is conserved along the magnetic lines of force (independent of "time").

$$\frac{\mathrm{d}\psi}{\mathrm{d}\zeta} = \frac{\boldsymbol{B}\cdot\nabla\psi}{\boldsymbol{B}\cdot\nabla\zeta} = 0 \tag{3.39}$$

$$\frac{\mathrm{d}\phi}{\mathrm{d}\zeta} = \frac{\boldsymbol{B}\cdot\nabla\phi}{\boldsymbol{B}\cdot\nabla\zeta} = 0 \tag{3.40}$$

$$\frac{\mathrm{d}\alpha}{\mathrm{d}\zeta} = \frac{\boldsymbol{B}\cdot\nabla\alpha}{\boldsymbol{B}\cdot\nabla\zeta} = 0\,. \tag{3.41}$$

Geometrically, the magnetic field $\boldsymbol{B}$ is perpendicular to gradients of both ϕ and α ($\boldsymbol{B}\cdot\nabla\phi = 0$, $\boldsymbol{B}\cdot\nabla\alpha = 0$) as seen from Figure 3.10 (b). The magnetic field line trajectory lies on constant ϕ surface. The conserved quantity along the trajectory is called the "first integral" in the dynamical system, and we call such a case integrable if the first integral exists.

In the derivation of the flux coordinate system, no geometrical symmetry is assumed for the torus plasma. But if we assume the existence of force equilibrium, double periodicity of the torus leads to "hidden symmetry" and hence becomes integrable. In analytical mechanics and the gauge field theory of elementary particles, classic methodology to find the conservation law from a cyclic coordinate is extended to "Noether's theorem" (see Section 4.1) which is independent of the choice of coordinates [17].

Salon: Hidden Symmetry in Algebraic Equation [18, 19]

It is well known that an nth ($n \geq 5$) order algebraic equation does not have a general solution as proved by Norwegian mathematician N. H. Abel (1802–1829). If there is a solution through root of power and arithmetic operations, there should be symmetry against the exchange of solutions as investigated by French mathematician J. L. Lagrange (1736–1813). This hidden symmetry in the algebraic equation led to the foundation of group theory by French mathematician Evariste Galois (1811–1832). By using group theory, he identified the solvable condition of 5th order algebraic equations.

3.6 Flux Coordinates: Hamada and Boozer Coordinates

The flux coordinates in Section 3.5 impose only one constraint $\theta_{\rm m} = \theta + \lambda$ to the two arbitrary angle variables θ and ζ. So, there is one more arbitrary factor to add to the angle variables. In fact, Equation 3.35 is invariant under the coordinate transformation $\theta_{\rm m1} = \theta_{\rm m} + \eta(\phi, \theta_{\rm m}, \zeta)$ and $\zeta_1 = \zeta + q(\phi)\eta(\phi, \theta_{\rm m}, \zeta)$ for arbitrary function $\eta(\phi, \theta_{\rm m}, \zeta)$. Thus freedom remains for the combination $(\theta_{\rm m}, \zeta)$. Using this arbitrariness, Hamada and Boozer coordinates are defined in this section.

Discussion of the flow functions in Section 3.5 can be extended to the current density vector as well as the magnetic field. Using flux coordinates (u, θ, ζ) with u as a label of the magnetic surface, $\boldsymbol{B}$ and $\boldsymbol{J}$ tangent to the magnetic surface can be expressed by the tangent vectors on u plane, $\partial \boldsymbol{x}/\partial\theta$ and $\partial \boldsymbol{x}/\partial\zeta$. Using the dual relation ($\partial \boldsymbol{x}/\partial u^i = J\nabla u^j \times \nabla u^k$), $\boldsymbol{B}$ and $\boldsymbol{J}$ are given by

$$\boldsymbol{a} = a_2 \nabla\zeta \times \nabla u + a_3 \nabla u \times \nabla\theta \quad (\boldsymbol{a} = \boldsymbol{B}, \boldsymbol{J})\,. \tag{3.42}$$

For the equilibrium state, the current density $\boldsymbol{J}$ is incompressible as well as the magnetic field, $\nabla \cdot \boldsymbol{a} = 0$ ($\boldsymbol{a} = \boldsymbol{B}, \boldsymbol{J}$). From the vector formula $\nabla \cdot (\phi \boldsymbol{a}) = \phi \nabla \cdot \boldsymbol{a} + \boldsymbol{a} \cdot \nabla\phi$ and $\nabla \cdot (\nabla F \times \nabla G) = 0$, we obtain $\partial a_2/\partial\theta + \partial a_3/\partial\zeta = 0$. So flow field $\boldsymbol{a}$ has the stream function h on a magnetic surface,

$$a_2 = \frac{\partial h}{\partial \zeta}\,, \quad a_3 = \frac{\partial h}{\partial \theta}\,. \tag{3.43}$$

Since a_2 and a_3 are periodic functions of (θ, ζ), stream functions for the magnetic field ($h = b$) and current density ($h = j$) are given by

$$\begin{aligned} b(u, \theta, \zeta) &= b_2(u)\theta + b_3(u)\zeta + \tilde{b}(u, \theta, \zeta) \\ j(u, \theta, \zeta) &= j_2(u)\theta + j_3(u)\zeta + \tilde{j}(u, \theta, \zeta)\,. \end{aligned} \tag{3.44}$$

Here $\tilde{b}(u, \theta, \phi)$, $\tilde{j}(u, \theta, \phi)$ are periodic functions of θ and ζ. Coordinate transformations to remove them are given by $\theta_h = \theta + \theta_1$ and $\zeta_h = \zeta + \zeta_1$ where θ_1 and ζ_1

are given by

$$\theta_1 = \frac{\tilde{b} j_3 - \tilde{j} b_3}{b_2 j_3 - b_3 j_2}, \quad \zeta_1 = \frac{-\tilde{b} j_2 + \tilde{j} b_2}{b_2 j_3 - b_3 j_2}. \tag{3.45}$$

The flux coordinates thus obtained (u, θ_h, ζ_h) are called (in a broad sense) Hamada coordinates. Coefficients of the stream functions of the magnetic field and current density, $b_2(u)$, $b_3(u)$, $j_2(u)$, $j_3(u)$ are related to the toroidal magnetic flux within the magnetic surface $2\pi\phi(u) = \int \boldsymbol{B} \cdot \mathrm{d}\boldsymbol{a}_\zeta$, poloidal flux $2\pi\psi(u) = -\int \boldsymbol{B} \cdot \mathrm{d}\boldsymbol{a}_\theta$, the toroidal current flux $2\pi f(u) = \int \boldsymbol{J} \cdot \mathrm{d}\boldsymbol{a}_\zeta$, and the poloidal current flux $2\pi g(u) = \int \boldsymbol{J} \cdot \mathrm{d}\boldsymbol{a}_\theta$ by the relationships $\psi'(u) = -b_3(u)$, $\phi'(u) = b_2(u)$, $g'(u) = j_3(u)$, $f'(u) = j_2(u)$ as follows,

$$\boldsymbol{B} = \nabla\phi \times \nabla(\theta_h - \zeta_h/q), \tag{3.46}$$

$$\boldsymbol{J} = \nabla f \times \nabla(\theta_h - \zeta_h/q_J). \tag{3.47}$$

Here, $q = \mathrm{d}\phi/\mathrm{d}\psi(u)$ and $q_J = -\mathrm{d}f(u)/\mathrm{d}g(u)$. $\boldsymbol{B}$ and $\boldsymbol{J}$ have to be linearly independent to enable coordinate transformation to the Hamada coordinates, $b_2 j_3 - b_3 j_2 \neq 0$ $(q \neq q_J)$. Defining surface functions by $\alpha = \theta_h - \zeta_h/q$ and $\alpha_J = \theta_h - \zeta_h/q_J$, $\boldsymbol{B}$ and $\boldsymbol{J}$ can be expressed by $\boldsymbol{B} = \nabla\phi \times \nabla\alpha$ and $\boldsymbol{J} = \nabla f \times \nabla\alpha_J$, which leads to $\boldsymbol{B} \cdot \nabla\alpha = 0$ and $\boldsymbol{J} \cdot \nabla\alpha_J = 0$. Both magnetic field and current density are given by straight lines in Hamada coordinates. This Hamada coordinates were derived by a Japanese physicist, Shigeo Hamada, (1931–2001; Figure 3.11 (a)) in 1962 [20]. Hamada called it the natural coordinate system and they are now called "Hamada coordinates." Consider the expression of magnetic field and current density in the flux coordinates (ϕ, θ_m, ζ). Using Equation 3.44, $\boldsymbol{B}$ and $\boldsymbol{J}$ are given by,

$$\boldsymbol{B} = \nabla\phi \times \nabla\theta_\mathrm{m} + q^{-1}\nabla\zeta \times \nabla\phi \tag{3.48}$$

$$\mu_0 \boldsymbol{J} = -\frac{\partial h}{\partial \zeta}\nabla\zeta \times \nabla\phi + \frac{\partial h}{\partial \theta_\mathrm{m}}\nabla\phi \times \nabla\theta_\mathrm{m} \tag{3.49}$$

$$h = f'(\phi)\theta_\mathrm{m} + g'(\phi)\zeta + \nu(\phi, \theta_\mathrm{m}, \zeta). \tag{3.50}$$

Here, we replace the notation to $j_2 = f'(\phi)$, $j_3 = g'(\phi)$ and $\tilde{j} = \nu$ to match notation of Boozer [13]. Substituting Equations 3.48, 3.49 and $\nabla P = \mathrm{d}P/\mathrm{d}\phi\, \nabla\phi$ into Equation 3.28, we obtain a relation of stream function $\partial h/\partial\zeta + q^{-1}\, \partial h/\partial\theta_m = -\mu_0 J \mathrm{d}P/\mathrm{d}\phi$. Taking flux surface average $(2\pi)^{-2} \int \mathrm{d}\zeta \mathrm{d}\theta_m$ and using the relation $\mathrm{d}V/\mathrm{d}\phi = \int J \mathrm{d}\theta_m \mathrm{d}\zeta$ (volume enclosed by ϕ is given by $V = \int J\, \mathrm{d}\phi \mathrm{d}\theta_m \mathrm{d}\zeta$), we obtain $g'(\phi) + f'(\phi)/q = \mu_0 V'(\phi) P'(\phi)$. Taking difference of two equations, we obtain

$$\left(\frac{\partial}{\partial\zeta} + \frac{1}{q}\frac{\partial}{\partial\theta_m}\right)\nu = \mu_0 \left(\frac{\mathrm{d}V}{\mathrm{d}\phi} - (2\pi)^2 J\right)\frac{\mathrm{d}P}{\mathrm{d}\phi}. \tag{3.51}$$

Case for $\nu = 0$ corresponds to Hamada coordinates. Therefore, Jacobian of Hamada coordinates is given by $J = (2\pi)^{-2}\mathrm{d}V/\mathrm{d}\phi$ and is a flux function. If we change coordinate ϕ to $v = V/4\pi^2$, Jacobian of Hamada coordinates is given by $J = 1$.

From Equation 3.49, $\boldsymbol{B}$ satisfying $\nabla \times \boldsymbol{B} = \mu_0 \boldsymbol{J}$ is given by,

$$\boldsymbol{B} = g(\phi)\nabla\zeta + f(\phi)\nabla\theta_\mathrm{m} - \nu(\phi, \theta_\mathrm{m}, \zeta)\nabla\phi + \nabla F(\phi, \theta_\mathrm{m}, \zeta). \tag{3.52}$$

Figure 3.11 (a) Shigeo Hamada who invented Hamada coordinates (with kind permision of Nihon University) and (b) Alan Boozer who invented Boozer coordinates (with kind permission of Prof. Boozer)

Here, F is the magnetic scalar potential satisfying $\nabla \times \boldsymbol{B} = 0$. The case where $\nu = 0$ in Equation 3.51 corresponds to the Hamada coordinates. US physicist A. Boozer (Figure 3.11 (b)) found another set of flux coordinates in 1981 [21]. In the Boozer coordinates, the gauge term ∇F in Equation 3.51 is eliminated by the transformation. Here, we show that such a coordinate transformation exists by using the remaining freedom of coordinate transformation. Boozer coordinates are given by the coordinate transformation $(\theta_b, \zeta_b) = (\theta_\mathrm{m} + \eta, \zeta + q(\phi)\eta)$ and Equation 3.52 reads,

$$\boldsymbol{B} = g(\phi)\nabla\zeta_b + f(\phi)\nabla\theta_b + \beta_*\nabla\phi, \quad \boldsymbol{B} = \nabla\phi \times \nabla\alpha\ , \tag{3.53}$$

$$\eta(\phi, \theta_\mathrm{m}, \zeta) = \frac{F(\phi, \theta_\mathrm{m}, \zeta)}{g(\phi)q(\phi) + f(\phi)}\ , \quad \alpha = \theta_b - \zeta_b/q\ , \tag{3.54}$$

$$\beta_* = \eta(\phi, \theta_\mathrm{m}, \zeta)(q(\phi)g'(\phi) + f'(\phi)) - \nu(\phi, \theta_\mathrm{m}, \zeta)\ . \tag{3.55}$$

Two-way expressions of $\boldsymbol{B}$ in Equation 3.53 is especially useful to simplify the particle orbit equation (see Chapter 4). The magnetic field in Boozer coordinates $(\phi, \theta_b, \zeta_b)$ can be transformed into the following form,

$$\boldsymbol{B} = \nabla\chi + \beta\nabla\phi\ , \qquad \boldsymbol{B} = \nabla\phi \times \nabla\alpha\ , \tag{3.56}$$

$$\chi = g(\phi)\zeta_b + f(\phi)\theta_b\ , \quad \beta = \beta_* - g'(\phi)\zeta_b - f'(\phi)\theta_b\ . \tag{3.57}$$

Corresponding to this form, (ϕ, α, χ) are called Boozer–Grad coordinates and are one of the variants of Clebsch coordinates (coordinates using the two Euler potentials ϕ and α are called Clebsch coordinates). It is important to note that $\boldsymbol{B}$ is expressed in two ways (covariant form and Clebsch form) in two Boozer coordinates.

3.7 Ergodicity: A Field Line Densely Covers the Torus

Figure 3.12 shows the magnetic surface of the torus plasma in the cylindrical coordinate system (R, ζ, Z). We use the flux coordinate system (ϕ, θ_m, ζ) in which $\boldsymbol{B}$ is expressed by $\boldsymbol{B} = \nabla\phi \times \nabla(\theta_m - \zeta/q)$ and its trajectory along toroidal direction $\mathrm{d}\theta_m/\mathrm{d}\zeta = 1/q(\phi)$ is a straight line with a gradient $1/q$. We choose the toroidal angle of cylindrical coordinates for ζ of flux coordinates.

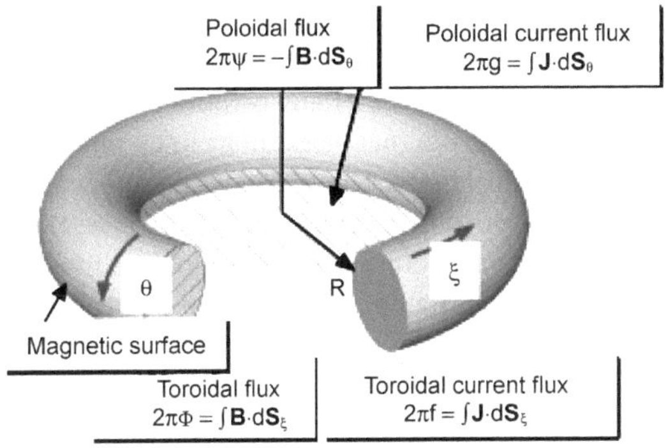

Figure 3.12 The definition of magnetic field and current density fluxes in a torus

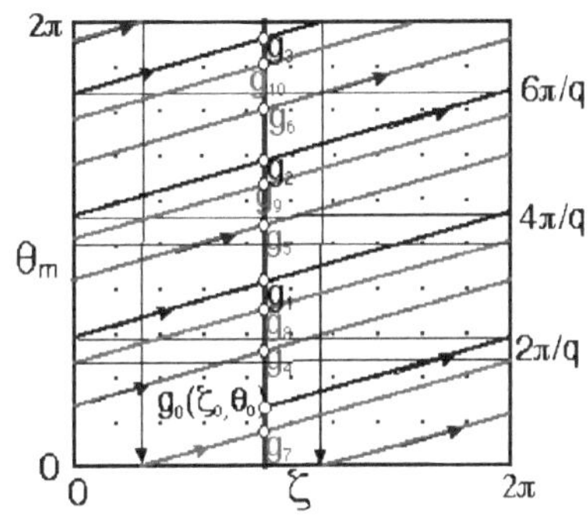

Figure 3.13 Locus of the magnetic field in the flux coordinates (ϕ, θ_m, ζ) and the mapping series of $g_0(\zeta_0, \theta_0), g_1, g_2, \ldots$

Consider the magnetic field starting from the point on the magnetic surface $(\zeta, \theta) = (\zeta_0, \theta_0)$, poloidal rotation angle per one toroidal rotation is given by $\Delta\theta = 2\pi/q$ and field line returns to the point $\theta = 2\pi/q + \theta_0$ at $\zeta = \zeta_0$. If $\theta \geq 2\pi$, 2π is subtracted so that θ is within $[0, 2\pi)$. Repeating this procedure, the sequence of points $g\theta_0, g^2\theta_0, g^3\theta_0, g^4\theta_0, \ldots, g^j\theta_0$ are drawn on the $\zeta = \zeta_0$ plane. The poloidal angle of the sequence of points $\{g^j\theta_0\}$ is given by $\theta^j = 2\pi j/q + \theta_0$. This mapping to some plane ($\zeta = \zeta_0$ in this case) is called "Poincaré mapping." Let real semi open set $\Theta = [0, 2\pi)$, this mapping is the mapping from Θ to Θ itself ($g : \Theta \to \Theta$, $g\theta_0 = \theta_0 + 2\pi/q$ for $\theta_0 \in \Theta$). Now, when q is a rational number, *i.e.*, integers m and n exist to satisfy $q = m/n$, rotation by mapping g^m is given by $\theta^m = 2\pi m/q + \theta_0 = 2n\pi + \theta_0$ and returns to the original position (ζ_0, θ_0) ("identity mapping"). See Figure 3.13.

However, if q is an irrational number, it can be proved by using "reduction to absurdity (reductio ad absurdum)," originating from Aristotle (384–322 BC), that the magnetic field line will not return to the original point after any toroidal circulations. In fact, if we assume that it returns to the original position, it contradicts the assumption of an irrational number, and it goes around the torus infinitely. Then, it can be shown by Poincaré mapping that the sequence of points $\{g^j\theta_0\}$ will densely cover the poloidal circumference. The magnetic field line is a 1-dimensional line. The line is a 1-dimensional set and the width of the line is zero according to Euclid's definition. If we place two lines side by side, there will still be a gap between them and we cannot form a continuous surface. However, we can reduce the gap infinitely. Then, we can form a magnetic field line originating from (ζ_0, θ_0) passes at any closest distance of any position of the torus.

Considering the magnetic field lines on the torus surface, the magnetic field is said to "densely cover the torus" if there always exists a field line in any neighborhood (closer distance if we set arbitrary $\varepsilon > 0$) of any point on the torus. In general, the set A (set of points of a magnetic field line, in this case) is said to densely cover the set B (torus magnetic surface, in this case) if there exists a point of the set A at any neighborhood of any point in set B.

The fact that the magnetic field densely covers the torus when q is irrational can be proved by using reduction to absurdity [5]. Consider an arbitrary poloidal angle θ_0 at $\zeta = \zeta_0$ and its neighborhood U. Since mapping points $\{g^i\theta_0\}$ continue indefinitely (do not return to any previous point), the mapping series $\{g^iU\}$ also continues indefinitely. If there is no intersection among mapping series, the poloidal length of the mapping series becomes infinity and contradicts a finite poloidal circumference. Therefore, this mapping series of the neighborhood should have a common set. This means that there are integers $k \geq 1$ and $l \geq 1$ $(k > l)$ such that $g^kU \cap g^lU \neq \emptyset$, then $g^{k-l}U \cap U \neq \emptyset$, and $\theta = g^{k-l}\theta_0$ should be in the neighborhood of θ_0. Since the choice of ζ_0 is arbitrary, there is a field line point $(\zeta_0, g^{k-l}\theta_0)$ in the neighborhood of arbitrary point (ζ_0, θ_0). "Densely covered" is also termed "ergodically covered." This stems from the "ergodic hypothesis" in the phase space to derive the "principle of equal weight" by L. Boltzmann.

In the force equilibrium of the plasma, the safety factor q continuously changes with different magnetic surfaces, and the range of q is a real closed interval. In this real closed interval, the number of irrational number is uncountable infinity. On the other hand, the number of the rational number is countable infinity, and the so-called "measure" is zero (see the salon).

Salon: Wonder of Infinity [22]

The German mathematician Georg Cantor (1845–1918; Figure 3.14), a famous founder of set theory, investigated the "number" or "number line" which relates the point in a line to a number. For example, the number of natural numbers is infinity, but they can be counted as $1, 2, 3, \ldots$ and are said to be countable infinity ("denumerable"). Counting infinity is different from counting the finite number. In 1874, Cantor showed that a set of rational numbers has the same number as a set of natural numbers using a "diagonal argument" (the "same number," to be precise, means there is one-to-one mapping between a set of natural numbers and a set of rational numbers). Natural numbers are discrete and rational numbers are dense on the number line. Rational numbers always exist in any neighborhood of any point in the number line (a set of real numbers), by which the set of rational numbers is said to be dense everywhere in the set of real numbers. It is the nature of infinity that such different sets of rational and natural numbers have "equal numbers."

In 1874, Cantor also showed that real numbers are uncountable using "reduction to absurdity" In fact, assuming real numbers of set [0, 1] are countable, he expressed a real number by an infinite decimal, arranged in a series in a vertical column with numbering $(1, 2, 3, \ldots, n)$ from the top. Then, he picked up each digit at n decimal places to form a new number (the diagonal number) and added 1 to each digit of the new number. It is easy to prove that this number is not included in the series since it differs at least at the n decimal place, which is a contradiction (absurdity). The "infinity" of real numbers is in a higher order than the "infinity" of rational and natural numbers [23].

Real number R has a one-to-one correspondence with a number line, and a non-negative real number, "length," can be defined. R^2 and R^3 have one-to-one correspondences with plane and space, respectively. And the corresponding nonnegative real numbers "area" and "volume" can be defined. A set of rational numbers is denumerable and its "length" is zero even if it densely covers the number line. Such length, area, and volume are generalized to a concept of "measure" [23]. This is a nonnegative real number set to meet the complete additivity.

Figure 3.14 G. Cantor is a founder of set theory who investigated the nature of "infinity" in depth

3.8 Apparent Symmetry: Force Equilibrium of Axisymmetric Torus

Let us consider the axisymmetric torus, which is a major object of present fusion research. In the cylindrical coordinate system, (R, ζ, Z), the ζ is a cyclic coordinate and $\partial/\partial\zeta = 0$ (Figure 3.15). If we define a flux function by using the ζ component of the vector potential A as $\psi = RA_\zeta(R, Z)$, B_R and B_Z in the poloidal cross section are given as follows:

$$\begin{aligned} RB_R &= -\frac{\partial\psi}{\partial Z}, \\ RB_Z &= \frac{\partial\psi}{\partial R}. \end{aligned} \tag{3.58}$$

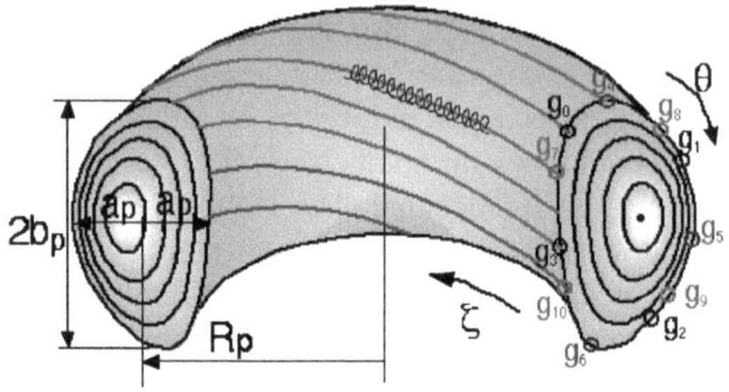

Figure 3.15 Locus of the magnetic field in cylindrical coordinates (R, ζ, Z)

Equation 3.58 satisfies the basic nature of $\boldsymbol{B}$, the incompressibility condition $\nabla \cdot \boldsymbol{B} = 0$ $((1/R)\partial(RB_R)/\partial R + \partial B_Z/\partial Z = 0)$ $(\nabla \cdot \boldsymbol{B} = 0 \to \boldsymbol{B}_p = \nabla\zeta \times \nabla\psi)$. Also, $\boldsymbol{B} \cdot \nabla\psi = 0$ can be checked easily. Then the magnetic field trajectory is on the $\psi =$ constant surface. In terms of terminology in dynamical system, the system has a first integral and the orbit is "integrable." The constant ψ surface is called a "magnetic surface," or "magnetic surface ψ." In terms of Hamilton dynamics in Section 3.4, the system is independent of "time" ζ and the Hamiltonian ψ becomes an invariant.

Axisymmetry guarantees that B_R and B_Z can be expressed by the first integral, but it does not impose any constraint on the toroidal magnetic field B_ζ. The constraint on B_ζ comes from the equilibrium condition $\boldsymbol{J} \times \boldsymbol{B} = \nabla P$. In fact, $\boldsymbol{B} \cdot \nabla P = 0$ reads $\partial(\psi, P)/\partial(R, Z) = 0$ and $P = P(\psi)$ is derived. Then, $\boldsymbol{J} \cdot \nabla P = 0$ reads $\partial(RB_\zeta P)/\partial(R, Z) = 0$ and $RB_\zeta = F(P) = F(\psi)$ is derived. Such functions of flux function ψ only are called a magnetic "flux function." From the above, the following relationships can be obtained:

$$\boldsymbol{B} = \nabla\zeta \times \nabla\psi + F\nabla\zeta \,, \tag{3.59}$$

$$\boldsymbol{J} = \mu_0^{-1}\left[\nabla F \times \nabla\zeta + \Delta^*\psi\nabla\zeta\right] . \tag{3.60}$$

Here, $\Delta^* = R\partial/\partial R(R^{-1}\partial/\partial R) + \partial^2/\partial R^2$ is called the "Grad–Shafranov operator." $F(\psi)$ plays the role of stream function for $\boldsymbol{J}_p = -\mu_0^{-1}\nabla\zeta \times \nabla F$. Substituting Equations 3.59 and 3.60 into $\boldsymbol{J} \times \boldsymbol{B} = \nabla P$ yields,

$$\Delta^*\psi = -\mu_0 R^2 P'(\psi) - FF'(\psi) . \tag{3.61}$$

This nonlinear elliptic partial differential equation is called the "Grad–Shafranov equation" [24, 25]. The functional form of $P(\psi)$ and $F(\psi)$ cannot be determined by the force equilibrium (determined by the transport equations of current and temperature/density). In general, the Grad–Shafranov equation is solved numerically by giving the functional form of $P(\psi)$ and $F(\psi)$. The Grad-Shafranov equation can be derived using the variational principle $\delta S = 0$ [25].

$$S = \int L(\psi, \psi_R, \psi_Z, R)\,\mathrm{d}R\,\mathrm{d}Z . \tag{3.62}$$

Here, $\psi_R = \partial\psi/\partial R$ and $\psi_Z = \partial\psi/\partial Z$. The Lagrangian L is given by,

$$L = R\left(\frac{B_p^2}{2\mu_0} - \frac{B_\zeta^2}{2\mu_0} - P\right) \tag{3.63}$$

where, $B_p = |\nabla\psi|/R$ and $B_z = F(\psi)/R$. The Euler–Lagrange equation derived from variational principle $\delta S = 0$ is

$$\frac{\partial L}{\partial\psi} - \frac{\partial}{\partial R}\frac{\partial L}{\partial\psi_R} - \frac{\partial}{\partial Z}\frac{\partial L}{\partial\psi_Z} = 0\,. \tag{3.64}$$

And the Equation 3.61 can be obtained. $B_p^2/2\mu_0$ plays the role of the kinetic energy of the Lagrangian, and $B_\zeta^2/2\mu_0 + P$ plays the role of the effective potential energy. It may be natural to question why the roles of the toroidal and poloidal magnetic field energies are different? In this variational principle, the toroidal magnetic field and pressure are already given by $B_\zeta = F(\psi)/R$ and $P = P(\psi)$ and the problem is reduced to obtain a solution of "motion" of ψ.

The "flux surface average" of a physical quantity $\langle A\rangle$ is defined by the volume integral in an infinitesimal small shell in $(\psi, \psi + \mathrm{d}\psi)$. Using $\mathrm{d}V = J\,\mathrm{d}\psi\,\mathrm{d}\theta\,\mathrm{d}\zeta$,

$$\langle A\rangle = \frac{\int_\psi^{\psi+\mathrm{d}\psi} AJ\,\mathrm{d}\psi\,\mathrm{d}\theta\,\mathrm{d}\zeta}{\int_\psi^{\psi+\mathrm{d}\psi} J\,\mathrm{d}\psi\,\mathrm{d}\theta\,\mathrm{d}\zeta} = \frac{\int_0^{2\pi}\frac{A\,\mathrm{d}\theta}{\boldsymbol{B}_p\cdot\nabla\theta}}{\int_0^{2\pi}\frac{\mathrm{d}\theta}{\boldsymbol{B}_p\cdot\nabla\theta}}\,. \tag{3.65}$$

Here, $J = 1/(\nabla\zeta\times\nabla\psi)\cdot\nabla\theta = 1/\boldsymbol{B}_p\cdot\nabla\theta$ is used. The flux surface average annihilates the differential operator $\boldsymbol{B}\cdot\nabla = J^{-1}\partial/\partial\theta$. The differential equation along the magnetic field appears frequently in the magnetic confinement theory and was named the "magnetic differential equation" by the famous MHD stability theoretician W. Newcomb.

$$\boldsymbol{B}\cdot\nabla h = S\,. \tag{3.66}$$

Here, h and S are single-valued. In a closed magnetic configuration, integrability of the magnetic field sets a constraint on h and S, called the "solvable condition." Equation 3.66 in the flux coordinates (ϕ, θ, ζ) becomes

$$(q\nabla\psi\times\nabla\theta - \nabla\psi\times\nabla\zeta)\left[\frac{\partial h}{\partial\psi}\nabla\psi + \frac{\partial h}{\partial\theta}\nabla\theta + \frac{\partial h}{\partial\zeta}\nabla\zeta\right] = S\,. \tag{3.67}$$

Using $\boldsymbol{B}\cdot\nabla\psi = 0$, axisymmetric condition $\partial h/\partial\zeta = 0$, and $J^{-1} - \nabla\psi\times\nabla\theta\cdot\nabla\zeta = \boldsymbol{B}\cdot\nabla\theta = \boldsymbol{B}_p\cdot\nabla\theta$,

$$\frac{\partial h}{\partial\theta} = \frac{S}{\boldsymbol{B}_p\cdot\nabla\theta}\,. \tag{3.68}$$

Since h is single-valued, it must be the same value at $\theta = 0$ and 2π. Integration of Equation 3.67 for θ gives the following "solvable condition."

$$\int_0^{2\pi}\frac{S}{\boldsymbol{B}_p\cdot\nabla\theta}\mathrm{d}\theta = 0\,. \tag{3.69}$$

This means $\langle S\rangle = 0$ and application of the flux surface averaging operator to the differential operator $\boldsymbol{B}\cdot\nabla = J^{-1}\partial/\partial\theta$ gives $\langle \boldsymbol{B}\cdot\nabla\rangle \equiv 0$. This is the origin of the naming of the annihilator of $\boldsymbol{B}\cdot\nabla = J^{-1}\partial/\partial\theta$.

Note: Symmetry and Invariant of the Dynamical System [15]

For the case where the system has symmetry and some position coordinate q_s is not included in Lagrangian $L(q_i, \dot{q}_i)$ (means $\partial L/\partial q_s = 0$, but assume $\dot{q}_s$ is included in L since q_s is not a dynamical variable if both are not included). Then, the Lagrange equation (see Section 4.1)

$$\frac{\mathrm{d}}{\mathrm{d}t}\left(\frac{\partial L}{\partial \dot{q}_s}\right) - \frac{\partial L}{\partial q_s} = 0 \tag{3.70}$$

reads

$$\frac{\mathrm{d}}{\mathrm{d}t}\left(\frac{\partial L}{\partial \dot{q}_s}\right) = 0\,. \tag{3.71}$$

Thus, generalized momentum $p_s = \partial L/\partial \dot{q}_s$ is conserved (invariant). Such a coordinate is called a "cyclic" coordinate. In the system of rotational symmetry (axisymmetric system), ζ is not included in L and the generalized angular momentum $p_\zeta = \partial L/\partial\dot{\zeta}$ is conserved. Symmetry in dynamical systems is closely related to the existence of the invariant (integrability).

3.9 3-dimensional Force Equilibrium: Search for Hidden Symmetry

A typical example of force equilibrium without apparent geometrical symmetry is the stellarator concept originated by Spitzer [2], see Figure 3.16. The 3-dimensional equilibrium may not have global equilibrium in some cases, in contrast to the tokamak equilibrium. There are few mathematical theories of 3-dimensional equilibrium except for KAM theory, which treats slight symmetry breaking. The variational principle may be useful to examine 3-dimensional equilibrium since it is independent of the coordinate system. In 1958, H. Grad derived a variational principle ($\delta S = 0$) equivalent to the plasma equilibrium condition $\boldsymbol{J}\times\boldsymbol{B} = \nabla P$ by defining the action integral S with a variable to $\boldsymbol{B}$ and P satisfying $\nabla\cdot\boldsymbol{B} = 0$ and $\boldsymbol{B}\cdot\nabla P = 0$ [24,27].

$$S(\boldsymbol{B}, P) = \int_v L\,\mathrm{d}V = \int_v \left[\frac{B^2}{2\mu_0} - P\right]\mathrm{d}V\,, \tag{3.72}$$

$$\delta S(\boldsymbol{B}, P) = \int_v \left[\frac{1}{\mu_0}\boldsymbol{B}\cdot\delta\boldsymbol{B} - \delta P\right]\mathrm{d}V\,,$$

where V is the plasma volume and $\boldsymbol{B}\cdot\boldsymbol{n} = 0$ at the plasma surface. The Lagrangian $L = B^2/2\mu_0 - P$ is the difference between magnetic pressure and plasma pressure,

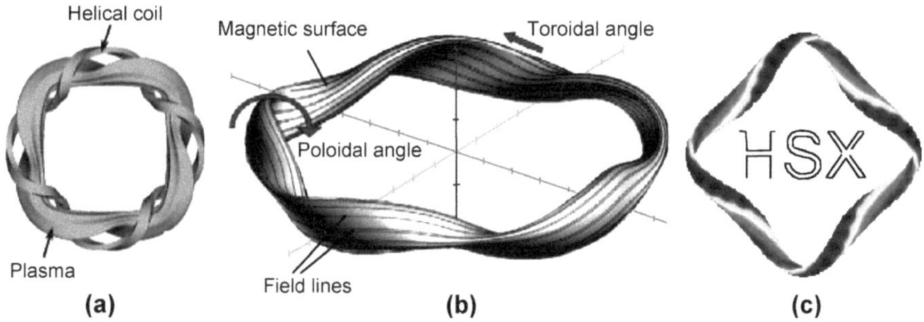

Figure 3.16 Typical 3-dimensional toroidal equilibrium configurations with periodic symmetry. (a) Heliotron-J (by kind permission of IAE Kyoto University) with 4-fold symmetry, (b) Wendelstein 7-X (by courtesy of the Max Planck Institute for plasma physics) with 5-fold symmetry and (c) HSX (with kind permission by Prof. D. Anderson, Wisconsin University) with 4-fold symmetry

implying magnetic field energy is "kinetic energy," plasma pressure is "potential energy." The magnetic field satisfying $\boldsymbol{B} \cdot \nabla P = 0$ and $\nabla \cdot \boldsymbol{B} = 0$ are given by,

$$\boldsymbol{B} = \nabla P \times \nabla \omega \, . \tag{3.73}$$

Here, the flow function ω is expected to be a multi-valued function containing angle variables from the discussion of surface function in Section 3.5. Under this strong constraint of $\boldsymbol{B}$ on the constant pressure surface, action integral S becomes a function of ω and P. Using Equation 3.73,

$$\boldsymbol{B} \cdot \delta \boldsymbol{B} = (\nabla \omega \times \boldsymbol{B}) \cdot \nabla \delta P + (\boldsymbol{B} \times \nabla P) \cdot \nabla \delta \omega \, . \tag{3.74}$$

Applying vector formula $\nabla \cdot (\phi \boldsymbol{a}) = \phi \nabla \cdot \boldsymbol{a} + \boldsymbol{a} \cdot \nabla \phi$ to Equation 3.74 and transforming $\nabla \cdot (\delta P \nabla \omega \times \boldsymbol{B} + \delta \omega \boldsymbol{B} \times \nabla P)$ into surface integral by the Gauss's theorem and set zero by the boundary conditions $\delta \psi = \delta \omega = 0$, the remaining term of volume integral of $\boldsymbol{B} \cdot \delta \boldsymbol{B}$ is given by,

$$\boldsymbol{B} \cdot \delta \boldsymbol{B} = -\left[\nabla \cdot (\nabla \omega \times \boldsymbol{B})\right] \delta P - \left[\nabla \cdot (\boldsymbol{B} \times \nabla P)\right] \delta \omega \, . \tag{3.75}$$

Then the following form of δS is obtained by using the vector formula $\nabla \cdot (\boldsymbol{a} \times \boldsymbol{b}) = \boldsymbol{b} \cdot \nabla \times \boldsymbol{a} - \boldsymbol{a} \cdot \nabla \times \boldsymbol{b}$,

$$\delta S(\omega, P) = \int \left[(\boldsymbol{J} \cdot \nabla \omega - 1)\delta P - \boldsymbol{J} \cdot \nabla P \delta \omega\right] \mathrm{d}V \, . \tag{3.76}$$

Then,

$$\boldsymbol{J} \cdot \nabla \omega = 1 \, , \tag{3.77}$$

$$\boldsymbol{J} \cdot \nabla P = 0 \tag{3.78}$$

are obtained as the Euler equations to extremize S. Since $\boldsymbol{J} \times \boldsymbol{B} - \nabla P = (\boldsymbol{J} \cdot \nabla \omega - 1)\nabla P - (\boldsymbol{J} \cdot \nabla P)\nabla \omega = 0$, the variational principle $\delta S = 0$ is equivalent to $\boldsymbol{J} \times \boldsymbol{B} = \nabla P$. As is clear from the proof of the above, the plasma pressure P plays a role of "potential energy" and strongly constrains the magnetic field ($\boldsymbol{B} \cdot \nabla P = 0$). Magnetic energy plays the role of "kinetic energy" and the variational principle becomes an extremal problem on the stream function ω under the strong constraint of P.

If a small displacement $\delta \boldsymbol{x}$ of the plasma induces a pressure change δP and the change in the stream function $\delta\omega$, we obtain $\delta P = -\delta \boldsymbol{x} \cdot \nabla P$ and $\delta\omega = -\delta \boldsymbol{x} \cdot \nabla\omega$. The variational principle, Equation 3.76, can be rewritten in the following form given by Kruskal–Krusrud [28]:

$$\delta S(\omega, P) = -\int \delta \boldsymbol{x} \cdot [\boldsymbol{J} \times \boldsymbol{B} - \nabla P]\,\mathrm{d}V\,. \tag{3.79}$$

From $\nabla \cdot \boldsymbol{J} = 0$ and $\boldsymbol{J} \cdot \nabla P = 0$, $\boldsymbol{J}$ is given by,

$$\boldsymbol{J} = \nabla P \times \nabla\omega_J\,. \tag{3.80}$$

Here, ω_J is a flow function for $\boldsymbol{J}$. Using $\boldsymbol{B} \cdot \nabla P = 0$, we see $\boldsymbol{J} \times \boldsymbol{B} = (\nabla P \times \nabla\omega_J) \times \boldsymbol{B} = -(\boldsymbol{B} \cdot \nabla\omega_J)\nabla P$ to reach,

$$\boldsymbol{B} \cdot \nabla\omega_J = -1\,. \tag{3.81}$$

Substituting Equations 3.73 and 3.80 into the force equilibrium $\boldsymbol{J} \times \boldsymbol{B} = \nabla P$, we obtain the following equation, equivalent to Equation 3.78,

$$(\nabla\omega_J \times \nabla\omega) \cdot \nabla P = 1\,. \tag{3.82}$$

"3-dimensional force equilibrium" with no apparent geometric symmetry, must have a coordinate transformation from the flux coordinates with hidden symmetry to the real coordinate system (such as cylindrical coordinate system). The equilibrium problem is understood as finding "inverse mapping" $\boldsymbol{x} = \boldsymbol{x}(u, \theta, \zeta)$ from the flux coordinates (u, θ, ζ) to the cylindrical coordinates (R, ζ, Z). The variational principle has merit in the simplicity of its coordinate-independent formulation as an extremal problem of a single scalar function rather than vector differential equations.

Such a variational principle for 3-dimensional plasma equilibrium is implemented in Hirshman's VMEC code [28]. Introducing the virtual time "t" into Equation 3.79 and changing δS to $\mathrm{d}S/\mathrm{d}t$, we have,

$$\frac{\mathrm{d}S}{\mathrm{d}t} = -\int [\boldsymbol{J} \times \boldsymbol{B} - \nabla P] \cdot \frac{\partial\delta\boldsymbol{x}}{\partial t}\,\mathrm{d}V\,. \tag{3.83}$$

This is an evolution of the equation from the state of $\boldsymbol{F} = \boldsymbol{J} \times \boldsymbol{B} - \nabla P \neq 0$ to the state of $\boldsymbol{F} = 0$ with virtual displacement $\delta\boldsymbol{x}$. Applying Cauchy-Schwartz inequality, $|\int \boldsymbol{A}^*\boldsymbol{B}\,\mathrm{d}u|^2 \leq \int |\boldsymbol{A}|^2\,\mathrm{d}u \int |\boldsymbol{B}|^2\,\mathrm{d}u$ to Equation 3.83, we have

$$\left|\int [\boldsymbol{J} \times \boldsymbol{B} - \nabla P] \cdot \frac{\partial\delta\boldsymbol{x}}{\partial t}\,\mathrm{d}V\right|^2 \leq \int |\boldsymbol{J} \times \boldsymbol{B} - \nabla P|^2\,\mathrm{d}V \int \left|\frac{\partial\delta\boldsymbol{x}}{\partial t}\right|^2\,\mathrm{d}V\,. \tag{3.84}$$

Here, the equality holds only when $\partial\delta\boldsymbol{x}/\partial t = c(\boldsymbol{J} \times \boldsymbol{B} - \nabla P)$ is satisfied (c can be set 1 since $\delta\boldsymbol{x}$ is the virtual displacement). Convergence of this equation can be accelerated by adding the second order time derivative (second order Richardson scheme). By setting unknown constants in the coordinate transformation equation from flux coordinates $\{u_i\} = (\phi, \theta_m, \zeta)$ to cylindrical coordinates (R, ζ, Z), we treat this problem as an extremal problem of $(\delta S)^2$. The toroidal angular variable ζ in the flux coordinate is chosen to be same as that of the cylindrical coordinates. Poloidal angle θ is determined by the condition of fast convergence of the Fourier expansion in the plasma surface, the unknown functions are $\boldsymbol{x} = (R, \lambda, Z)$. Assuming F_R, F_λ, F_Z are virtual forces which are zero in equilibrium, each Fourier component F_j^{mn} $(j = R, \lambda, Z)$ should be zero in equilibrium. However, equilibrium with separatrix cannot be reconstructed if we expand in Fourier series. Minimization of the action integral is possible numerically, but it must be noted that this does not mean that the 3-dimensional equilibrium is obtained [30].

References

1. Sakharov A (1990) Memoirs. The English Agency.
2. Spitzer Jr. L (1958) Physics of Fluids, 1, 253.
3. Iiyoshi A et al. (1990) Fusion Technology 17 169; Flux surface figure is courtesy of Professor T. Watanabe.
4. Diacu F, Holmes P (1996) Celestial Encounters: The Origin of Chaos and Stability. Princeton University Press.
5. Arnold VI (1978) Mathematical Methods of Classical Mechanics. Springer-Verlag New York Inc.
6. Richtenberg AJ and Lieberman MA (1992) Regular and Chaotic Dynamics. Springer-Verlag New York Inc.
7. Poincaré H (1881) Journal de Mathematiques pures et appliqués. 3, No. 7, 375–422.
8. Poincaré H (1882) Journal de Mathematiques pures et appliqués. 3, No. 8, 251–296.
9. Poincaré H (1885) Journal de Mathematiques pures et appliqués, 4, No. 1, 167–244.
10. Poincaré H (1886) Journal de Mathematiques pures et appliqués, 4, No. 2, 151217.
11. Massey WS (1967) Algebraic Topology: An Introduction. Springer-Verlag.
12. O'Shea D (2007) The Poincaré Conjecture. Walker & Company.
13. Boozer A (2004) Review of Modern Physics 76, 1071.
14. D'haeseleer WD, Hutchon WNG, Callen JD, Shohet JL (1991) Flux Coordinates and Magnetic Field Structure. Springer-Verlag Berlin Heidelberg.
15. Goldstein H (1950) Classical Mechanics. Addison-Wesley.
16. Cary JR and Littlejohn RG (1983) Annals of Physics, 151, 1.
17. Lederman L and Hill CT (2008) Symmetry and the Beautiful Universe. Prometheus Books.
18. Ronan M (2006) Symmetry and the Monster. Oxford University Press.
19. Tignol JP (2001) Galois' Theory of Algebraic Equations. World Scientific Publishing.
20. Hamada S (1962) Nuclear Fusion, 2, 23; Hamada S (1959) Prog. Theor. Phys., 22, 145.
21. Boozer A (1981) Physics of Fluids 24,1999.
22. Aczel AD (2000) The Mystery of the Aleph Mathematics. Four Walls Eight Windows Inc.
23. Kolmogorov AN, Fomin SV (1999) Elements of the Theory of Functions and Functional Analysis. Dover, Chapter I, Chapter V.
24. Grad H and Rubin H (1958) Proc. 2nd UN Int. Conf. Peaceful Use of Atomic Energy, Geneva, 31, 190.
25. Shafranov VD (1958) Soviet Physics JETP, 26, 400.
26. Lao LL, Hirshman SP, Wieland RM (1981) Physics of Fluids, 24, 1431.
27. Grad H (1964) Physics of Fluids, 7, 1159.
28. Kruskal MD and Krusrud RM (1958) Physics of Fluids, 1, 265.
29. Hirshman SP, Whitson JC (1983) Physics of Fluids, 26, 3553.
30. Courant R, Hilbert D (1989) Methods of Mathematical Physics Vol. 1. Wiley-Interscience, Chap. 4, § 1,4.

Chapter 4
Charged Particle Motion: Lagrange–Hamilton Orbit Dynamics

The motion of light and objects in nature follows an orbit in which the path integral of the "action" becomes extremal. Fermat's well-known optics principle tells us that light draws an orbit such that the time required becomes minimum between fixed point A and B. For example, Snell's law describing refraction of light in media with different refractive indexes can be derived from Fermat's principle (Figure 4.1). Nature is governed beautifully by the variational principle.

The variational principle for the object motion is expressed by Hamilton's principle. The complex charged particle motion, a combination of Larmor and drift motions in the confinement magnetic field, can be simplified using the above variational principle.

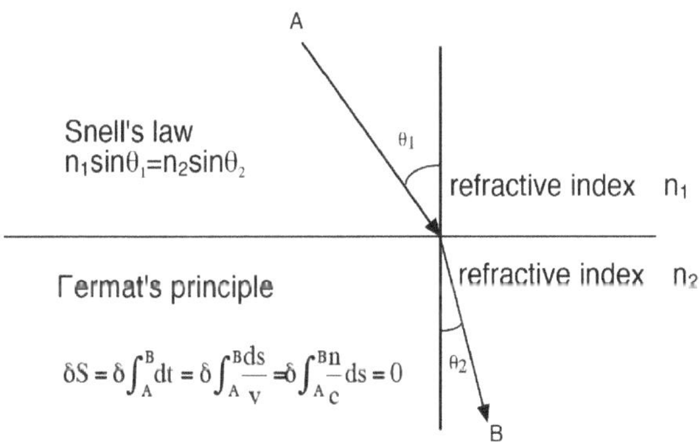

Figure 4.1 Relationship between Snell's law and Fermat's principle

4.1 Variational Principle: Hamilton's Principle

Many natural phenomena follow the "action principle" [1–3]. The most famous example is "Fermat's principle" in geometric optics derived by the French mathemati-

Figure 4.2 (a) P. Fermat and (b) W. Hamilton

cian P. Fermat (1601–1665; Figure 4.2 (a)): "Light chooses the path that it can travel in the minimum time." This principle gives an image of "preestablished harmony" as if the light knew the path to travel before it reaches the final point. This principle implies more meaning to "time" than just being different from spatial coordinates through its direction from past, present, and future (time arrow).

The action principle holds for object motion as well as light motion. The motion of a charged particle in plasma follows the principle devised by Hamilton (1805–1865; Figure 4.2 (b)), in that it chooses a path $\boldsymbol{x}(t)$ such that following action integral defined by the Lagrangian function $L(\boldsymbol{x}, \mathrm{d}\boldsymbol{x}/\mathrm{d}t, t) = T - V$ (T: kinetic energy, V: potential energy) take the extreme,

$$S(\boldsymbol{x}) = \int_{t_1}^{t_2} \mathrm{d}t \; L(\boldsymbol{x}(t), \dot{\boldsymbol{x}}(t), t) \,. \tag{4.1}$$

Hamilton's principle: If the orbit from P_1 at t_1 to P_2 at t_2 is realized by the equation of motion, this orbit is extremal to the action integral $S[x]$ among various curves between two points (P_1, t_1), (P_2, t_2).

The above variational principle does not regard $\dot{\boldsymbol{x}} = \mathrm{d}\boldsymbol{x}/\mathrm{d}t$ as an independent variable, and is called the variational principle in Lagrange form. Using generalized coordinates q_i instead of $\boldsymbol{x}$ and $\delta\dot{q}_i = \mathrm{d}\delta q_i/\mathrm{d}t$ (see Figure 4.3 (a)) the variation of action integral δS becomes,

$$\delta S = \int_{t_1}^{t_2} \mathrm{d}t \sum_i \left[\frac{\mathrm{d}}{\mathrm{d}t}\left(\frac{\partial L}{\partial \dot{q}_i}\right) - \frac{\partial L}{\partial q_i} \right] \delta q_i \,. \tag{4.2}$$

Thus, the following Euler–Lagrange equation (Lagrange equation of motion) must be satisfied for the path $q_i(t)$ to extremize action integral S [1].

$$\frac{\mathrm{d}}{\mathrm{d}t}\left(\frac{\partial L}{\partial \dot{q}_i}\right) - \frac{\partial L}{\partial q_i} = 0 \,, \quad (i = 1, \ldots, n) \tag{4.3}$$

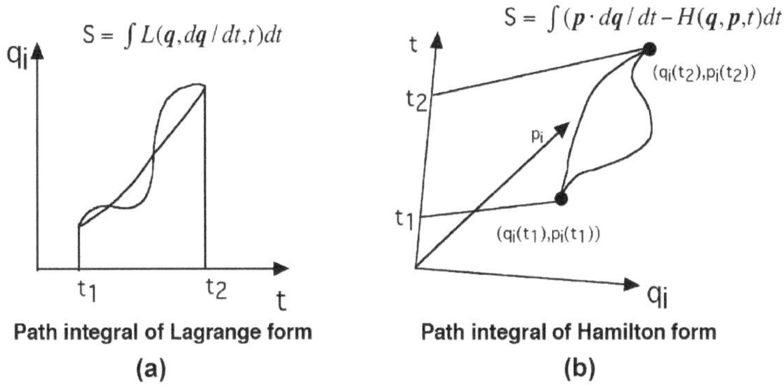

Figure 4.3 Action integrals of (a) Lagrange form in position space and (b) Hamilton forms in phase space. System chooses orbit where action integral takes extremal among possible various paths

For a given equation of motion, Lagrangian function does not always exist (for example, no Lagrangian function exists for $m\,\mathrm{d}^2\boldsymbol{r}/\mathrm{d}t^2 = a(\boldsymbol{r} \times \dot{\boldsymbol{r}})$). Also, the Lagrangian for a given equation of motion is not unique since addition of total derivative $\mathrm{d}G(\boldsymbol{q}, t)/\mathrm{d}t$ to $L(\boldsymbol{q}, \dot{\boldsymbol{q}}, t)$, namely $L + \mathrm{d}G/\mathrm{d}t$ also gives the same equation of motion. In fact,

$$S = \int_{t_1}^{t_2} \mathrm{d}t \left[L(\boldsymbol{q}, \dot{\boldsymbol{q}}, t) + \frac{\mathrm{d}G(\boldsymbol{q}, t)}{\mathrm{d}t} \right] = \int_{t_1}^{t_2} L(\boldsymbol{q}, \dot{\boldsymbol{q}}, t)\mathrm{d}t + G(\boldsymbol{q}(t_2), t_2) - G(\boldsymbol{q}(t_1), t_1) . \tag{4.4}$$

Considering $\delta \boldsymbol{q}(t_1) = \delta \boldsymbol{q}(t_2) = 0$, $\delta S(L) = \delta S(L + \mathrm{d}G/\mathrm{d}t)$ easily follows. Under the variational principle in Lagrange form, initial values of generalized coordinates q_i and generalized speeds $\dot{q}_i$ are both specified to define the motion, but $\dot{q}_i$ is treated as the time derivative of q_i by definition and is not an independent variable, thus $\delta \dot{q}_i$ is to be converted to δq_i through partial integration.

There is another variational principle in Hamilton form, which is different from the variational principle in Lagrange form above. The Hamiltonian is defined by $H(\boldsymbol{q}, \boldsymbol{p}, t) = \sum p_i \dot{q}_i - L(\boldsymbol{q}, \dot{\boldsymbol{q}}, t)$, where generalized momentum is defined by $p_i \equiv \partial L / \partial \dot{q}_i$. Extending the variational principle from position space to phase space $(\boldsymbol{q}, \boldsymbol{p})$ and treating q_i and p_i as independent variables (see Figure 4.3 (b)). Then the action integral is,

$$S(\boldsymbol{x}, \boldsymbol{p}) = \int_{t_1}^{t_2} \left[\sum p_i \dot{q}_i - H(\boldsymbol{q}, \boldsymbol{p}, t) \right] \mathrm{d}t . \tag{4.5}$$

Taking the variation, the following equation is obtained.

$$\delta S(\boldsymbol{p}, \boldsymbol{q}, t) = \int_{t_1}^{t_2} \sum_i \left[\left(\dot{q}_i - \frac{\partial H}{\partial p_i} \right) \delta p_i - \left(\dot{p}_i + \frac{\partial H}{\partial q_i} \right) \delta q_i \right] \mathrm{d}t . \tag{4.6}$$

Unlike the variation on the Lagrangian, variations δp_i and δq_i are independent. So, the following Hamilton equation is obtained.

$$\frac{\mathrm{d}q_j}{\mathrm{d}t} = \frac{\partial H}{\partial p_j}\,, \quad \frac{\mathrm{d}p_j}{\mathrm{d}t} = -\frac{\partial H}{\partial q_j}\,. \tag{4.7}$$

The variational principle in the Hamilton form treats δp_i as independent variation as well as δq_i, in contrast to the case for $\delta \dot{q}_i$. Both δq_i and δp_i are independent with equal capacity. By extending the independent variables from n to $2n$, the equation of motion becomes the first order differential equation.

Note: Noether's theorem

As mentioned in Chapter 3, it is important to find the constants of motion in a dynamical system. A general method to find the constant of motion was derived by a German mathematician A. E. Noether (1882–1935; Figure 4.4). Consider the following transformation of canonical variable with infinitesimal parameters ε

$$q_i' = q_i + \varepsilon S_i(q, \dot{q}, t)\,. \tag{4.8}$$

If the change in Lagrangian $\delta L = L(\boldsymbol{q}', \dot{\boldsymbol{q}}', t) - L(\boldsymbol{q}, \dot{\boldsymbol{q}}, t)$ satisfies $\delta L = \varepsilon\,\mathrm{d}W(\boldsymbol{q}, t)/\mathrm{d}t$ (W is gauge term and L essentially unchanged, the constant of motion is

$$I = \sum_{i=1}^{n} \frac{\partial L(\boldsymbol{q}, \dot{\boldsymbol{q}}, t)}{\partial \dot{q}_i} S_i - W(\boldsymbol{q}, t)\,. \tag{4.9}$$

Proof: $\delta L = \sum(\partial L/\partial q_i \delta q_i + \partial L/\partial \dot{q}_i \delta \dot{q}_i) = \sum[\mathrm{d}\{\partial L/\partial \dot{q}_i \delta q_i\}/\mathrm{d}t - \{\mathrm{d}(\partial L/\partial \dot{q}_i)/\mathrm{d}t - \partial L/\partial q_i\}\delta q_i]$. Since the second term of right-hand side is zero from Equation 4.3, substitution of Equation 4.8 gives $\sum[\mathrm{d}\{(\partial L/\partial q_i)S_i\}/\mathrm{d}t = \mathrm{d}W/\mathrm{d}t$, and Equation 4.9 holds.

Figure 4.4 Emmy Noether

4.2 Lagrange–Hamilton Mechanics: Motion in an Electromagnetic Field

The motion of a charged particle in an electromagnetic field follows Hamilton's Principle and the action integral is given as the time integral of the Lagrangian which is extreme under the boundary conditions $\delta\boldsymbol{x}(t_1) = \delta\boldsymbol{x}(t_2) = 0$.

$$S[\boldsymbol{x}(t)] = \int_{t_1}^{t_2} L_a(\boldsymbol{x}, \dot{\boldsymbol{x}}, t)\,\mathrm{d}t \tag{4.10}$$

$$L_a(\boldsymbol{x}, \dot{\boldsymbol{x}}, t) = \frac{1}{2} m_a \dot{\boldsymbol{x}}^2 + e_a \boldsymbol{A}(\boldsymbol{x}, t) \cdot \dot{\boldsymbol{x}} - e_a \Phi(\boldsymbol{x}, t)\,. \tag{4.11}$$

Here, velocity dependent generalized potential $V = e\Phi(\boldsymbol{x}, t) - e\boldsymbol{v} \cdot \boldsymbol{A}(\boldsymbol{x}, t)$ is used in the Lagrangian. Φ is the electrostatic potential. Lagrange's equation of motion is given by [3],

$$\frac{\mathrm{d}}{\mathrm{d}t}\frac{\partial L_a}{\partial \dot{\boldsymbol{x}}} - \frac{\partial L_a}{\partial \boldsymbol{x}} = 0\,. \tag{4.12}$$

Then, we calculate,

$$\begin{aligned} \frac{\partial L_a}{\partial \dot{\boldsymbol{x}}} &= m_a \dot{\boldsymbol{x}} + e_a \boldsymbol{A}\,, \\ \frac{\partial L_a}{\partial \boldsymbol{x}} &= -e_a \frac{\partial \Phi}{\partial \boldsymbol{x}} + e_a \frac{\partial (\boldsymbol{A} \cdot \dot{\boldsymbol{x}})}{\partial \boldsymbol{x}} \end{aligned} \tag{4.13}$$

and using $\nabla(\boldsymbol{A} \cdot \boldsymbol{v}) = \boldsymbol{v} \cdot \nabla \boldsymbol{A} + \boldsymbol{v} \times \nabla \times \boldsymbol{A}$ (∇ is the derivative under constant $\boldsymbol{v} = \dot{\boldsymbol{x}} = \mathrm{d}\boldsymbol{x}/\mathrm{d}t$; $\nabla = \partial/\partial \boldsymbol{x}$ is the partial derivative in Equation 4.12). So,

$$\frac{\partial L_a}{\partial \boldsymbol{x}} = -e_a \frac{\partial \Phi}{\partial \boldsymbol{x}} + e_a (\dot{\boldsymbol{x}} \cdot \nabla) \boldsymbol{A} + e_a \dot{\boldsymbol{x}} \times (\nabla \times \boldsymbol{A})\,. \tag{4.14}$$

Substituting of Equation 4.14 into Equation 4.12 and considering $\mathrm{d}\boldsymbol{A}/\mathrm{d}t = \partial \boldsymbol{A}/\partial t + \boldsymbol{v} \cdot \nabla \boldsymbol{A}$ and $\boldsymbol{E} = -\nabla\Phi - \partial \boldsymbol{A}/\partial t$, we obtain the following Lorentz equation.

$$m_a \frac{\mathrm{d}^2 \boldsymbol{x}}{\mathrm{d}t^2} = e_a [\boldsymbol{E} + \boldsymbol{v} \times \boldsymbol{B}]\,. \tag{4.15}$$

When the Maxwell and Lorentz equations are combined they have a "Gauge invariance" under the gauge transformation $\boldsymbol{A} \to \boldsymbol{A} + \nabla G$, $\Phi \to \Phi - \partial G/\partial t$ for arbitrary function $G(\boldsymbol{x}, t)$. Substituting this gauge transformation into Equation 4.11, the Lagrangian becomes $L + \mathrm{d}G/\mathrm{d}t$, which means that the equation of motion does not change from that discussed in Section 4.1.

As described in the Note of Section 3.8, symmetry is key for integrability. This is also true for charged particle motion. Generally speaking, generalized momentum p_j is conserved when the Lagrangian L has a translational symmetry on coordinate q_j. Tokamak is an axisymmetric system and the toroidal angle ζ becomes a cyclic coordinate. So, p_ζ is conserved.

$$p_\zeta = \frac{\partial L}{\partial \dot{\zeta}} = m_a\, R^2 \dot{\zeta} + e_a\, R A_\zeta = \text{constant}\,. \tag{4.16}$$

Charges in particle motion can be written in the Hamilton form. Generalized momentum and the Hamiltonian H are defined by $\boldsymbol{p} = \partial L_a / \partial \boldsymbol{x}$, $H = \boldsymbol{p} \cdot \boldsymbol{x} - L(\boldsymbol{x}, \boldsymbol{x}, t)$ and are given by using Equations 4.14 and 4.11 as follows,

$$\boldsymbol{p} = m_a \boldsymbol{v} + e_a \boldsymbol{A}\,, \tag{4.17}$$

$$H(\boldsymbol{p}, \boldsymbol{x}, t) = \frac{1}{2} m_a \dot{\boldsymbol{x}}^2 + e_a \Phi(\boldsymbol{x}, t) = \frac{1}{2m_a}(\boldsymbol{p} - e_a \boldsymbol{A})^2 + e_a \Phi(\boldsymbol{x}, t)\,. \tag{4.18}$$

The first term of Equation 4.17 is the momentum of mass and the second term is the momentum of electromagnetic field. In Hamilton form, $\boldsymbol{p}$ and $\boldsymbol{x}$ are independent variables. The Hamilton equation for charged particle motion is given as follows,

$$\begin{aligned} \frac{\mathrm{d}\boldsymbol{x}}{\mathrm{d}t} &= \frac{\partial H}{\partial \boldsymbol{p}}\,, \\ \frac{\mathrm{d}\boldsymbol{p}}{\mathrm{d}t} &= -\frac{\partial H}{\partial \boldsymbol{x}}\,. \end{aligned} \tag{4.19}$$

This equation is equivalent to Equations 4.12 and 4.15.

Note: Relativistic dynamics of charged particles [3]

We must use the relativistic expression of the Hamiltonian and Lagrangian when charged particle velocity approaches to the speed of light and relativistic effects becomes important.

Free particle action must be invariant for all inertial systems (invariance for Lorentz transformation). The action form satisfying this necessary condition is an integration of the world interval s ($\mathrm{d}s \equiv (c^2\,\mathrm{d}t^2 - \mathrm{d}x^2 - \mathrm{d}y^2 - \mathrm{d}z^2)^{1/2}$), $S = C \int \mathrm{d}s$. From the condition to match the first term of Equation 4.11, $C = -m_{a0}c$ is obtained. Free particle action $S = -m_{a0}c \int \mathrm{d}s$ is given by,

$$S_{\text{Free Particle}} = -m_{a0}c \int_1^2 \mathrm{d}s = -m_{a0}c^2 \int_{t_1}^{t_2} \sqrt{1 - \frac{v^2}{c^2}}\,\mathrm{d}t\,. \tag{4.20}$$

The action describing the interaction of the particles and the field does not change since Maxwell's equation is invariant under Lorentz transformation. However, the generalized potential is interpreted as a path integral of covariant components of 4-dimensional potential $\{A_\mu\} = (\Phi, -\boldsymbol{A})$ on the world line.

$$S_{\text{Field–Particle}} = -e_a \int_1^2 A_\mu \mathrm{d}x^\mu = -\int_{t_1}^{t_2} e_a \Phi\,\mathrm{d}t + \int_{t_1}^{t_2} e_a \boldsymbol{A} \cdot \mathrm{d}\boldsymbol{x}\,. \tag{4.21}$$

Here, the second term on the right-hand side of Equation 4.21 agrees with action Equation 3.23. From the above, the relativistic Lagrangian is given by the following equation.

$$L_a(\boldsymbol{x}, \boldsymbol{v}, t) = -m_{a0}c^2\sqrt{1-\left(\frac{v}{c}\right)^2} + e_a(\boldsymbol{v}\cdot\boldsymbol{A} - \Phi) \,. \tag{4.22}$$

Relativistic generalized momentum $\boldsymbol{p}(= \partial L/\partial \boldsymbol{v})$ is given by,

$$\boldsymbol{p} = \frac{m_{a0}\boldsymbol{v}}{\sqrt{1-(v/c)^2}} + e_a\boldsymbol{A} \,. \tag{4.23}$$

Hamiltonian $H(= \boldsymbol{v}\cdot\partial L/\partial \boldsymbol{v} - L)$ is given by,

$$H(\boldsymbol{p}, \boldsymbol{q}, t) = \frac{m_{a0}c^2}{\sqrt{1-(v/c)^2}} + e_a\Phi(\boldsymbol{x}, t) \,. \tag{4.24}$$

Using the relation $(H - e_a\Phi)^2 = m_{a0}^2c^2 + (\boldsymbol{p} - e_a\boldsymbol{A})^2$,

$$H(\boldsymbol{p}, \boldsymbol{q}, t) = \sqrt{m_{a0}c^4 + c^2(\boldsymbol{p} - e_a\boldsymbol{A})^2} + e_a\Phi(\boldsymbol{x}, t) \,. \tag{4.25}$$

So far, only two actions of the free particle and the interaction of the particle and field have been considered. In addition to these actions, the action of the field itself S_{Field} exists. As long as we are concerned with the motion of charged particles in an electromagnetic field, this term is not required, but S_{Field} is essential for the dynamics of the electromagnetic field. Since the superposition principle holds for an electromagnetic field, field action should be a quadratic form of the field (electric field and magnetic field). Then, the following form of S_{Field} is obtained,

$$S_{\text{Field}} = \int_{t_1}^{t_2} \left[\frac{\varepsilon_0 \boldsymbol{E}^2}{2} - \frac{\boldsymbol{B}^2}{2\mu_0}\right] \mathrm{d}V\,\mathrm{d}t \,. \tag{4.26}$$

4.3 Littlejohn's Variational Principle: Orbital Dynamics of the Guiding Center

Charged particle orbit in an electromagnetic field is the superposition of the fast ($\sim$ 1 GHz) gyro-motion ($\boldsymbol{\rho}(t)$) and the slow ($\sim$ 100 KHz) drift motion of the guiding center ($\boldsymbol{r}(t)$). In plasma confinement, the behavior of the average slow drift motion is important and its equation of motion, called the guiding center equation, plays an important role. Here, we assume that time variation of the field is negligible. The guiding center equation first derived by H. Alfven [4] has an important drawback that it does not satisfy Liouville theorem that all Hamilton systems conserve phase space volume. A guiding center equation satisfying Liouville theorem can be obtained by the variational principle using the gyro-averaged (average gyro motion)

Figure 4.5 Guiding center variables ($\boldsymbol{r}(t)$: guiding center; $\boldsymbol{b}$: unit vector in the direction of $\boldsymbol{B}$; $\boldsymbol{e}_x, \boldsymbol{e}_y$: unit vectors in gyro plane)

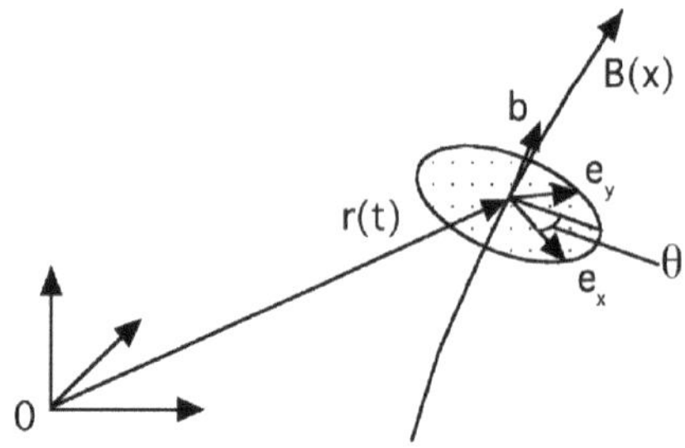

Lagrangian from Equation 4.11 for the non-relativistic case. As mentioned in Section 4.1, there are Hamilton and Lagrange forms in the variational principle. Littlejohn [5] extended the Hamilton form to the non-canonical variables $\boldsymbol{z} = \boldsymbol{z}(\boldsymbol{q}, \boldsymbol{p}, t)$ in phase space. This is Littlejohn's variational principle. Let non-canonical phase space variable $\boldsymbol{z} = (\boldsymbol{x}, \boldsymbol{v})$, the Lagrangian is given by,

$$L(\boldsymbol{x}, \boldsymbol{v}, t) = (e_a \boldsymbol{A} + m_a \boldsymbol{v}) \cdot \dot{\boldsymbol{x}} - H(\boldsymbol{x}, \boldsymbol{v}, t) , \quad (4.27)$$
$$H(\boldsymbol{x}, \boldsymbol{v}, t) = \frac{1}{2} m_a \boldsymbol{v}^2 + e_a \Phi(\boldsymbol{x}, t) .$$

Here, $\boldsymbol{v}$ becomes $\dot{\boldsymbol{x}}$ as a result, but they are treated as different quantities. As shown in Figure 4.5, $\boldsymbol{x}(t)$ is the position of charged particles, $\boldsymbol{r}(t)$ is the position of the guiding center, $\boldsymbol{\rho}(t) = \rho[\boldsymbol{e}_x \cos\theta + \boldsymbol{e}_y \sin\theta]$ is the component of gyro motion, $\rho = v_\perp/\Omega$ is the gyro radius, $\boldsymbol{e}_x$ and $\boldsymbol{e}_y$ are unit vectors which form an orthogonal set with the unit vector of the magnetic field ($\boldsymbol{b} \cdot (\boldsymbol{e}_x \times \boldsymbol{e}_y) = 1$). We obtain $\boldsymbol{x}(t) = \boldsymbol{r}(t) + \boldsymbol{\rho}(t)$ and $\boldsymbol{v} = v_\parallel \boldsymbol{b} - v_\perp \boldsymbol{c}$ ($\boldsymbol{c} = -\boldsymbol{e}_x \sin\theta + \boldsymbol{e}_y \cos\theta$). Then we evaluate gyro average from the right-hand side of Equation 4.27. Here, the independent variables of L are $(\boldsymbol{r}, v_\parallel, v_\perp, \theta)$. Considering $\boldsymbol{A} \cdot \dot{\boldsymbol{x}} \simeq [\boldsymbol{A}(\boldsymbol{r}) + (\partial \boldsymbol{A}/\partial x)\rho\cos\theta + (\partial \boldsymbol{A}/\partial y)\rho\sin\theta] \cdot \dot{\boldsymbol{r}} + \rho\dot{\theta}(-\boldsymbol{e}_x \sin\theta + \boldsymbol{e}_y \cos\theta)]$, we obtain,

$$\langle \boldsymbol{A} \cdot \dot{\boldsymbol{x}} \rangle = \boldsymbol{A}(\boldsymbol{r}) \cdot \dot{\boldsymbol{r}} + \frac{1}{2}\rho^2 \dot{\theta} \left(\frac{\partial A_y}{\partial x} - \frac{\partial A_x}{\partial y} \right) = \boldsymbol{A}(\boldsymbol{r}) \cdot \dot{\boldsymbol{r}} + \frac{1}{2} B \rho^2 \dot{\theta} . \quad (4.28)$$

Since $\langle \boldsymbol{v} \cdot \dot{\boldsymbol{x}} \rangle = \langle (v_\parallel \boldsymbol{b} - v_\perp(-\boldsymbol{e}_x \sin\theta + \boldsymbol{e}_y \cos\theta))(\dot{\boldsymbol{r}} + \rho\dot{\theta}(-\boldsymbol{e}_x \sin\theta + \boldsymbol{e}_y \cos\theta)) \rangle$ $= v_\parallel \boldsymbol{b} \cdot \dot{\boldsymbol{r}} - v_\perp \rho \dot{\theta}$, gyro-averaged Lagrangian L is given by,

$$L(\boldsymbol{r}, \dot{\boldsymbol{r}}, v_{//}, \mu, \dot{\theta}, t) = e_a \boldsymbol{A}^*(\boldsymbol{r}, v_{//}, t) \cdot \dot{\boldsymbol{r}} - \frac{m_a}{e_a} \mu \dot{\theta} - H(\boldsymbol{r}, v_\parallel, \mu, t) , \quad (4.29)$$

$$\boldsymbol{A}^* = \boldsymbol{A} + (m_a/e_a) v_\parallel \boldsymbol{b} , \quad \mu = \frac{m_a v_\perp^2}{2B} , \quad (4.30)$$

$$H(\boldsymbol{r}, v_{//}, \mu, t) = \frac{1}{2} m_a v_{//}^2 + \mu B(\boldsymbol{r}) + e_a \Phi(\boldsymbol{r}, t) . \quad (4.31)$$

Here, the variables of L are changed from $(\boldsymbol{r}, v_\parallel, v_\perp, \theta)$ to $(\boldsymbol{r}, v_\parallel, \mu, \theta)$, where $\mu = m_a v_\perp^2/B$ is the magnetic moment. Also, $\boldsymbol{A}^* = \boldsymbol{A} + (m_a/e_a) v_\parallel \boldsymbol{b}$ is the modified vector potential of Morozov–Solovev [6] and H is the Hamiltonian of the guiding center. The variational principle to give the guiding center motion is given by

$$\delta S = \delta \int L \, \mathrm{d}t = 0 \, . \tag{4.32}$$

Since the Lagrangian L does not contain gyro phase θ, Euler–Lagrange equation for θ reads $\mathrm{d}/\mathrm{d}t(\partial L/\partial \dot{\theta}) = 0$ and the conservation of magnetic moment $\mathrm{d}\mu/\mathrm{d}t = 0$ is obtained. In addition, since L does not contain $\dot{v}_\parallel$, the Euler–Lagrange equation for $v_\parallel$ reads $\partial L/\partial v_\parallel = 0$ and $v_\parallel = \boldsymbol{b} \cdot \dot{\boldsymbol{r}}$ is obtained. Also, L is free from $\dot{\mu}$, the Euler–Lagrange equation for μ reads $\partial L/\partial \mu = 0$ and $\dot{\theta} = -\Omega$ is obtained. Finally, the Euler–Lagrange equation for the guiding center $\boldsymbol{r}$ is obtained as follows,

$$\frac{\mathrm{d}}{\mathrm{d}t}\left(\frac{\partial L}{\partial \dot{\boldsymbol{r}}}\right) - \frac{\partial L}{\partial \boldsymbol{r}} = 0 \, , \tag{4.33}$$

$$L(\boldsymbol{r}, \dot{\boldsymbol{r}}, v_\parallel, \mu, \dot{\theta}, t) = e_a \boldsymbol{A}^*(\boldsymbol{r}, v_{//}, t) \cdot \dot{\boldsymbol{r}} - H(\boldsymbol{r}, v_\parallel, \mu, t) \, ,$$

where the second term of right-hand side of Equation 4.29 is dropped since that is constant for variable $\boldsymbol{r}$. This Lagrangian is called the Taylor Lagrangian.

Considering $\partial L/\partial \dot{\boldsymbol{r}} = e_a \boldsymbol{A} + m_a v_\parallel \boldsymbol{b}$, $\mathrm{d}/\mathrm{d}t = \partial/\partial t + \dot{\boldsymbol{r}} \cdot \nabla$, and $\nabla(\boldsymbol{A} \cdot \boldsymbol{C}) = \boldsymbol{C} \cdot \nabla \boldsymbol{A} + \boldsymbol{C} \times \nabla \times \boldsymbol{A}$ for a constant vector $\boldsymbol{C}$, we see,

$$\frac{\partial L}{\partial \boldsymbol{r}} = e_a(\dot{\boldsymbol{r}} \cdot \nabla \boldsymbol{A}^* + \dot{\boldsymbol{r}} \times \nabla \times \boldsymbol{A}^*) - \mu \nabla B - e_a \nabla \Phi \, , \tag{4.34}$$

$$\frac{\mathrm{d}}{\mathrm{d}t}\frac{\partial L}{\partial \dot{\boldsymbol{r}}} = m_a \frac{\mathrm{d}v_\parallel}{\mathrm{d}t}\boldsymbol{b} + e_a\left(\frac{\partial \boldsymbol{A}}{\partial t} + \dot{\boldsymbol{r}} \cdot \nabla \boldsymbol{A}\right) + m_a v_\parallel \left(\frac{\partial \boldsymbol{b}}{\partial t} + \dot{\boldsymbol{r}} \cdot \nabla \boldsymbol{b}\right) . \tag{4.35}$$

From $\dot{\boldsymbol{r}} \cdot \nabla \boldsymbol{A}^* = \dot{\boldsymbol{r}} \cdot \nabla \boldsymbol{A} + (m_a v_\parallel/e_a)\dot{\boldsymbol{r}} \cdot \nabla \boldsymbol{b}$, the following guiding center equation of motion is obtained.

$$m_a \frac{\mathrm{d}v_\parallel}{\mathrm{d}t}\boldsymbol{b} = e_a \dot{\boldsymbol{r}} \times \boldsymbol{B}^* + e_a \boldsymbol{E}^* - \mu \nabla B \, , \tag{4.36}$$

$$\boldsymbol{B}^* = \nabla \times \boldsymbol{A}^* = \boldsymbol{B} + (m_a v_\parallel/e_a)\nabla \times \boldsymbol{b} \, , \tag{4.37}$$

$$\boldsymbol{E}^* = -\frac{\partial \boldsymbol{A}^*}{\partial t} - \nabla \Phi = \boldsymbol{E} - (m_a v_\parallel/e_a)\frac{\partial \boldsymbol{b}}{\partial t} \, . \tag{4.38}$$

From Equation 4.36 $\cdot \boldsymbol{B}^*$ and Equation 4.36 $\times \boldsymbol{b}$, the following guiding center equation is obtained

$$\frac{\mathrm{d}v_\parallel}{\mathrm{d}t} = -\frac{1}{B_\parallel^*}\boldsymbol{B}^* \cdot (\mu \nabla B - e_a \boldsymbol{E}^*) \tag{4.39}$$

$$\frac{\mathrm{d}\boldsymbol{r}}{\mathrm{d}t} = \frac{1}{B_\parallel^*}\left[v_\parallel \boldsymbol{B}^* + \boldsymbol{b} \times ((\mu/e_a)\nabla B - \boldsymbol{E}^*)\right] \tag{4.40}$$

where

$$B_\parallel^* = \boldsymbol{b} \cdot \boldsymbol{B}^* = B + (m_a \, v_\parallel/e_a)\boldsymbol{b} \cdot \nabla \times \boldsymbol{b} \, . \tag{4.41}$$

When the system is time independent, energy is conserved and $v_\parallel$ becomes space-dependent, and $m_a v_\parallel \nabla v_\parallel = -\nabla(\mu B + e_a \Phi)$ is obtained from Equation 4.31. Substituting this into Equation 4.39 gives,

$$\frac{\mathrm{d}\boldsymbol{r}}{\mathrm{d}t} = \frac{v_\parallel}{\boldsymbol{b}\cdot\boldsymbol{B}^*}\nabla\times\left(\boldsymbol{A} + \frac{m_a v_\parallel}{e_a}\boldsymbol{b}\right) . \tag{4.42}$$

Morozov–Solovev equation [6]

Under the approximation $\boldsymbol{b}\cdot\boldsymbol{B}^* = B$, we have,

$$\frac{\mathrm{d}\boldsymbol{r}}{\mathrm{d}t} = \frac{v_\parallel}{B}\nabla\times(\boldsymbol{A} + \rho_\parallel \boldsymbol{B}) \tag{4.43}$$

where

$$\rho_\parallel = \frac{m_a v_\parallel}{e_a B} . \tag{4.44}$$

4.4 Orbital Dynamics: Hamilton Orbit Dynamics in Flux Coordinates

If the plasma is in force equilibrium ($\boldsymbol{J}\times\boldsymbol{B} = \nabla P$), we have to develop orbit theory in flux coordinates. The use of Boozer–Grad coordinates (ϕ, α, χ) and Boozer coordinates (ϕ, θ, ζ) described in Section 3.7 is efficient to analyze particle orbit because field is given by both co- and contra-variant forms [7]. First, let us derive the orbit equation under a static field using the Morosov–Solovev approximate expression.

$$\frac{\mathrm{d}\boldsymbol{r}}{\mathrm{d}t} = \frac{v_\parallel}{B}\nabla\times(\boldsymbol{A} + \rho_\parallel \boldsymbol{B}) . \tag{4.45}$$

In the Boozer–Grad coordinates (ϕ, α, χ), $\boldsymbol{B} = \nabla\phi\times\nabla\alpha = \nabla\times(\phi\nabla\alpha)$ ($\boldsymbol{A} = \phi\nabla\alpha$ except gauge term) as well as $\boldsymbol{B} = \nabla\chi + \beta\nabla\phi$.

$$\begin{aligned}\frac{\mathrm{d}\boldsymbol{r}}{\mathrm{d}t} &= \frac{v_\parallel}{B}\nabla\times\left[\phi\nabla\alpha + \rho_\parallel\nabla\chi + \rho_\parallel\beta\nabla\phi\right] \\ &= \frac{v_\parallel}{B}\left[\nabla\phi\times\nabla\alpha\left(1 - \frac{\partial\rho_\parallel\beta}{\partial\alpha}\right) + \nabla\alpha\times\nabla\chi\frac{\partial\rho_\parallel}{\partial\alpha} + \nabla\chi\times\nabla\phi\left(\frac{\partial\rho_\parallel\beta}{\partial\chi} - \frac{\partial\rho_\parallel}{\partial\phi}\right)\right] .\end{aligned} \tag{4.46}$$

Considering the orthogonal relation $\nabla u^i \cdot \partial\boldsymbol{x}/\partial u^j = \delta_{ij}$ (Equation 3.5), the Jacobian of the Boozer–Grad coordinates $J = 1/\nabla\phi\times\nabla\alpha\cdot\nabla\chi = 1/B^2$, and following vector relation,

$$\frac{\mathrm{d}\boldsymbol{r}}{\mathrm{d}t} = \frac{\partial\boldsymbol{r}}{\partial\phi}\frac{\mathrm{d}\phi}{\mathrm{d}t} + \frac{\partial\boldsymbol{r}}{\partial\alpha}\frac{\mathrm{d}\alpha}{\mathrm{d}t} + \frac{\partial\boldsymbol{r}}{\partial\chi}\frac{\mathrm{d}\chi}{\mathrm{d}t} . \tag{4.47}$$

We obtain the following orbit equations in the Boozer–Grad coordinates (ϕ, α, χ).

$$\frac{\mathrm{d}\phi}{\mathrm{d}t} = \frac{\mathrm{d}\boldsymbol{r}}{\mathrm{d}t} \cdot \nabla\phi = v_\parallel B \frac{\partial \rho_\parallel}{\partial \alpha} , \tag{4.48}$$

$$\frac{\mathrm{d}\alpha}{\mathrm{d}t} = \frac{\mathrm{d}\boldsymbol{r}}{\mathrm{d}t} \cdot \nabla\alpha = v_\parallel B \left(\frac{\partial \rho_\parallel \beta}{\partial \chi} - \frac{\partial \rho_\parallel}{\partial \phi} \right) , \tag{4.49}$$

$$\frac{\mathrm{d}\chi}{\mathrm{d}t} = \frac{\mathrm{d}\boldsymbol{r}}{\mathrm{d}t} \cdot \nabla\chi = v_\parallel B \left(1 - \frac{\partial \rho_\parallel \beta}{\partial \alpha} \right) . \tag{4.50}$$

Here, the evolution of $\rho_\parallel$ can be obtained either from Equation 4.51 under constant H and μ or from Equation 4.52 as follows,

$$H = (e_a^2 B^2 / 2m_a) \rho_\parallel^2 + \mu B + e_a \Phi , \tag{4.51}$$

$$\frac{\mathrm{d}\rho_\parallel}{\mathrm{d}t} = \frac{\mathrm{d}\boldsymbol{r}}{\mathrm{d}t} \cdot \nabla\rho_\parallel = v_\parallel B \left[\frac{\partial \rho_\parallel}{\partial \chi} - \rho_\parallel \left(\frac{\partial \beta}{\partial \chi} \frac{\partial \rho_\parallel}{\partial \alpha} - \frac{\partial \beta}{\partial \alpha} \frac{\partial \rho_\parallel}{\partial \chi} \right) \right] . \tag{4.52}$$

A more accurate orbit equation satisfying the Liouville theorem can be obtained from Equation 4.33. Eliminating the second term in the right-hand side of Equation 4.29, which is independent of the guiding center motion, we have the following Taylor Lagrangian [8],

$$L(\boldsymbol{r}, \dot{\boldsymbol{r}}) = e_a \boldsymbol{A} \cdot \dot{\boldsymbol{r}} + (m_a/2)(\dot{\boldsymbol{r}} \cdot \boldsymbol{b})^2 - \mu B(\boldsymbol{r}) - e_a \Phi(\boldsymbol{r}, t) . \tag{4.53}$$

This Taylor Lagrangian in Boozer–Grad coordinates is expressed as follows,

$$L = e_a \phi \dot{\alpha} + \frac{m_a}{2B^2} (\dot{\chi} + \beta \dot{\phi})^2 - \mu B - e_a \Phi \tag{4.54}$$

where we use

$$\begin{aligned} e_a \boldsymbol{A}(\boldsymbol{r}, t) \cdot \dot{\boldsymbol{r}} &= e_a \phi \nabla\alpha \cdot \tfrac{\partial \boldsymbol{r}}{\partial \alpha} \dot{\alpha} = e_a \phi \dot{\alpha} , \\ \boldsymbol{b} \cdot \dot{\boldsymbol{r}} &= \tfrac{1}{B} \left(\nabla\chi \cdot \tfrac{\partial \boldsymbol{r}}{\partial \chi} \dot{\chi} + \beta \nabla\phi \cdot \tfrac{\partial \boldsymbol{r}}{\partial \phi} \dot{\phi} \right) = (\dot{\chi} + \beta \dot{\phi})/B . \end{aligned} \tag{4.55}$$

Using (4.55), canonical momentum $P_\alpha = \partial L / \partial \dot{\alpha}$ and $P_\chi = \partial L / \partial \dot{\chi}$ conjugate to α and χ, and the Hamiltonian H is obtained as follows,

$$P_\alpha = e_a \phi , \tag{4.56}$$

$$P_\chi = \frac{m_a}{B^2} (\dot{\chi} + \beta \dot{\phi}) = e_a \rho_\parallel , \tag{4.57}$$

$$H = \frac{B^2}{2m_a} P_\chi^2 + \mu B + e_a \Phi . \tag{4.58}$$

The Hamilton equation is given by,

$$\frac{\mathrm{d}\alpha}{\mathrm{d}t} = \frac{\partial H}{\partial P_\alpha} , \quad \frac{\mathrm{d}P_\alpha}{\mathrm{d}t} = -\frac{\partial H}{\partial \alpha} , \tag{4.59}$$

$$\frac{\mathrm{d}\chi}{\mathrm{d}t} = \frac{\partial H}{\partial P_\chi} , \quad \frac{\mathrm{d}P_\chi}{\mathrm{d}t} = -\frac{\partial H}{\partial \chi} . \tag{4.60}$$

The above Hamilton equations have independent variables P_α, P_χ, α, χ and P_α and P_χ agrees with ϕ and $\rho_\parallel$, respectively except numerical constants from Equations 4.56 and 4.57. So, the following orbit equations for ϕ, $\rho_\parallel$, α and χ are obtained,

$$\frac{\mathrm{d}\alpha}{\mathrm{d}t} = \frac{\partial H}{e_a \partial \phi} = \Phi'(\phi) + \left(\frac{\mu}{e_a} + \frac{e_a \rho_\parallel^2 B}{m_a}\right)\frac{\partial B}{\partial \phi}\,, \tag{4.61}$$

$$\frac{\mathrm{d}\chi}{\mathrm{d}t} = \frac{\partial H}{e_a \partial \rho_\parallel} = \frac{e_a \rho_\parallel B^2}{m_a}\,, \tag{4.62}$$

$$\frac{\mathrm{d}\phi}{\mathrm{d}t} = -\frac{\partial H}{e_a \partial \alpha} = -\left(\frac{\mu}{e_a} + \frac{e_a \rho_\parallel^2 B}{m_a}\right)\frac{\partial B}{\partial \alpha}\,, \tag{4.63}$$

$$\frac{\mathrm{d}\rho_\parallel}{\mathrm{d}t} = -\frac{\partial H}{e_a \partial \chi} = -\left(\frac{\mu}{e_a} + \frac{e_a \rho_\parallel^2 B}{m_a}\right)\frac{\partial B}{\partial \chi}\,. \tag{4.64}$$

Similar, but more complicated, orbit equations can be obtained for Boozer coordinates (ϕ, θ, ζ) [9, 10].

Canonical Momentum and Hamiltonian in the Magnetic Coordinates

Even in a non-integrable magnetic field, canonical formulation is possible using the magnetic coordinates $(\phi,\ \theta,\ \zeta)$ introduced in Section 3.4. The Taylor Lagrangian L in magnetic coordinates is given by,

$$L = e_a(\phi\dot{\theta} - \psi\dot{\zeta}) + \frac{m_a}{2B^2}(B_\phi\dot{\phi} + B_\theta\dot{\theta} + B_\zeta\dot{\zeta})^2 - \mu B - e_a\Phi\,. \tag{4.65}$$

Canonical momentum $P_\theta = \partial L/\partial\dot{\theta}$, $P_\zeta = \partial L/\partial\dot{\zeta}$ conjugate to θ, ζ and the Hamiltonian H are given by

$$P_\theta = e_a(\phi + \rho_\parallel B_\theta)\,, \tag{4.66}$$

$$P_\zeta = e_a(-\psi + \rho_\parallel B_\zeta)\,, \tag{4.67}$$

$$H = \frac{e_a^2}{2m_a}\rho_\parallel^2 B^2 + \mu B + e_a\Phi\,. \tag{4.68}$$

Here, following relations are satisfied,

$$\rho_\parallel = \frac{m_a v_\parallel}{e_a B} = \frac{m_a}{e_a B^2}(B_\phi\dot{\phi} + B_\theta\dot{\theta} + B_\zeta\dot{\zeta})\,, \tag{4.69}$$

$$\boldsymbol{B} = B_\phi\nabla\phi + B_\theta\nabla\theta + B_\zeta\nabla\zeta\,. \tag{4.70}$$

4.5 Periodicity and Invariants: Magnetic Moment and Longitudinal Adiabatic Invariant

As in analytical mechanics, the adiabatic invariant is conserved in the case of periodic motion. Adiabatic invariant J is given by,

$$J = \oint \boldsymbol{p} \cdot d\boldsymbol{q} \,. \tag{4.71}$$

Here, the integration is performed around the closed orbit $C(t)$ of the periodic motion. In fact, using the following Hamilton equations of motion,

$$\frac{\mathrm{d}\boldsymbol{q}}{\mathrm{d}t} = \frac{\partial H}{\partial \boldsymbol{p}} \,, \quad \frac{\mathrm{d}\boldsymbol{p}}{\mathrm{d}t} = -\frac{\partial H}{\partial \boldsymbol{q}} \tag{4.72}$$

and setting s as parameter along the orbit, the time derivative of J is given by [11],

$$\begin{aligned}\frac{\mathrm{d}J}{\mathrm{d}t} &= \oint \left[\frac{\mathrm{d}\boldsymbol{p}}{\mathrm{d}t}\frac{\mathrm{d}\boldsymbol{q}}{\mathrm{d}s} + \boldsymbol{p}\frac{\mathrm{d}}{\mathrm{d}s}\left(\frac{\mathrm{d}\boldsymbol{q}}{\mathrm{d}t}\right)\right]\mathrm{d}s = \oint \left[-\frac{\partial H}{\partial \boldsymbol{q}}\frac{\mathrm{d}\boldsymbol{q}}{\mathrm{d}s} + \boldsymbol{p}\frac{\mathrm{d}}{\mathrm{d}s}\left(\frac{\partial H}{\partial \boldsymbol{p}}\right)\right]\mathrm{d}s \\ &= \oint \frac{\mathrm{d}}{\mathrm{d}s}\left(\boldsymbol{p}\frac{\partial H}{\partial \boldsymbol{p}} - H\right)\mathrm{d}s = \left[\boldsymbol{p}\frac{\partial H}{\partial \boldsymbol{p}} - H\right]_A^B = 0 \,. \end{aligned} \tag{4.73}$$

Position A agrees with B for periodic motion and J is conserved. This property of conservation of the adiabatic invariant originates from the fact that the motion is described by the Hamiltonian.

Conservation of magnetic moment: The adiabatic invariant associated with gyro motion as the periodic motion of charged particles is given by considering Equation 4.17

$$J = \oint p\,dq = \oint [m_a v_\perp + e_a A_\perp]\,dq_\perp = 2\pi m_a v_\perp \rho - e_a B \pi \rho^2 = \mu(2\pi m_a/e_a) \,. \tag{4.74}$$

Here, the closed integral of $\boldsymbol{A}$ is rewritten to flux using the Stokes theorem. Conservation of J leads to the conservation of the magnetic moment $\mu = m_a v_\perp^2/2B$ as given in Section 4.3.

Motion along the magnetic field has some limitation due to conservation of the magnetic moment (see Figure 4.6). The magnetic field strength B changes periodically along the magnetic field in toroidal configuration. Parallel velocity is given by $v_\parallel = (2(E-\mu B)/m_a)^{0.5}$ and $v_\parallel = 0$ at $B = E/\mu$. A particle cannot move to a location where B is higher than E/μ, and is reflected. The reflected particle moves in the opposite direction and is reflected again at the position satisfying $B = E/\mu$. This effect is called the "magnetic mirror effect" and the particles contained in the weak magnetic field strength are "trapped particles."

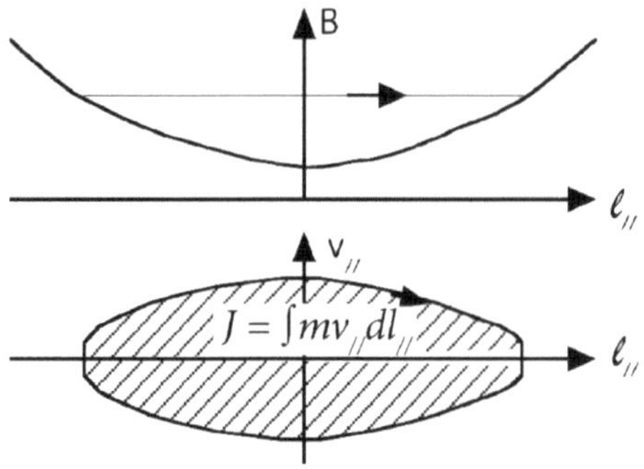

Figure 4.6 Magnetic field strength and parallel velocity $v_\parallel$ as a function of length along magnetic field $l_{//}$. Adiabatic invariant J is shaded area in $(l_{//}, v_\parallel)$ plane

Longitudinal adiabatic invariant: Since trapped particle exhibits periodic motion along the magnetic field, new invariant, called longitudinal adiabatic invariant appears.

$$J = \oint \boldsymbol{p} \cdot \mathrm{d}\boldsymbol{q} = \oint p_\parallel \mathrm{d}l_\parallel = \oint m_a(v_\parallel - e_a A_\parallel)\mathrm{d}l_\parallel = m_a \oint v_\parallel \mathrm{d}l_\parallel \,. \tag{4.75}$$

Here, the term $A_\parallel$ disappears due to periodicity, while the term $v_\parallel$ remains since it changes sign after reflection. For Boozer–Grad coordinates (ϕ, α, χ), Equation 4.75 reads,

$$J(H, \mu, \phi, \alpha) = m_a \oint v_\parallel \mathrm{d}l_\parallel = \oint p_\chi \,\mathrm{d}\chi = e_a \oint \rho_\parallel \,\mathrm{d}\chi \,. \tag{4.76}$$

From Equation 4.31, $v_\parallel$ is given by,

$$m_a v_\parallel = \pm\sqrt{2m_a(H - \mu B - e_a \Phi)} \,. \tag{4.77}$$

So, we have $\partial v_\parallel/\partial H = 1/m_a v_\parallel$. With this, an important relation of the global motion of the trapped particle orbit (sometimes called "Banana Orbit" due to its shape) is obtained from the partial derivatives of J with respect to H, ϕ and α. The partial derivative of J with respect to H is bounce period τ_b of the trapped particle orbit. In fact,

$$\frac{\partial J}{\partial H} = \oint \frac{\partial m_a v_\parallel}{\partial H} \mathrm{d}l_\parallel = \oint \frac{\mathrm{d}l_\parallel}{v_\parallel} = \tau_b \,. \tag{4.78}$$

Secondly, since $\mathrm{d}\phi/\mathrm{d}t = v_\parallel B \partial\rho_\parallel/\partial\alpha$ from Equation 4.48, we have

$$\frac{\partial J}{\partial \alpha} = e_a \oint \frac{\partial \rho_\parallel}{\partial \alpha} \,\mathrm{d}\chi = e_a \oint \frac{\mathrm{d}\phi/\mathrm{d}t}{v_\parallel B} B \,\mathrm{d}l_\parallel = e_a \oint (\mathrm{d}\phi/\mathrm{d}t)\,\mathrm{d}t = e_a \Delta\phi \,. \tag{4.79}$$

Namely, trapped particles move radially by $\Delta\phi = (\partial J/\partial\alpha)/e_a$ in every circulating motion of trapped particle orbit. Thirdly, since $\mathrm{d}\alpha/\mathrm{d}t = v_\parallel B(\partial\beta\rho_\parallel/\partial\chi - \partial\rho_\parallel/\partial\phi)$

from Equation 4.49, we have

$$\begin{aligned}\frac{\partial J}{\partial \phi} &= e_a \oint \frac{\partial \rho_\parallel}{\partial \phi}\,\mathrm{d}\chi = e_a \oint \left(\frac{\partial \beta \rho_\parallel}{\partial \chi}\,\mathrm{d}\chi - \frac{\mathrm{d}\alpha/\mathrm{d}t}{v_\parallel B} B\,\mathrm{d}l_\parallel \right) \\ &= -e_a \oint (\mathrm{d}\alpha/\mathrm{d}t)\,\mathrm{d}t = -e_a \Delta\alpha\,. \end{aligned} \tag{4.80}$$

Namely, trapped particles move in α direction by $\Delta\alpha = (\partial J/\partial\phi)/e_a$ in every circulating motion of trapped particle orbit. This motion is called "Precession motion." This precession motion in ϕ and α directions is given below from Equations 4.78–4.80.

$$\begin{aligned}\frac{\mathrm{d}\phi}{\mathrm{d}t} &= \frac{1}{e_a}\frac{\partial J/\partial\alpha}{\partial J/\partial H}\,, \\ \frac{\mathrm{d}\alpha}{\mathrm{d}t} &= -\frac{1}{e_a}\frac{\partial J/\partial\phi}{\partial J/\partial H}\,. \end{aligned} \tag{4.81}$$

Using Equation 4.81, we can prove the conservation of longitudinal adiabatic invariant again.

$$\frac{\mathrm{d}J}{\mathrm{d}t} = \frac{\partial J}{\partial\phi}\frac{\mathrm{d}\phi}{\mathrm{d}t} + \frac{\partial J}{\partial\alpha}\frac{\mathrm{d}\alpha}{\mathrm{d}t} = \frac{1}{e_a}\left(\frac{\partial J}{\partial\phi}\frac{\partial J}{\partial\alpha} - \frac{\partial J}{\partial\alpha}\frac{\partial J}{\partial\phi}\right) \Big/ \frac{\partial J}{\partial H} = 0\,. \tag{4.82}$$

Note: Chaotic Motion and Confinement in Dipole Field [14,15]

The simplest form of magnetic field is the dipole field ($\psi = M \sin^2\theta/r$ with polar coordinates (r, θ, ζ)) since there is no monopole for a magnetic field. In this case, there are three invariants in the particle motion, namely, magnetic moment $J_1 = \mu$, longitudinal adiabatic invariant $J_2 = J$ which gives bounce motion in the vertical direction, and $J_3 = e\psi$ if the gyro radius is small enough. However, conservation may be broken down if the gyro radius is comparable with special scale length. Especially if the bounce motion resonates with cyclotron motion, charged particle motion becomes chaotic.

Laboratory dipole experiments with a levitated superconducting ring show the formation of a stable electron vortex [15].

4.6 Coordinate Invariance: Non-canonical Variational Principle and Lie Transformation

One of the benefits of the variational principle is that it is applicable for any coordinates. The equation of motion with new coordinates can be obtained simply by rewriting the Lagrangian with the new coordinates and taking the variation with respect to the new coordinates. This is because the Lagrangian is a scalar. This property is called coordinate invariance. As we discussed in Section 4.3, non-canonical coordinates are useful for examining the dynamics of the charged particle orbit in plasma. Here, we describe the variational principle in non-canonical coordinates. The coordinate invariance is also a property of the differential form, which was invented by the French mathematician Elie J. Cartan (1869–1951) to describe the differential equation in a coordinate-independent form and has commonality with the variational principle. Dropping $\int$ from the action $S = \int L\,\mathrm{d}t$, $L\,\mathrm{d}t$ is called a "differential 1 form" and we write it as $L\,\mathrm{d}t = \gamma$. Below, we discuss expression of the differential form [16]. The Lagrangian L is given in a canonical form as $L = \boldsymbol{p}\cdot\dot{\boldsymbol{q}} - H(\boldsymbol{q},\boldsymbol{p},t)$ (in the differential 1 form $\gamma = \boldsymbol{p}\cdot\mathrm{d}\boldsymbol{q} - H\,\mathrm{d}t$). This can be expressed in arbitrary 6D coordinates $\boldsymbol{z} = \boldsymbol{z}(\boldsymbol{p},\boldsymbol{q})$ as follows,

$$L(\boldsymbol{z},\dot{\boldsymbol{z}},t) = \sum_{i=1}^{6} \gamma_i \dot{z}^i - h\,. \tag{4.83}$$

Equation 4.83 becomes $\gamma = \gamma_\mu\,\mathrm{d}z^\mu = \gamma_i\,\mathrm{d}z^i - h\,\mathrm{d}t$ in differential form (Poincaré–Cartan fundamental 1 form). Here, we take the sum on $\mu = 0, 6$ ($z^0 = t$, $\gamma_0 = -h$), $i = 1, 6$ (Einstein's rule). Except for the independence of variables, $\boldsymbol{z}$ can be non-canonical variables of position and velocity in general. Assuming the relationship between the new coordinates with canonical variables is given by $\boldsymbol{q} = \boldsymbol{q}(\boldsymbol{z},t)$, $\boldsymbol{p}(\boldsymbol{z},t)$, γ_i and h are given as follows,

$$\gamma_i(\boldsymbol{z},t) = \boldsymbol{p}\cdot\frac{\partial \boldsymbol{q}}{\partial z^i}\,, \tag{4.84}$$

$$h(\boldsymbol{z},t) = H(\boldsymbol{q}(\boldsymbol{z},t),\boldsymbol{p}(\boldsymbol{z},t),t) - \boldsymbol{p}\cdot\frac{\partial \boldsymbol{q}}{\partial t}\,. \tag{4.85}$$

Since $\delta L = \delta(\gamma_i(\mathrm{d}z^i/\mathrm{d}t) - h) = [(\partial\gamma_j/\partial z^i - \partial\gamma_i/\partial z^j)(\mathrm{d}z^j/\mathrm{d}t) - (\partial h/\partial z^i + \partial\gamma_i/\partial t)]\delta z^i + \mathrm{d}(\gamma_i\delta z^j)/\mathrm{d}t$, the Euler–Lagrange equation to satisfy $\delta S = 0$ becomes $\omega_{ij}(\mathrm{d}z^j/\mathrm{d}t) = \partial h/\partial z^i + \partial\gamma_i/\partial t$, where $\omega_{ij} = \partial\gamma_j/\partial z^i - \partial\gamma_i/\partial z^j$. From the definition of γ_i (Equation 4.84), we obtain $\omega_{ij} = [z^i, z^j] \equiv (\partial\boldsymbol{p}/\partial z^i)\cdot(\partial\boldsymbol{q}/\partial z^j) - (\partial\boldsymbol{p}/\partial z^j)\cdot(\partial\boldsymbol{q}/\partial z^i)$. Here, $[z^i, z^j]$ is called the Lagrange brackets [1]. Known in analytical mechanics, the inverse matrix of ω_{ij}, π_{ij} is given by $\pi_{ij} = \{z^i, z^j\} \equiv (\partial z^i/\partial\boldsymbol{q})\cdot(\partial z^j/\partial\boldsymbol{p}) - (\partial z^i/\partial\boldsymbol{p})\cdot(\partial z^j/\partial\boldsymbol{q})$ [1]. Here, $\{z^i, z^j\}$ is called the Poisson bracket. Then, the equation of motion in non-canonical coordinates z^i is given by $\mathrm{d}z^i/\mathrm{d}t = \pi_{ij}(\partial h/\partial z^j + \partial\gamma_j/\partial t)$. If γ_j does not explicitly depend on t, $\partial\gamma_j/\partial t = 0$ and we obtain $\mathrm{d}z^i/\mathrm{d}t = \pi_{ij}\partial h/\partial z^j = \{z^i, z^j\}\partial h/\partial z^j$ and can be written as follows,

$$\frac{\mathrm{d}z^i}{\mathrm{d}t} = \{z^i, h\}\,. \tag{4.86}$$

When the non-canonical coordinates $\boldsymbol{z} = \{z^\mu\} = \{t, z^i\}$ are transformed to other non-canonical coordinates $\bar{\boldsymbol{z}} = \{\bar{z}^\mu\} = \{t, \bar{z}^i\}$ Lagrangian differential 1 form is given by $\gamma = \gamma_\mu\mathrm{d}z^\mu = \Gamma_\nu\mathrm{d}\bar{z}^\nu$ ($\mu, \nu = 0, \ldots, 6$). The transformation law of γ due to the coordinate transformation is given by,

$$\Gamma_\mu = \frac{\partial z^\nu}{\partial \bar{z}^\mu}\gamma_\nu\,. \tag{4.87}$$

Figure 4.7 Sophus Lie who developed Lie group and Lie algebra

Norwegian mathematician M. S. Lie (1842–1899; Figure 4.7) formulated the Lie transform, which is useful for obtaining the orbit of charged particles when a small perturbation is applied to the plasma as given in the note.

Note: Lie Transformation [13]

Consider the coordinate transformation from $z = \{z^\mu\}$ to a new coordinates $\bar{z} = \{\bar{z}^\mu\}$, given by $\partial \bar{z}^\mu(z,\varepsilon)/\partial\varepsilon = g^\mu(\bar{z})$ (n1), where ε is a smallness parameter (here, $z^0 = \bar{z}^0 = t$ is invariant ($g^0 \equiv 0$)). An important point here is g^μ does not depend on ε explicitly and is called the **Lie transform**. The vector $\{g^\mu\}$ is called the generation vector of transformation. If we define the inverse as $z^\mu = z^\mu(\bar{z},\varepsilon)$, the identity relation $z^\mu(\bar{z}^\mu(z,\varepsilon),\varepsilon) = z^\mu$ is obtained. Differentiating with respect to ε and considering Equation n1, the relation to be satisfied by $z^\mu(\bar{z},\varepsilon)$ is obtained as $\partial z^\mu(\bar{z},\varepsilon)/\partial\varepsilon = -g^\nu(\bar{z})\partial z^\mu(\bar{z},\varepsilon)/\partial\bar{z}^\nu$ (n2).

Lie transform of scalar function: We study ε dependence of scalar functions brought by the Lie transform. Let $s(z)$ be a scalar function defined in the coordinates $\{z^\mu\}$, and $S(\bar{z},\varepsilon)$ be that in $\{\bar{z}^\mu\}$ coordinates to satisfy $S(\bar{z},\varepsilon) = s(z)$. Here, explicit ε dependence of S comes from the ε dependence of the coordinate transformation. Differentiation of $S(z(z,\varepsilon),\varepsilon) = s(z)$ by ε gives $\partial S(\bar{z},\varepsilon)/\partial\varepsilon = -g^\mu(\bar{z})\partial S(\bar{z},\varepsilon)/\partial\bar{z}^\mu$ (n3). Define an operator by $L = g^m\partial/y^m$ with arbitrary coordinate y^m. Here y^m can be z^m or $\bar{z}^\mu$. Then, Equation n3 becomes $\partial S(y,\varepsilon)/\partial\varepsilon = -LS(y,\varepsilon)$. Consider the Taylor expansion of $S(y,\varepsilon)$ noting that $\partial^n S(y,\varepsilon)/\partial\varepsilon^n|_{\varepsilon=0} = (-L)^n S(y,0)$ and $S(y,0) = s(y)$, we obtain $S(y,\varepsilon) = \exp(-\varepsilon L)s(y)$ (n4). Equation n4 can be extended if s has ε dependence. Let $s(z,\varepsilon) = s_0(z)+\varepsilon s_1(z)+(\varepsilon^2/2)s_2(z)+\ldots$ and continue similar discussion for $S_n(\bar{z},\varepsilon) = s_n(z(\bar{z},\varepsilon))$, then we obtain $S_n(y,\varepsilon) = \exp(-\varepsilon L)s_n(y)$ is obtained and finally $S(y,\varepsilon) = \exp(-\varepsilon L)s(y,\varepsilon)$ (n5).

Lie transform of differential form: We start by writing Equation 4.87 for the Lie transform as $\Gamma_\mu(\bar{z},\varepsilon) = (\partial z^\nu(\bar{z},\varepsilon)/\partial\bar{z}^\mu)\gamma_\nu(z(\bar{z},\varepsilon))$ (n6). Similar to the scalar case, we assume that γ is independent of ε. Differentiating Equation n6 by ε and noting Equation n2 and the circulation law $(\partial\gamma_\mu/\partial z^\lambda)(\partial z^\lambda/\partial\bar{z}^\nu) = \partial\gamma_\mu/\partial\bar{z}^\nu$, we have

$$\begin{aligned}
&\partial\Gamma_\mu(\bar{z},\varepsilon)/\partial\varepsilon \\
&= -\left\{\partial/\partial\bar{z}^\mu\left[g^\lambda(\bar{z})\partial z^\nu(\bar{z},\varepsilon)/\partial\bar{z}^\lambda\right]\right\}\gamma_\nu(z(\bar{z},\varepsilon)) \\
&\quad - g^\lambda(\bar{z})(\partial z^\nu(\bar{z},\varepsilon)/\partial\bar{z}^\mu)\partial\gamma_\nu(z(\bar{z},\varepsilon)/\partial\bar{z}^\lambda) \\
&= -\partial/\partial\bar{z}^\mu\left[g^\lambda(\partial z^\nu/\partial\bar{z}^\lambda)\gamma_\nu\right] \\
&\quad - g^\lambda\left[(\partial z^\nu/\partial\bar{z}^\mu)(\partial\gamma_\nu/\partial\bar{z}^\lambda) - (\partial z^\nu/\partial\bar{z}^\lambda)(\partial\gamma_\nu/\partial\bar{z}^\mu)\right] \\
&= -\partial/\partial\bar{z}^\mu\left[g^\lambda\Gamma_\lambda\right] - g^\lambda\left[\partial/\partial\bar{z}^\lambda\left((\partial z^\nu/\partial\bar{z}^\mu)\gamma_\nu\right) - \partial/\partial\bar{z}^\mu\left((\partial z^\nu/\partial\bar{z}^\lambda)\gamma_\nu\right)\right].
\end{aligned}$$

So,

$$\begin{aligned}
\partial\Gamma_\mu(\bar{z},\varepsilon)/\partial\varepsilon = -g^\lambda(\bar{z})&\left[\partial\Gamma_\mu(\bar{z},\varepsilon)/\partial\bar{z}^\lambda - \partial\Gamma_\lambda(\bar{z},\varepsilon)/\partial\bar{z}^\mu\right] \\
&- \partial\left[g^\nu(\bar{z})\Gamma_\nu(\bar{z},\varepsilon)\right]/\partial\bar{z}^\mu .
\end{aligned}$$

Since the second term of the right-hand side of this equation is a gauge term, we consider it in the end. This equation is written only with $\bar{z}$ so the following equation holds for arbitrary coordinate $\boldsymbol{y}$,

$$\partial\Gamma_\mu(\boldsymbol{y},\varepsilon)/\partial\varepsilon = -g^\lambda(\boldsymbol{y})\left[\partial\Gamma_\mu(\boldsymbol{y},\varepsilon)/\partial y^\lambda - \partial\Gamma_\lambda(\boldsymbol{y},\varepsilon)/\partial y^\mu\right]. \tag{4.88}$$

Defining the operator for any differential form ω as $(L\omega_\mu = g^\lambda(\partial\omega_\mu/\partial y^\lambda - \partial\omega_\lambda/\partial y^\mu)$, we obtain $\partial\Gamma_\mu(\bar{z},\varepsilon)/\partial\varepsilon = -L\Gamma_\mu$. Like scalars, we consider the Taylor expansion, we obtain $\Gamma(\boldsymbol{y},\varepsilon) = \exp(-\varepsilon L)\gamma(\boldsymbol{y},\varepsilon)$ (n7) as in Equation n5. To execute multiple Lie transforms, we set $T_n(\varepsilon) = \exp(-\varepsilon^n L_n)$ $(L_n\omega)_\mu = g_n^\lambda(\partial\omega_\mu/\partial y^\lambda - \partial\omega_\lambda/\partial y^\mu)$ (n8) and define the transformation $T = \ldots T_3T_2T_1$. Then the transformation law for the Lagrange differential 1 form is obtained considering the gauge term $\mathrm{d}S$ as follows,

$$\Gamma = T\gamma + \mathrm{d}S . \tag{4.89}$$

4.7 Lie Perturbation Theory: Gyro Center Orbit Dynamics

Littlejohn [12] and Cary [13] introduced Lie perturbation theory to the guiding center orbit theory and Brizard and Hahm [17] applied it to the orbit theory in the fluctuating fields for application to the gyrokinetic theory. The guiding center orbit equation derived in Section 4.3 is applicable if the field is stationary. But, it has a weak point that it is not valid if fluctuations exist with wavelength around the gyro radius of the plasma ($k_\perp \rho_i = O(1)$). With such low-frequency fluctuation in plasma ($\omega/\Omega_i \approx k_\parallel/k_\perp = O(\varepsilon)$), even the magnetic moment derived in Section 4.5 is no longer a conserved quantity. However, if the fluctuation level is small ($e\tilde{\Phi}/T_e = O(\varepsilon)$), it is possible to find new coordinates where the magnetic moment is conserved. Here, we treat the case of electrostatic perturbation $\delta\varphi$ with equilibrium electrostatic potential $\Phi = 0$ [18]. Set ε as a smallness parameter characterizing fluctuation in the plasma, the Lagrangian differential 1 form $L\,\mathrm{d}t$ in the coordinates $\boldsymbol{z} = \{z^\mu\} = \{t, z^i\}$ is given by,

$$L\,\mathrm{d}t = \gamma = \gamma_i(\boldsymbol{z}, \varepsilon)\mathrm{d}z^i(\bar{z}, \varepsilon) - h(\boldsymbol{z}, \varepsilon)\mathrm{d}t \ . \tag{4.90}$$

Our target is to transform this without ε dependence in new coordinates $\bar{\boldsymbol{z}} = \{\bar{z}^\mu\} = \{t, \bar{z}^i\}$.

$$L\,\mathrm{d}t = \Gamma = \Gamma_i(\bar{z})\mathrm{d}\bar{z}^i - H(\bar{z})\mathrm{d}t + \mathrm{d}S(\bar{z}) \ . \tag{4.91}$$

Here, we expand the transformation law of differential 1 form associated with Lie transform $\Gamma = T\gamma + \mathrm{d}S$ introduced in the previous section's note noting that $T = \ldots \exp(-\varepsilon^2 L_2)\exp(-\varepsilon L_1) = 1 - \varepsilon L_1 + \varepsilon^2((1/2)L_1^2 - L_2) + \ldots$ Relations in each order of ε can be derived as follows,

$$\varepsilon^0 \text{ order: } \Gamma_0 = \mathrm{d}S_0 + \gamma_0 \ , \tag{4.92}$$

$$\varepsilon^1 \text{ order: } \Gamma_1 = \mathrm{d}S_1 - L_1\gamma_0 + \gamma_1 \ , \tag{4.93}$$

$$\varepsilon^2 \text{ order: } \Gamma_2 = \mathrm{d}S_2 - L_2\gamma_0 + \gamma_2 - L_1\gamma_1 + (1/2)L_1^2\gamma_0 \ . \tag{4.94}$$

Although we choose Γ without ε dependence, ε expansion $\Gamma = \Gamma_0 + \Gamma_1 + \Gamma_2 \ldots$ is assumed to show such a solution exists. Lagrangian differential 1 form γ in an electromagnetic field is obtained by using Equation 4.27 and $\boldsymbol{z} = (t, \boldsymbol{x}, \boldsymbol{v})$ as

$$\gamma(t, \boldsymbol{x}, \boldsymbol{v}) = (e_a \boldsymbol{A}(\boldsymbol{x}, t) + m\boldsymbol{v}) \cdot \mathrm{d}\boldsymbol{x} - \left[m_a v^2/2 + e_a\varphi(\boldsymbol{x}, t)\right] \mathrm{d}t \ . \tag{4.95}$$

Zero, first, and second order Lagrangian differential 1 forms for the guiding center γ_0, γ_1, and γ_2 for a electrostatic perturbation are obtained from Equation 4.29 as follows,

$$\begin{aligned}\gamma_0(t, \boldsymbol{r}, v_\parallel, \mu, \theta) &= (e_a \boldsymbol{A} + m_a v_\parallel \boldsymbol{b}) \cdot \mathrm{d}\boldsymbol{r} - (m_a/e_a)\mu \mathrm{d}\theta \\ &\quad - [m_a\, v_\parallel^2/2 + \mu B(\boldsymbol{r})]\,\mathrm{d}t \ ,\end{aligned} \tag{4.96}$$

$$\gamma_1 = -e_a\varphi(\boldsymbol{r} + \boldsymbol{\rho}, t)\,\mathrm{d}t \tag{4.97}$$

$$\gamma_2 = 0 \ . \tag{4.98}$$

Since γ_1 has only t component ($\gamma_{1t} = -h_1 = -e_a\varphi, \gamma_{1i} = 0\ (i = 1, 6)$), we would like to absorb the perturbation only by Γ_{1t} with $\Gamma_{1i} = 0\ (i = 1, 6)$. Taking the i component of Equation 4.93 we obtain

$$0 = 0 - (L_1\gamma_0)_i + (\mathrm{d}S_1)_i \ . \tag{4.99}$$

Note Equation n8 from the note in Section 4.6 gives $(L_1\gamma_0)_i = g_1^j(\partial\gamma_{0i}/\partial z^j - \partial\gamma_{0j}/\partial z^i) = \omega_{ij}g_1^j$ and substitution to Equation 4.99 gives $\omega_{ij}g_1^j = \partial S_1/\partial z^i$. Similar to the case of Equation 4.94 from Equation 4.86, the generation vector g_1^i is obtained as follows,

$$g_1^i = \{S_1, z^i\}\,. \tag{4.100}$$

Substituting $\Gamma_{1t} = -H_1$, $\gamma_{1t} = -e_a\varphi$, $(L_1\gamma_0)_t = g_1^j(\partial\gamma_{00}/\partial z^j - \partial\gamma_{0j}/\partial z^0) = -\{S_1, z^j\}\partial h_0/\partial z^j = -\{S_1, h_0\}$ (note $\partial\gamma_{0j}/\partial t = 0$) into the t component of Equation 4.93

$$-H_1 = -e_a\varphi + \{S_1, h_0\} + \frac{\partial S_1}{\partial t} \quad \left(H_1 = h_1 - \frac{\mathrm{d}S_1}{\mathrm{d}t}\right)\,. \tag{4.101}$$

Here, we seek the solution $\Gamma_{1t} = -H_1$ that has no gyro phase dependence. Let gyro angle θ average expressed as $\langle\rangle$ and take a gyro angle average of (4.101). The second and third terms of the right-hand side are 0 and we obtain $\langle H_1\rangle = H_1 = \langle e_a\varphi\rangle$. Defining $e_a\tilde{\varphi} = e_a(\varphi - \langle\varphi\rangle)$ and substituting it into Equation 4.99, we obtain an equation to determine S_1 as $0 = -e_a\tilde{\varphi} + \{S_1, h_0\} + \partial S_1/\partial t = -e_a\tilde{\varphi} + \mathrm{d}S_1/\mathrm{d}t$. The main term of the Poisson brackets is $\{S_1, h\} \sim \Omega_a \partial S_1/\partial\theta$, $\partial S_1/\partial t \sim \omega/\Omega_a = O(\varepsilon)$ and we obtain the following equation,

$$S_1 = -e_a\int \varphi\,\mathrm{d}t \approx -\frac{e_a}{\Omega_a}\int \varphi \mathrm{d}\theta\,. \tag{4.102}$$

As for the second order in ε, the following relation is obtained from the i component of Equation 4.94 as follows,

$$\Gamma_{2i} = (\mathrm{d}S_2)_i - (L_2\gamma_0)_i + \gamma_{2i} - (L_1\gamma_1)_i + (1/2)(L_1[\mathrm{d}S_1 + \gamma_1 - \Gamma_1])_i\,. \tag{4.103}$$

Considering $L_1\mathrm{d}S_1 \equiv 0$, $\Gamma_{2i} = \Gamma_{1i} = \gamma_{2i} = \gamma_{1i} = 0\ (i = 1, 6)$, we obtain and

$$g_2^i = \{S_2, z^i\}\,. \tag{4.104}$$

Similarly, the t component of the ε^2 relation in Equation 4.94 gives the following results, noting $\gamma_{1t} = -e_a\varphi$, $\gamma_{2t} = -h_2 = 0$,

$$\Gamma_{2t} = (\mathrm{d}S_2)_t - (L_2\gamma_0)_t + \gamma_{2t} - (L_1\gamma_1)_t + (1/2)\left(L_1[\mathrm{d}S_1 + \gamma_1 - \Gamma_1]\right)_t\,. \tag{4.105}$$

Namely,

$$-H_2 = \frac{\partial S_2}{\partial t} + \{S_2, h_0\} + 0 + \frac{1}{2}\{S_1, h_1\} + \frac{1}{2}\{S_1, H_1\}\,. \tag{4.106}$$

Thus, to seek a solution that $\Gamma_{2t} = -H_2$ has no gyro angle dependence, we take a gyro angle average of Equation 4.106, noting that H_1 has no gyro angle dependence,

$$H_2 = \langle H_2\rangle = -\frac{1}{2}\langle\{S_1, h_1\}\rangle\,. \tag{4.107}$$

As described above, the Lie transformed Hamiltonian $H = H_0 + H_1 + H_2$ is determined while the coordinate system after Lie transform $\bar{z} = \{\bar{z}_\mu\} = \{t, \bar{z}_i\}$ is given by,

$$\bar{z}^\mu = z^\mu + \varepsilon \left. \frac{\partial \bar{z}^\mu}{\partial \varepsilon} \right|_{\varepsilon=0} + O(\varepsilon^2) = z^\mu + \varepsilon g(z^\mu) + O(\varepsilon^2) \,, \tag{4.108}$$

$$\bar{z}^\mu = z^\mu + \varepsilon \{S_1, z^\mu\} + O(\varepsilon^2) \,. \tag{4.109}$$

References

1. Goldstein H (1950) Classical Mechanics. Addison-Wesley.
2. Courant R, Hilbert D (1989) Methods of Mathematical Physics Vol. 1 Wiley-Interscience.
3. Landau LD, Lifshitz EM (1994) Classical Theory of Fields 4th edn. Pergamon Press.
4. Alfven H. (1940) Ark. Mat., Astron. Fys. 27A, 22.
5. Littlejohn RG (1983) J. Plasma Phys 29, 111–125.
6. Morozov AI, Solovev LS (1966) Rev. Plasma Phys., 2, 201.
7. Boozer A (2004) Rev. Mod. Phys., 76, 1071–1141.
8. Taylor JB (1964) Phys. Fluids, 7, 767–773.
9. Fowler RH, et al. (1985) Phys. Fluids 28, 338.
10. Boozer A and Kuo-Petravic G (1981) Phys. Fluids, 24, 851–859.
11. Kulsrud RM (2005) Plasma Physics for Astrophysics. Princeton University Press.
12. Littlejohn RH (1982) J. Math. Phys., 23, 742–747.
13. Cary JR, Littlejohn RG (1983) Ann. Phys., 151, 1–34.
14. Hasegawa A (2005) Physica Scr. T116, 72–74.
15. Yoshida, Z, et al. (2010) Phys. Rev. Lett., 104, 235004.
16. Flanders H (1989) Differential Forms with Applications to the Physical Sciences. Dover Books.
17. Brizard AJ, Hahm TS (2007) Rev. Mod. Phys., 79, 421–468.
18. Hahm TS (1988) Phys. Fluids, 31, 2670–2673.

Chapter 5
Plasma Kinetic Theory: Collective Equation in Phase Space

The motion of a large number of charged particles in plasma could be determined completely if the initial conditions are known since an individual particle follows Newton's equations of motion. The flow of the probability distribution function of the system in $6N$ phase space consisting of the position and momentum of N particles shows incompressibility (Liouville theorem). This property leads to an important theorem of the isolated dynamical system "Poincare's recurrence theorem," which guarantees that the system will return to be arbitrarily close to the initial state. The kinetics equation represented by the Boltzmann equation is derived from reversible mechanics equations, but is often irreversible. In the Boltzmann equation, a statistical assumption "Stosszahl Ansatz" leads to a collision term exhibiting the arrow of time. Thus, there is a fundamental difference between the reversible dynamical equation and the kinetic equation.

In the kinetic equation for high temperature plasma, a strange phenomenon (called Landau damping) occurs where the oscillating electric field damps with time even when collisions are negligible through the mechanism of "phase mixing" in the velocity space, since the operator of the kinetic equation $\boldsymbol{v} \cdot \partial f / \partial \boldsymbol{x}$ has a continuous spectrum. In this chapter, the basics of plasma kinetic equations including Coulomb collisions, the drift kinetic equation, and the gyro kinetic equations are introduced based on the orbit theories described in Chapter 4.

5.1 Phase Space: Liouville Theorem and Poincaré Recurrence Theorem

Plasma consists of many electrons and ions and the state of plasma motion is determined by their position and velocity. Once the initial values are determined, they are uniquely governed by the dynamical equation. The information necessary for each ion/electron is the position (x, y, z) and speed (v_x, v_y, v_z). Then, to specify the state of, for example, 10^{23} particles, a set of 6×10^{23} variables is necessary. This set of variables is regarded as "space" called "phase space" and the trajectory in the phase space is considered. Visualization of a four or higher dimensional space is not possible, but it is easier to imagine the motion of one point in the 6×10^{23} dimensional virtual space (phase space) than to imagine the motion of 10^{23} particles in 3-dimensional space [1]. When N particles move according to Newton's equations of motion, the point representing the system state draws the trajectory according to the following Hamilton equation in the $6N$-dimensional phase space $\boldsymbol{Z} = (\boldsymbol{q}_{3N}, \boldsymbol{p}_{3N})$ (called Γ space in statistical mechanics).

$$\begin{aligned} \frac{\mathrm{d}q_j}{\mathrm{d}t} &= \frac{\partial H}{\partial p_j}, \\ \frac{\mathrm{d}p_j}{\mathrm{d}t} &= -\frac{\partial H}{\partial q_j}. \end{aligned} \qquad (j = 1, 3N) \tag{5.1}$$

According to the above considerations, the system state is determined as a single point in the phase space, according to Newton's equation, if the initial values are given. US physicist J. W. Gibbs (1839–1903) who constructed statistical mechanics [2] introduced the concept of "ensemble" (he considered that a measurable macroscopic state includes a large number of microscopic states and this set is called an "ensemble"). To define the probability of the system in a microscopic state in the ensemble, he introduced a probability density D. In other words, the description of the system was changed from a "deterministic view" to a non-deterministic "probabilistic view." The macroscopically identical system is assumed to have a smooth distribution in phase space and the distribution is given by the probability distribution function. By this "smoothness," the possibility of the direction of time is introduced. Combining the continuity equation in phase space and the Hamilton equation, the phase space flow of the probability density is shown to be incompressible. Arbitrary volume element Ω in the phase space changes its shape with time but conserves its volume. Phase space flow $\boldsymbol{v}$ of the probability density D in the $6N$-dimensional phase space satisfies $\nabla \cdot \boldsymbol{v} = 0$. In fact, using Equation 5.1,

$$\nabla \cdot \boldsymbol{v} = \sum_{j=1}^{3N} \left[\frac{\partial \dot{p}_j}{\partial p_j} + \frac{\partial \dot{q}_j}{\partial q_j} \right] = \sum_{j=1}^{3N} \left[\frac{\partial}{\partial p_j} \left(-\frac{\partial H}{\partial q_j}\right) + \frac{\partial}{\partial q_j} \frac{\partial H}{\partial p_j} \right] = 0 \,. \tag{5.2}$$

Substituting $\nabla \cdot \boldsymbol{v} = 0$ into the continuity equation in $6N$-dimensional phase space $\partial D/\partial t + \nabla \cdot (D\boldsymbol{v}) = 0$, we obtain $\mathrm{d}D/\mathrm{d}t = \partial D/\partial t + \boldsymbol{v} \cdot \nabla D = 0$.

$$\frac{\mathrm{d}D}{\mathrm{d}t} = \frac{\partial D}{\partial t} + \sum_{j=1}^{3N} \left[\frac{\partial D}{\partial q_j} \frac{\partial H}{\partial p_j} - \frac{\partial D}{\partial p_j} \frac{\partial H}{\partial q_j} \right] = \frac{\partial D}{\partial t} + \{D, H\} = 0 \,. \tag{5.3}$$

Here, $\{D, H\}$ is called the Poisson bracket. The total derivative, $\mathrm{d}D/\mathrm{d}t$ is the time derivative of probability density along the $6N$-dimensional phase space flow, so probability density is conserved along the phase space flow. This is called the Liouville theorem [3]. Incompressibility of the phase space flow leads to an interesting

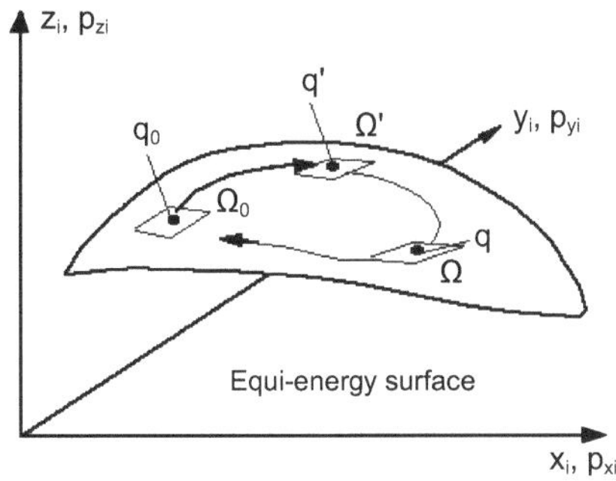

Figure 5.1 Phase space motion of dynamical system. System state is expressed by a point in the phase space. Starting from initial point q_0, q is always on the equi-energy surface

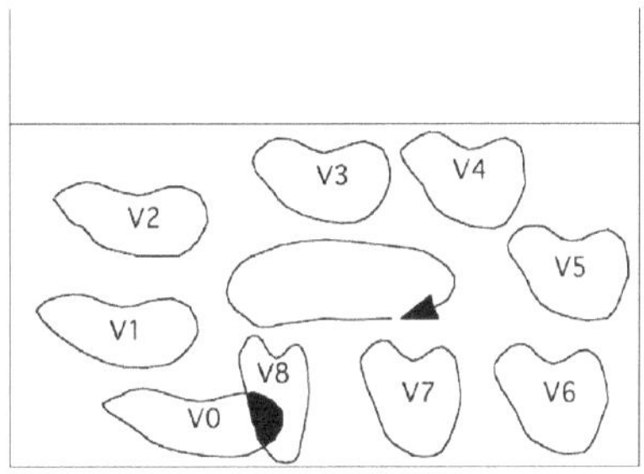

Figure 5.2 Water convection in a water tank. Initial volume element V_0 move as time goes $V_1, \ldots, V_8$. The incompressibility of water means all volumes are same

property of the "Poincaré recurrence theorem" [4] that an isolated dynamical system will return to be arbitrarily close to the starting point as time passes. This theorem plays an important role in the paradoxical discussion in deriving the laws of thermodynamics with an arrow of time from the time reversible dynamical equation [5].

Total energy is conserved in the isolated dynamical system consisting of N particles. The $6N$-dimensional phase space trajectory is shown in Figure 5.1 on the equi-energy surface. The Poincaré recurrence theorem claims that the system will return to an arbitrary close point from the initial point q_0 on the phase space (arbitrary neighborhood of q_0).

The Poincaré recurrence theorem is proved using reduction to absurdity. That is, contradiction occurs if the state does not return to be arbitrary close from the starting point. The motion on the phase space "behaves like a incompressible fluid," so the concept of the proof can be explained by using water as typical example of an incompressible fluid [6]. Figure 5.2 shows water convection in a finite volume water tank. The water occupying region V_0 at initial time t_0 moves to the new region V_1 at time t_1. The shape of V_1 may be different from that of V_0 but has same volume. Let $V_2, V_3, \ldots$ be regions at later time $t_2, t_3, \ldots$ If they never overlap, the volume of water becomes infinite and contradicts the initial assumption of the finite volume water tank. This shows that the initial assumption was wrong.

This Poincaré recurrence theorem predicts that in Figure 5.3 all gas molecules are in a left box (Figure 5.3 (a)) will expand to the full box after opening the shutter (Figure 5.3 (b)), but that a state arbitrary close to (Figure 5.3 (a)) can be realized some day (Figure 5.3 (c)).

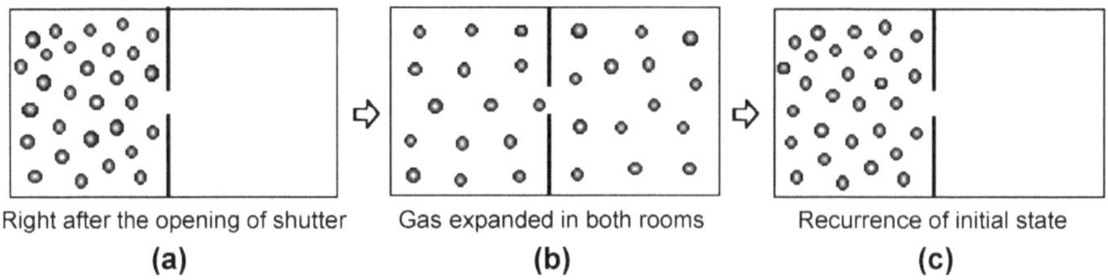

Figure 5.3 Molecular diffusion and prediction from the recurrence theorem

Note: The Poincaré Recurrence Theorem [7]

A mathematically more accurate definition and proof of the Poincaré theorem are given in Arnold [7].

Poincaré recurrence theorem: Let S be a bounded region in phase space and g a volume (measure) conserving one-to-one mapping from S to $S(gS = S)$. Then, in any neighborhood V of an arbitrary point in S, there is a point q that returns to the neighborhood V, $q \in V$. Namely, there exists $n > 0$ such that $g^n q \in V$.

Proof: Consider an infinite series of the mapping of neighborhood V, gV, g^2V, ..., $g^nV, \ldots$. Since g is volume (measure) conserving mapping, these mappings have the same volume. If they do not intersect, volume (measure) of S becomes infinite. This is inconsistent with the assumption that S is a bounded region. Therefore, there is an intersection among mappings. Let g^kV and g^mV $(k > m \geq 0)$ intersect, this means $g^kV \cap g^mV \neq \emptyset$. Here $\cap$ and $\emptyset$ represent the intersection and empty sets, respectively. Then, we obtain $g^{(k-m)}V \cap V \neq \emptyset$. Therefore, selecting q from the intersection, we have $q \in V$ and $g^{(k-m)}q \in V$ and $n = k - m$.

5.2 Dynamics and Kinetics: Individual Reversible and Collective Non-reversible Equations

The concept of the "velocity distribution function" was introduced by British physicist J. C. Maxwell (1831–1879; Figure 5.4 (a)) in 1860 [8]. Instead of specifying the status of all the particles in a deterministic dynamical equation, the smooth function f is defined so that $f(\boldsymbol{x}, \boldsymbol{v}, t)\mathrm{d}\boldsymbol{x}\mathrm{d}\boldsymbol{v}$ is the number of particles in position interval $\boldsymbol{x} \sim \boldsymbol{x} + \mathrm{d}\boldsymbol{x}$ and velocity interval $\boldsymbol{v} \sim \boldsymbol{v} + \mathrm{d}\boldsymbol{v}$. This smooth distribution function is constructed from the original discrete distribution function through some statistical operation. The deterministic reversible equation is converted to an irreversible collective equation. The exact velocity distribution function F, considering plasma is a group of discrete particles, is given in the following form by using a delta func-

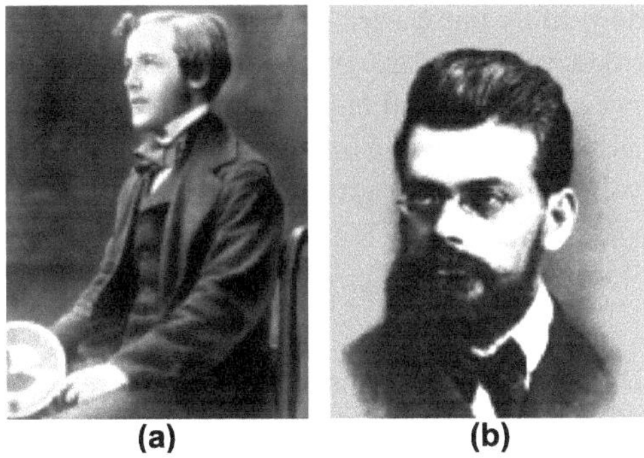

Figure 5.4 (a) James Clark Maxwell who invented the concept of velocity distribution function and (b) Ludwig Boltzmann who invented the Boltzmann equation and derived H-theorem to explain macroscopic irreversibility from microscopic law

tion, and it follows the "Klimontovich equation" in which particle density is conserved along the phase space $\boldsymbol{z} = (\boldsymbol{x}, \boldsymbol{v})$ (called γ space in statistical mechanics) trajectory [9].

$$F(\boldsymbol{x}, \boldsymbol{v}, t) = \sum_{i=1}^{N} \delta(\boldsymbol{x} - \boldsymbol{x}_i(t))\delta(\boldsymbol{v} - \boldsymbol{v}_i(t)) \,, \tag{5.4}$$

$$\frac{\mathrm{d}F}{\mathrm{d}t} = \frac{\partial F}{\partial t} + \boldsymbol{v} \cdot \frac{\partial F}{\partial \boldsymbol{x}} + \boldsymbol{a} \cdot \frac{\partial F}{\partial \boldsymbol{v}} = 0 \,. \tag{5.5}$$

Here, the acceleration $\boldsymbol{a} = (e/m)(\boldsymbol{E} + \boldsymbol{v} \times \boldsymbol{B})$ includes the force of the average electromagnetic field and the Coulomb collisions between charged particles. Taking the ensemble average of F to get a smooth velocity distribution function $f = \langle F \rangle_{\text{ensemble}}$ and we obtain the following "collision term" [10],

$$C(f) = \bar{\boldsymbol{a}} \cdot \frac{\partial f}{\partial \boldsymbol{v}} - \left\langle \boldsymbol{a} \cdot \frac{\partial F}{\partial \boldsymbol{v}} \right\rangle_{\text{ensemble}} \tag{5.6}$$

Also, the Boltzmann-type transport equation is obtained.

$$\frac{\mathrm{d}f}{\mathrm{d}t} = \frac{\partial f}{\partial t} + \boldsymbol{v} \cdot \frac{\partial f}{\partial \boldsymbol{x}} + \bar{\boldsymbol{a}} \cdot \frac{\partial f}{\partial \boldsymbol{v}} = C(f) \tag{5.7}$$

$$C(f) = -\left\langle \tilde{\boldsymbol{a}} \cdot \frac{\partial \tilde{F}}{\partial \boldsymbol{v}} \right\rangle_{\text{ensemble}} \,. \tag{5.8}$$

Here, $\boldsymbol{a}$ and F are divided into an average part and a microscopic fluctuating part ($\boldsymbol{a} = \bar{\boldsymbol{a}} + \tilde{\boldsymbol{a}}$, $F = f + \tilde{F}$). We find that the collision term is a correlation of the acceleration by the microscopic Coulomb field and the associated velocity space gradient of the fluctuating distribution function. Acceleration by mean-field is treated deterministically and collision by the microscopic Coulomb field is treated statistically. The "smoothness" of the distribution function plays an important role in explaining collisionless damping (Landau damping) by "phase mixing." The collision term in the plasma is discussed in Section 5.6. The collision term for molecular gas was derived by the Austrian physicist L. Boltzmann (1844–1906 Figure 5.4 (b)). He introduced "Stosszahl Ansatz" which states that that there will be no correlation between position and momentum of two colliding particles [1]. With this assumption, we can calculate the time variation of particle number into and out from phase space volume $\mathrm{d}\boldsymbol{x}\mathrm{d}\boldsymbol{v}$ through collisional short-range force using f as follows,

$$C(f) = \int [f(\boldsymbol{v}')f(\boldsymbol{v}_1') - f(\boldsymbol{v})f(\boldsymbol{v}_1)]\,|\boldsymbol{v}_1 - \boldsymbol{v}|\,\sigma\,\mathrm{d}\Omega \mathrm{d}\boldsymbol{v}_1 \,. \tag{5.9}$$

Here, $\boldsymbol{v}$ and $\boldsymbol{v}_1$ are the velocities before collision and $\boldsymbol{v}'$ and $\boldsymbol{v}_1'$ are the velocities after collision. Also, the first term on the right-hand side is incoming particles to the velocity interval $\mathrm{d}\boldsymbol{v}$ from the inverse collision of particles with $\boldsymbol{v}'$ and $\boldsymbol{v}_1'$, and the second term is outgoing particles from the velocity interval $\mathrm{d}\boldsymbol{v}$ from the collision. The Boltzmann equation is an irreversible equation constructed from reversible dynamics (see Salon).

Salon: Reversible Dynamical Equation and Irreversible Kinetic Equation

The Newton equation $m\,\mathrm{d}^2x/\mathrm{d}t^2 = F$ does not change by $t \to -t$ and is symmetric to time reversal. On the other hand, phenomena concerning heat are not time reversible, for example, when hot water gets cold or gas expands to a low-pressure region. Such phenomena are called irreversible processes. If the heat is from the microscopic motion of atom and molecule, question arises whether irreversible thermal phenomena such as the law of entropy increase can be explained by the reversible dynamical equation.

Boltzmann attacked this problem by applying a dynamical equation to the microscopic molecule. He constructed an equation (Boltzmann equation 5.7 and 5.9) governing the process whereby a non-equilibrium gas relaxes to equilibrium by collisions between molecules. The Boltzmann collision term shows that the Boltzmann H function $H = \int f \cdot \ln f\,\mathrm{d}\boldsymbol{v}$ decreases monotonically with time (the Boltzmann H theorem). He used a time-symmetric dynamical equation for the collision process, but the equation got a time arrow through the statistical operation of counting the number of colliding particles.

The German physicist E. Zermelo (1871–1953) pointed out that H theorem contradicts Poincaré recurrence theorem [5]. Boltzmann's H theorem is a statistical theorem and the very low probability of recurrence is neglected (or, the equivalently neglected case which takes a very long time from the deterministic dynamical equation). Boltzmann's equation eliminates the recurrence and describes evolution to the high probability state. The Japanese Nobel Prize in Physics winner S. Tomonaga discussed this paradox, which lies between the kinetics and dynamics, in detail [11]. This situation is the same for the Coulomb collision in plasma, and $H = \int f \cdot \ln f\,\mathrm{d}\boldsymbol{v}$ decreases monotonically with time.

Consider the small size of the realization probability of the Poincaré recurrence state from the example of particle diffusion in a box shown in Figure 5.3). The group of particles in the left box expands to both boxes but will revert to left box at some time according to the Poincaré recurrence theorem. Then, try to evaluate the probability. The probability of having a particle in either box is the same and is $1/2$. Assuming the probability of each particle is independent, the probability of all particles entering the left box is $(1/2)^N$ (N is the number of particles). If $N = 10^{23}$, the chance to realize the state predicted by the Poincaré recurrence is very small. This means that it take an enormous amount of time to realize the recurrent state. From the viewpoint of the initial value, there is a set of initial values, with which particles can return to the left box in a relatively short timescale (for example, the time reversal solution of diffusion). But, this has a very low probability among all possible initial values for the particles spread in the whole box.

Interestingly, it seems apparent that the probability of having a particle in either box is the same, but it is not self-evident that the probability that exists in any location in the whole box is the same. This problem is an example in a dynamical system called an "ergodic problem" and is explained by "Weyl's billiards" [12]. The arrow of time in a many-particle system is caused by a macroscopic manipulation (for example, the operation of opening the shutter), which leads to a change in the number of possible motion states, and the system tends to the state of dominant probability. The key here is that the macro-operation can increase the number of microscopic states, but rarely reduce it. British astronomer Arthur Eddington (1882–1944) coined the phrase "arrow of time" in 1927. Some physical phenomena are difficult to reverse in time and this phrase indicated that "time" has direction related to the occurring phenomena. The arrow of time is discussed in detail by Davies in [13].

5.3 Vlasov Equation: Invariants, Time-reversal Symmetry and Continuous Spectrum

"Collision" in terms of changes in the velocity distribution function due to the microscopic electric field of the Coulomb potential becomes negligible compared with the average force when the plasma temperature is higher and such a plasma state is called "collisionless plasma." In this case, the right-hand side of Equation 5.7 can be ignored. Russian physicist A. Vlasov (1908–1975; Figure 5.5 (a)) pointed out for the first time that the collision term can be ignored in high-temperature plasma and this equation is called Vlasov equation [14].

$$\frac{\mathrm{d}f_s}{\mathrm{d}t} = \frac{\partial f_s}{\partial t} + \boldsymbol{v} \cdot \frac{\partial f_s}{\partial \boldsymbol{x}} + \bar{\boldsymbol{a}} \cdot \frac{\partial f_s}{\partial \boldsymbol{v}} = 0 . \tag{5.10}$$

Here, $\sum \bar{\boldsymbol{a}} = (e_s/m_s)(\boldsymbol{E} + \boldsymbol{v} \times \boldsymbol{B})$ is the average acceleration excluding the microscopic Coulomb field. The f_s in Vlasov equation is the ensemble-averaged "smooth" distribution function. Equation 5.10 means that the density f_s is conserved for the observer moving with particle trajectory in phase space $\boldsymbol{z} = (\boldsymbol{x}, \boldsymbol{v})$

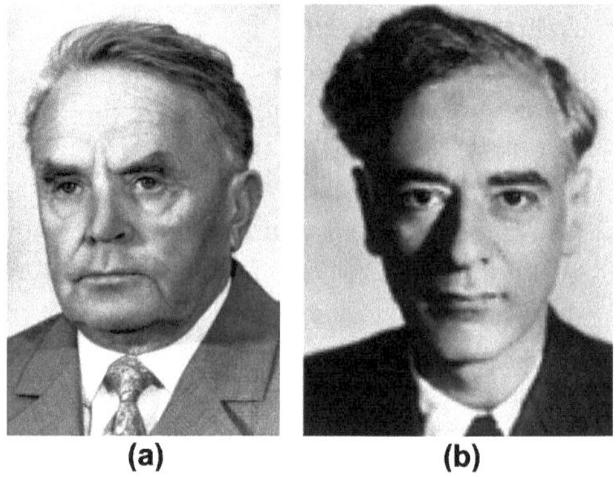

Figure 5.5 (a) A. A. Vlasov who identified Vlasov equation as basic equation for collisionless plasma (with kind permission of Physics-Uspekhi journal) and (b) L. D. Landau who found Landau damping as collisionless damping mechanism

(density variation along phase space flow, $\mathrm{d}f_s/\mathrm{d}t$, the Lagrange differential, is zero). From this property, $f(\boldsymbol{x}, \boldsymbol{v}, t) > 0 (t > 0)$ if $f(\boldsymbol{x}, \boldsymbol{v}, t = 0) > 0$ are met. In other words, the trajectory of particle motion in phase space (characteristic curve) contours of constant f_s. If no sink and source exists in phase space, the particle conservation law in phase space is given by $\partial f_s/\partial t + \partial/\partial \boldsymbol{z} \cdot (\boldsymbol{u} f_s) = 0$ ($\boldsymbol{u} = (\boldsymbol{v}, \bar{\boldsymbol{a}})$ and comparison with Vlasov equation leads to $(\partial/\partial \boldsymbol{z}) \cdot \boldsymbol{u} = 0$. This means that the phase space flow is incompressible.

In a collision-dominated gas in equilibrium, entropy is conserved in the isolated system. In collisionless plasma satisfying the Vlasov equation, H defined by $H = \int G(f_s)\mathrm{d}z$ is a conserved quantity ($\mathrm{d}H/\mathrm{d}t = 0$) for arbitrary function $G(f_s)$. In fact,

$$\frac{\mathrm{d}H}{\mathrm{d}t} = \int \frac{\partial G(f_s)}{\partial t} \mathrm{d}z = -\int G'(f_s) \boldsymbol{u} \cdot \frac{\partial f_s}{\partial \boldsymbol{z}} \mathrm{d}z = -\int \frac{\partial \boldsymbol{u} G(f_s)}{\partial \boldsymbol{z}} \mathrm{d}z = 0 \,. \quad (5.11)$$

Since G is an arbitrary function, the Vlasov equation has an infinite number of invariants. If we choose $G = f_s$, it gives the conservation of particles. Also, $G = -f_s \ln f_s$ gives the conservation of entropy in collisionless plasma. This property of the Vlasov equation is also called the generalized entropy conservation law [15].

The Vlasov equation has interesting properties, although we need to note that the equation is valid in the zero-collision limit. One of them is time-reversal symmetry, the Boltzmann equation does not have such symmetry. If $\boldsymbol{\psi} = (f_s, \boldsymbol{E}, \boldsymbol{B})$ is a solution of the Vlasov equation, $T\boldsymbol{\psi}(-t)$ is also a solution and is called the time-reversal solution (here, T is the "time reversal operator," and requires a reversal of the magnetic field) [16]. In the Boltzmann equation, the distribution function will converge to an equilibrium solution due to the property of the collision term. On the other hand, the solution of the Vlasov equation does not necessarily converge to an equilibrium solution due to its conservation property and time-reversal symmetry. If $\boldsymbol{\psi}(t)$ is the converging solution to equilibrium, the time reversal solution $T\boldsymbol{\psi}(-t)$ is the solution away from equilibrium.

The Vlasov equation has the structure of a wave equation. In addition to the wave frequency determined by the dispersion properties of the system, there is a wave with frequency $\omega_c = \boldsymbol{k} \cdot \boldsymbol{v}$ continuously changing with particle velocity $\boldsymbol{v}$ at fixed wave number $\boldsymbol{k}$, $(\exp(-\mathrm{i}\boldsymbol{k} \cdot \boldsymbol{v}t))$, which is called free streaming solution [17].

In fact, if we expand the electrostatic wave solution of the Vlasov equation without magnetic field as $f_a = F_{a0} + f_{a1}$ $(f_{a1} \ll F_{a0})$, the following Poisson equation and the linearized Vlasov equation are obtained,

$$\frac{\partial f_{a1}}{\partial t} + \boldsymbol{v} \cdot \frac{\partial f_{a1}}{\partial \boldsymbol{x}} = \frac{e_a}{m_a} \nabla\varphi \cdot \frac{\partial f_{a0}}{\partial \boldsymbol{v}} , \tag{5.12}$$

$$\varepsilon_0 \nabla^2 \varphi = -e_a \int_{-\infty}^{\infty} f_{a1} d\boldsymbol{v} . \tag{5.13}$$

For the case of $f_{a1} = f_{a1kw} \exp(\mathrm{i}\boldsymbol{k} \cdot \boldsymbol{x} - \mathrm{i}\omega t)$ and $\boldsymbol{\varphi} = \boldsymbol{\varphi}_{kw} \exp(\mathrm{i}\boldsymbol{k} \cdot \boldsymbol{x} - \mathrm{i}\omega t)$, we obtain

$$(\omega - \boldsymbol{k} \cdot \boldsymbol{v}) f_{a1\boldsymbol{k}\omega} = -\frac{e_a}{m_a} \varphi_{\boldsymbol{k}\omega} \boldsymbol{k} \cdot \frac{\partial f_{a0}}{\partial \boldsymbol{v}} . \tag{5.14}$$

Here we note that the homogeneous solution of (5.14) is free streaming solution $e^{(-\mathrm{i}\boldsymbol{k} \cdot \boldsymbol{v}t)}$. Since the general solution of $xf(x) = g(x)$ is given by $f(x) = g(x)P[x^{-1}] + \lambda\delta(x)$ (P is a principal value, $\delta(x)$ is the Dirac delta function), we obtain

$$f_{a1\boldsymbol{k}\omega} = \left[-\frac{e_a}{m_a} \boldsymbol{k} \cdot \frac{\partial f_{a0}}{\partial \boldsymbol{v}} P \frac{1}{\omega - \boldsymbol{k} \cdot \boldsymbol{v}} + \lambda\delta(\omega - \boldsymbol{k} \cdot \boldsymbol{v}) \right] \varphi_{\boldsymbol{k}\omega} . \tag{5.15}$$

If we inverse-Fourier transform the delta function term of Equation 5.15, we obtain a free streaming solution $f_{a1} = \exp[-\mathrm{i}\boldsymbol{k} \cdot (\boldsymbol{x} - \boldsymbol{v}t)]$. The range of $\boldsymbol{k} \cdot \boldsymbol{v}$ spans $-\infty$ to $+\infty$ and the waves have a "continuous spectrum." In fact, from the Poisson equation 5.13, we obtain following [17],

$$\left[1 + \frac{e_a}{\varepsilon_0 k^2 m_a} \int_{-\infty}^{\infty} \frac{P}{\omega - \boldsymbol{k} \cdot \boldsymbol{v}} \boldsymbol{k} \cdot \frac{\partial f_{a0}}{\partial \boldsymbol{v}} \mathrm{d}\boldsymbol{v} \right] + \frac{e_a}{\varepsilon_0 k^2} \lambda = 0 . \tag{5.16}$$

This equation gives the relationship between two unknowns, ω and λ for a given wave number k. This means that the angular frequency ω is arbitrary for a given k, namely, the spectrum of eigenvalues is continuous. This property of linear Vlasov equations originates from the fact that operator $\Lambda = \boldsymbol{k} \cdot \boldsymbol{v}(\mathrm{d}f/\mathrm{d}t = -\mathrm{i}\Lambda f)$ is a linear operator with a "continuous spectrum" [18, 19]. A non-damping wave can exist with a real ω and λ can be determined using Equation 5.16 which gives the necessary resonant particles in Equation 5.15. This wave is called the "Van Kampen mode" [16]. A free-streaming solution coupled to Maxwell distribution $f_{a1} = \exp[\mathrm{i}kut - (u/u_{\mathrm{th}})^2/2]$ oscillates in the velocity space more violently with time, as shown in Figure 5.6, and actual physical quantities, such as the electric field calculated from the velocity integral, tends to zero with time. This "structure extinction" is called "phase mixing" since it occurs due to the phase overlapping of the wave. Mathematically, this structure extinction is guaranteed by the "Riemann–Lebesgue theorem" [16, 20]. Collisionless damping of the electric field caused by the contin-

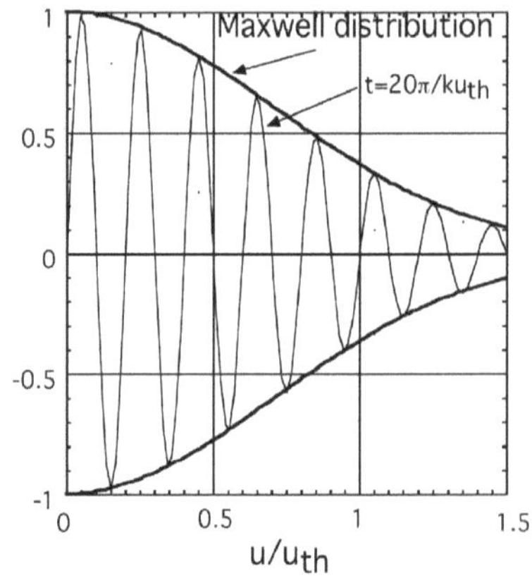

Figure 5.6 Maxwell distribution function ($\exp(-u^2/u_{\text{th}}^2)$) and the phase mixing of perturbed distribution function $f_1 = \exp(ikut - u^2/2u_{\text{th}}^2)$ ($t = 20\pi/ku_{\text{th}}$)

uous spectrum of velocity space is called "Landau damping" as described in detail in Section 5.4 and was developed by L. D. Landau (1908–1968; Figure 5.5 (b)).

5.4 Landau Damping: Irreversible Phenomenon Caused by Reversible Equation

The Vlasov equation is symmetric to time reversal, and in order to satisfy causality (cause prior to the results or "arrow of time"), the Laplace transform on time (or an equivalent method) can be used. This corresponds to analyzing the problem as an initial value problem by restricting $t \geq 0$, in contrast to the "spectral analysis" detailed in Section 5.3. The electron plasma oscillation is described by considering that the electric field is determined by the perturbed plasma density according to the Poisson equation. The linear response of the system is a solution that meets both the Poisson equation and the Fourier transformed linearized Vlasov equation in space.

$$\frac{\partial f_{e1\boldsymbol{k}}}{\partial t} + \mathrm{i}\boldsymbol{k}\cdot\boldsymbol{v} f_{e1\boldsymbol{k}} = -\mathrm{i}\frac{e}{m_e}\varphi_{\boldsymbol{k}}\boldsymbol{k}\cdot\frac{\partial f_{e0}}{\partial \boldsymbol{v}}\,, \tag{5.17}$$

$$\varepsilon_0 k^2 \varphi_k = -e\int_{-\infty}^{\infty} f_{e1\boldsymbol{k}}\mathrm{d}\boldsymbol{v}\,. \tag{5.18}$$

A standard way to solve this equation correctly as an initial value problem is the Laplace transform in time (integration at $t \geq 0$).

$$f_{e1\boldsymbol{k}\omega}(\boldsymbol{v}) = \frac{1}{2\pi}\int_0^{\infty} f_{e1\boldsymbol{k}}(\boldsymbol{v},t)e^{\mathrm{i}\omega t}\,\mathrm{d}t\,, \tag{5.19}$$

$$\varphi_{\boldsymbol{k}\omega} = \frac{1}{2\pi}\int_0^{\infty}\varphi_{\boldsymbol{k}}(t)e^{\mathrm{i}\omega t}\,\mathrm{d}t\,. \tag{5.20}$$

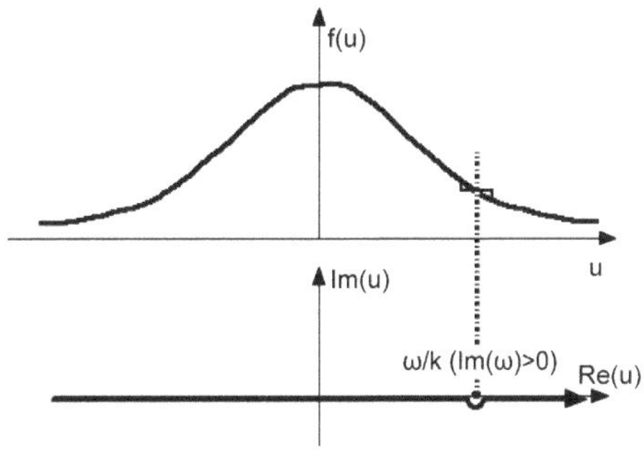

Figure 5.7 Maxwell distribution function ($\exp(-u^2/2u_{\text{th}}^2)$) and the phase velocity of the wave that interacts with resonant particles with almost same velocity. Integration path in complex u plane for Landau problem

The key here is the Laplace transformation, Equations 5.19 and 5.20 are defined so that $\text{Im}(\omega) = \omega_i$ (ω_i is a positive constant) is large enough to ensure the convergence of the integral. Causality is satisfied through this choice. The Fourier-Laplace transformation of Equations 5.17 and 5.18 are given as follows,

$$(\omega - \boldsymbol{k}\cdot\boldsymbol{v}) f_{e1\boldsymbol{k}\omega}(\boldsymbol{v}) = i f_{e1\boldsymbol{k}}(\boldsymbol{v}, t=0) + \frac{e}{m_e}\varphi_{\boldsymbol{k}\omega}\boldsymbol{k}\cdot\frac{\partial f_{e0}}{\partial \boldsymbol{v}}\,, \tag{5.21}$$

$$\mathrm{i}\varepsilon_0 k^2 \varphi_{\boldsymbol{k}\omega} = -e\int_{-\infty}^{\infty} f_{e1\boldsymbol{k}}(\boldsymbol{v},\omega)\mathrm{d}\boldsymbol{v}\,. \tag{5.22}$$

Substituting Equation 5.21 into 5.22 and eliminating $f_{e1\boldsymbol{k}}(\boldsymbol{v}, t)$, we obtain

$$\varphi_{\boldsymbol{k}\omega} = -\frac{\mathrm{i}e}{\varepsilon_0 k^2 K(\omega,\boldsymbol{k})}\int_{-\infty}^{\infty}\frac{f_{e1\boldsymbol{k}}(\boldsymbol{v}, t=0)}{\omega - \boldsymbol{k}\cdot\boldsymbol{v}}\mathrm{d}\boldsymbol{v}\,, \tag{5.23}$$

$$K(\boldsymbol{k},\omega) = 1 + \frac{\omega_{pe}^2}{n_e k^2}\int\frac{\boldsymbol{k}\cdot\partial f_{e0}/\partial\boldsymbol{v}}{\omega - \boldsymbol{k}\cdot\boldsymbol{v}}\mathrm{d}\boldsymbol{v} \tag{5.24}$$

where $\omega_{pe}^2 = e^2 n_e/\varepsilon_0 m_e$ and ω_{pe} is called plasma frequency.

The eigen modes are given by $K(\boldsymbol{k},\omega) = 0$, and this expression is called the "dispersion equation." In the integral of the dispersion equation, $u \equiv \boldsymbol{k}\cdot\boldsymbol{v}/k = \omega/k$ is a singular point (the denominator of the integral is zero: ω/k is the wave phase speed) and the integration method becomes the issue. Consider the initial value problem when a wave is excited in the plasma with a real wave number k. If the wave grows or decays, ω is complex. Therefore, the integral in Equation 5.24 has to be treated as a complex integration in the u plane. Vlasov took the Cauchy principal value. But Russian Nobel Prize winner L. D. Landau (1908–1968) realized that it should be treated as an initial value problem. This creates a term to circumvent singularity (the Landau damping term) [21]. In this case, the integration path of u is below the singular point since $\text{Im}(\omega) > 0$ (see Figure 5.7). Separating the dispersion function K into real and imaginary parts with $\omega = \omega_r + \mathrm{i}\omega_i$, we obtain the following equations:

$$K(\boldsymbol{k},\omega) = K_r(\boldsymbol{k},\omega_r) + \mathrm{i}[K_i(\boldsymbol{k},\omega_r) + \omega_i\frac{\partial K_r(\boldsymbol{k},\omega_r)}{\partial\omega_r}] = 0\,, \tag{5.25}$$

$$\omega_i = -\frac{K_i(\boldsymbol{k},\omega_r)}{\partial K_r(\boldsymbol{k},\omega_r)/\partial\omega_r}, \tag{5.26}$$

$$K_r(\boldsymbol{k},\omega) = 1 + \frac{\omega_{pe}^2}{n_e k^2} P \int \frac{\boldsymbol{k}\cdot\partial f_{e0}/\partial\boldsymbol{v}}{\omega_r - \boldsymbol{k}\cdot\boldsymbol{v}} \mathrm{d}\boldsymbol{v},\ K_i(\boldsymbol{k},\omega_r) = -\pi\frac{\omega_{pe}^2}{k^2}\frac{\partial f_{e0}}{\partial v}\Big|_{u=\omega_r/k}. \tag{5.27}$$

P indicates the Cauchy principal value integral. When electrons are Maxwellian and the wave phase velocity is much larger than the thermal velocity ($\omega_r/k \gg v_{te}$), we obtain $K_r(k,\omega) = 1 - (\omega_{pe}/\omega_r)^2 - 3(\omega_{pe}/\omega_r)^4 k^2\lambda_{\mathrm{D}}^2$. So,

$$\omega_r = \omega_{pe}(1 + 1.5k^2\lambda_{\mathrm{D}}^2),\ \omega_i = -\sqrt{\frac{\pi}{8}}\frac{\omega_{pe}}{k^3\lambda_{De}^3}\exp\left[-\left(\frac{1}{2k^2\lambda De^2}+\frac{3}{2}\right)\right]. \tag{5.28}$$

Since $\omega_i < 0$, the wave will damp. This damping occurs without any dissipation of energy due to collision and is "collisionless damping." Landau was the first to identify this phenomenon [21] and so this is called "Landau damping". The physical mechanism of Landau damping is intuitively simple. First, since the decay rate comes from the residue at $u = \omega/k$, it is caused by the particle having almost the same speed of wave phase velocity (called the "resonant particle"). These particles can exchange energy with the wave creating an almost DC electric field, since particles move with the waves. In Landau damping, the number of particles gaining energy from the wave is larger than that losing energy to the wave as seen from the relation $\omega_i \sim \mathrm{d}f/\mathrm{d}v$ in Equation 5.27. Landau damping can be compared to surfing. If the surfboard is not on the wave, the wave simply passes and surfboard cannot gain energy. However, if the speed of the surfboard is the same as the wave, the board is pushed by the wave, giving it energy.

The Vlasov equation describing collisionless plasma does not have irreversibility such as that due to the collision term in the Boltzmann equation, but, it has an arrow of time through the damping of the wave called "Landau damping." Irreversibility of the Boltzmann equation was created by the "Stosszahl Ansatz," while Landau damping originated from "phase mixing" in the processes described in Section 5.3. The inverse operator of the linear operator $L = \omega - \boldsymbol{k}\cdot\boldsymbol{v}$ is $L^{-1} = P[1/(\omega - \boldsymbol{k}\cdot\boldsymbol{v})] + \lambda\delta(\omega - \boldsymbol{k}\cdot\boldsymbol{v})$ (λ is an arbitrary constant). If we impose the condition that the velocity distribution function is "smooth" at $t = 0$, λ needs to take a specific value $\lambda = \mathrm{i}\pi$. An inverse Laplace transformation of Equation 5.23 gives,

$$\boldsymbol{E}_{\boldsymbol{k}}(t) = -\frac{e\boldsymbol{k}}{2\pi\varepsilon_0 k^2}\int_{-\infty}^{\infty}\mathrm{d}\boldsymbol{v}\, f_{e1\boldsymbol{k}}(\boldsymbol{v},t=0)\int_{-\infty+\mathrm{i}\omega_i}^{\infty+\mathrm{i}\omega_i}\frac{\exp(-\mathrm{i}\omega t)\mathrm{d}\omega}{K(\omega,\boldsymbol{k})(\omega-\boldsymbol{k}\cdot\boldsymbol{v})}. \tag{5.29}$$

The free streaming term $\exp(-\mathrm{i}\boldsymbol{k}\cdot\boldsymbol{v}t)$ is produced from the pole $\omega = \boldsymbol{k}\cdot\boldsymbol{v}$ of ω integration of Equation 5.29, and the density perturbation in velocity space oscillates more strongly with time. The resulting density perturbation n_1 and electric field $\boldsymbol{E}$ after integration in velocity space will damp with time due to this phase mixing. It might be thought that this collisionless damping by phase mixing would be inconsistent with the reversibility of the Vlasov equation (similar to the discussion with Loschmidt's "reversibility paradox" against the Boltzmann equation). For the solution $f_{e1k}(\boldsymbol{v}, t)$, the time reversal solution $f_{e1k}(\boldsymbol{x}, -\boldsymbol{v}, -t)$ is also a solution of the Vlasov equation and the density perturbation $n_1(\boldsymbol{x}, t)$ increases exponentially with time. However, this time reversal solution may have an initial value $f_{e1k}(\boldsymbol{x}, \boldsymbol{v}, t_1)$ and is not smooth since it includes $\exp(-i\boldsymbol{k} \cdot \boldsymbol{v}t_1)$. After t_1 passes, the density perturbation reaches a maximum where f_{e1k} is a smooth function. Then, the density will decay again with time due to phase mixing [16].

5.5 Coulomb Logarithm: Collective Behavior in the Coulomb Field

Now, let us consider collisions in the Coulomb field ignored in previous sections. Since plasma consists of charged particles (ions and electrons), a repulsion force acts between similarly charged particles and an attractive force acts between differently charged particles. Shielding of the electric field occurs in the same way as Debye shielding in electrolytes. Incidentally, this shielding phenomenon does not occur in many body systems under gravitational force, which have only attractive forces. When a potential is formed by density changes in the plasma, electrons and ions will follow a Boltzmann distribution $n_e = n_{e0}\exp(e\phi/kT)$ and $n_i = n_{i0}\exp(-eZ_i\phi/kT)$. Consider the potential around the ion. Assuming the thermal energy kT is much larger than the potential energy $e\phi$ ($e\phi \ll kT$), the solution of the Poisson equation for ϕ is obtained as follows,

$$\phi = \frac{e}{4\pi\varepsilon_0 r}\mathrm{e}^{-r/\lambda_\mathrm{D}} \,. \tag{5.30}$$

Here, $\lambda_\mathrm{D}^{-2} = \lambda_{De}^{-2} + \sum \lambda_{Di}^{-2}$, $\lambda_{De}^{2} = (\varepsilon_0 kT/e^2 n_e)^{0.5} (= 7.43 \times 10^3 [T_e\,(\mathrm{eV})/n_e(m^{-3})]^{0.5}\ [\mathrm{m}])$, $\lambda_{Di}^{2} = (\varepsilon_0 kT/e^2 Z_i^2 n_i)^{0.5}$. This shielding effect is called Debye shielding and ϕ is the Debye potential. For this relationship to be valid and statistically meaningful many particles must exist in the potential well (in the Debye sphere), $n\lambda_\mathrm{D}^3 \gg 1$ must be met. When this condition is met, the collective shielding effect of the Coulomb field works. This condition can be modified to $kT \gg e^2/4\pi\varepsilon_0 d$ ($d = n^{-1/3}$ is inter-electron distance), which means that kinetic energy is sufficiently larger than the potential energy between electrons (close to the ideal gas). Coulomb collisions in the plasma generally occur within the Debye radius, but the degree of scattering due to collision varies greatly depending on the value of the impact parameter. The impact parameter b in the center-of-mass system in Coulomb scattering is related to the scattering angle θ as follows (see Figure 5.8) [3],

$$b = b_0 \cot\left(\frac{\theta}{2}\right) \,. \tag{5.31}$$

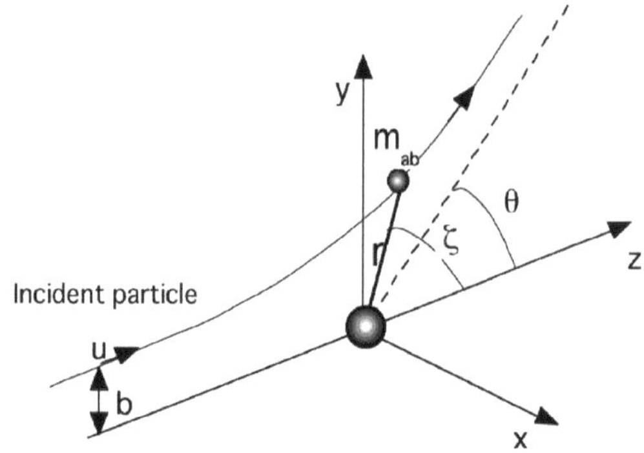

Figure 5.8 Scattering geometry for Coulomb collision. Scattering angle θ is defined as $\xi(r \to \infty)$

Here, $b_0 = e_a e_b/(4\pi e_0 m_{ab} u^2) = 7.2 \times 10^{-10} Z_a Z_b / E_r$, (eV) (m) is the impact parameter at 90 degrees scattering and is called the Landau parameter. Here, $m_{ab} = m_a m_b/(m_a + m_b)$ is reduced mass, u is the relative velocity, $E_r = m_{ab}u^2/2$ is the particle energy in the center-of-mass system. Substituting Equation 5.31 into the differential cross section $\sigma(\theta) = b(\mathrm{d}b/\mathrm{d}\theta)/\sin\theta$ scattered into differential solid angle $\mathrm{d}\Omega = 2\pi \sin\theta \mathrm{d}\theta$, we obtain following well-known Rutherford scattering cross section [3].

$$\sigma(\theta) = \frac{b_0^2}{\sin^4(\theta/2)} . \tag{5.32}$$

The velocity of particle species a, $\boldsymbol{v}_a$ is given by the velocity of the center of mass V and the relative velocity $\boldsymbol{u}_{ab} = \boldsymbol{v}_a - \boldsymbol{v}_b$ as $\boldsymbol{v}_a = \boldsymbol{V} + m_b \boldsymbol{u}_{ab}/(m_a + m_b)$ and the change in velocity of particle a is given by $\Delta \boldsymbol{v}_a = (m_{ab}/m_a)\Delta \boldsymbol{u}_{ab}$. Since the relative speed, u_{ab} is conserved for the elastic collisions, a change in u can be obtained as $\Delta \boldsymbol{u}_{ab} = u_{ab} \sin\theta \boldsymbol{n} - 2\sin^2(\theta/2)\boldsymbol{u}_{ab}$ using the formula for two isosceles triangle $(\sin^2(\theta/2) = b_0^2/(b_0^2 + b^2))$. Here, $\boldsymbol{n}$ is a unit vector perpendicular to $\boldsymbol{u}_{ab}$. Considering that the interaction occurs in the Debye area $(\pi\lambda_{\mathrm{D}}^2)$ and the particle flux of species b with velocity $\boldsymbol{v}_b$ passing through the Debye area in the time interval Δt is given by $\Delta\phi_b = \delta n_b(\boldsymbol{v}_b)u\Delta t$, the velocity change of species a by a collision with b is given by,

$$\Delta \boldsymbol{v}_a = \Delta\phi_b \frac{m_{ab}}{m_a} \int_0^{\lambda_{\mathrm{D}}} \Delta \boldsymbol{u}_{ab} 2\pi b \,\mathrm{d}b = -4\pi b_0^2 \Delta\phi_b \boldsymbol{u}_{ab} \frac{m_{ab}}{m_a} \int_0^{\lambda_{\mathrm{D}}} \frac{b}{b^2 + b_0^2} \mathrm{d}b . \tag{5.33}$$

Here, $b = 0$ corresponds to head-on collision $(\theta = \pi)$ and $b = \lambda_{\mathrm{D}}$ corresponds to the scattering angle $\theta_{\min} \sim b_0/\lambda_{\mathrm{D}}$. The component of $\Delta \boldsymbol{v}_a$ perpendicular to $\boldsymbol{u}_{ab}$ disappears due to rotational symmetry. The integral term is $(1/2)\ln(1+(\lambda_{\mathrm{D}}/b_0)^2) \sim \ln(\lambda_{\mathrm{D}}/b_0)$, and $\ln \Lambda \equiv \ln(\lambda_{\mathrm{D}}/b_0)$ is called Coulomb logarithm.

There are some subtleties in Debye shielding as the origin of the Coulomb logarithm. Consider the example of magnetic fusion plasma in Figure 5.9. A huge number of charged particles are contained in the Debye sphere $(n\lambda_{\mathrm{D}}^3 = 4 \times 10^7)$, but the integration of Debye potential in Equation 5.30 gives a charge number for the total

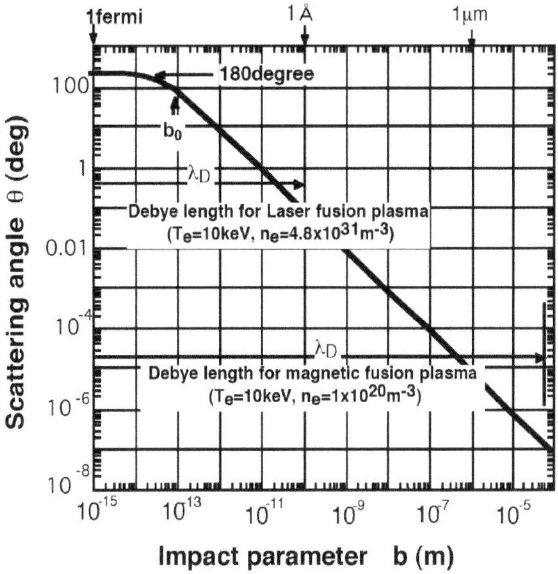

Figure 5.9 Relation between impact parameter b and scattering angle θ in the Coulomb scattering for relative energy $E_r = 10\,\text{keV}$ ($b_0 = 7.2 \times 10^{-14}\,\text{m}$). Parameters are magnetic fusion parameters ($T_e = 10\,\text{keV}$, $n_e = 1 \times 10^{20}\,\text{m}^{-3}$, $d = n_e^{-1/3} = 2 \times 10^{-7}\,\text{m}$, $\lambda_D = 7.4 \times 10^{-5}\,\text{m}$) and laser fusion parameters ($T_e = 10\,\text{keV}$, $n_e = 4.8 \times 10^{31}\,\text{m}^{-3}$, $d = n_e^{-1/3} = 2.8 \times 10^{-11}\,\text{m}$, $\lambda_D = 1.1 \times 10^{-10}\,\text{m}$)

electron cloud as only 1. This value is much smaller than the fluctuations in the number of electrons in the sphere ($(n\lambda_{De}^3)^{1/2} = 6000$). Thus, Van Kampen expressed the Debye sphere as "somewhat ghost-like existence." The Coulomb logarithm is a logarithmic integral, and the circumstances of collision depend on the particle distance. For $0 \leq b < 3b_0 = 2 \times 10^{-13}\,\text{m}$ ($0.2\pi < \theta \leq \pi$), the scattering between particles is large-angle scattering. The logarithmic integral in this regime is 1.15. For $3b_0 \leq b < \lambda_D/3.7 = 2 \times 10^{-5}\,\text{m}$, approximation of small-angle scattering is valid and the Debye shielding effect ($e^{-1/3.7} = 0.76$) can also be neglected. The logarithmic integral is 18.4 in this region. The logarithmic integral where Debye potential is effective (ghost region of Van Kampen), has the small value of 1.3, which validates the rough approximation to cut at Debye length. On the other hand, the two-body correlation or interaction with other particle cannot be ignored for $d(= n^{-1/3} = 2 \times 10^{-7}\,\text{m}) < b$ and the question of the validity of two-body collision remains. The two-body correlation is assumed to be negligible statistically. A brainstorming discussion on the Coulomb logarithm in a plasma is given by Van Kampen [16].

Here we discuss the difference between the collision process in plasma and the molecular collision considered by Boltzmann. In molecular collisions, interaction occurs only when a molecule reaches the molecular radius ($r_0 \sim 10^{-10}\,\text{m} = 1\,\text{Å}$). Since the molecular radius r_0 is much smaller than the inter-molecule distance $n^{-1/3}$ ($r_0 \ll n^{-1/3}$), the two-body correlation is expected to be small. On the other hand, since Coulomb force is a long-range force, collision time t_c is relatively long ($t_c \sim \lambda_D/v_{th} = 10^{-10}\,\text{s}$) and is longer than the time taken to approach next target for collision ($t_{mfp} \sim 1/n^{1/3}v_{th} = 10^{-13}\,\text{s}$). Namely, the relation t_c (collision time) $\ll \Delta t \ll t_{mfp}$ (mean free time) valid for molecular collision does not hold in plasmas. The general collision theory of a many-body system is discussed by Balescu [22].

5.6 Fokker–Planck Equation: Statistics of Soft Coulomb Collision

As mentioned in the previous section, the momentum change in Coulomb interaction is quite small and small-angle scattering is dominant except for the small region of the impact parameter b close to the Landau parameter b_0. Small-angle scattering process is called the Fokker–Planck process and the form of the collision term can be determined without detailed information about collision dynamics. If we assume the timescale Δt to be much longer than the correlation time of random force t_c ($\Delta t \gg t_c$), the force can be assumed to be statistically independent of the previous value. Under this circumstance, the state at $t + \Delta t$ is determined only by the state at t, independent of past history. This is the "Markov process." If we define $P(\boldsymbol{v}; \Delta\boldsymbol{v}, \Delta t)$ as the probability of a particle changing velocity $\boldsymbol{v}$ by $\Delta\boldsymbol{v}$ in the time interval Δt, the velocity distribution function $f(\boldsymbol{v}, t)$ is given as follows,

$$f_a(\boldsymbol{v}, t) = \int \mathrm{d}\Delta\boldsymbol{v}\, f_a(\boldsymbol{v} - \Delta\boldsymbol{v}, t - \Delta t) P(\boldsymbol{v} - \Delta\boldsymbol{v}; \Delta\boldsymbol{v}, \Delta t) \,. \tag{5.34}$$

Here, $P(\boldsymbol{v}; \Delta\boldsymbol{v}, \Delta t)$ is rapidly decreasing function with $|\Delta\boldsymbol{v}|$. Then, taking up to the second terms of the Taylor expansion, the following equation is obtained for $C(f_s) = \Delta f_s / \Delta t$.

$$C(f_a) = -\frac{\partial}{\partial \boldsymbol{v}} \cdot \left(\frac{\langle \Delta\boldsymbol{v} \rangle}{\Delta t} f_a \right) + \frac{\partial^2}{\partial \boldsymbol{v} \partial \boldsymbol{v}} : \left(\frac{\langle \Delta\boldsymbol{v} \Delta\boldsymbol{v} \rangle}{2\Delta t} f_a \right) . \tag{5.35}$$

This is called the Fokker–Planck collision term. If we take the first coordinate along $\boldsymbol{u}_{ab} = \boldsymbol{v}_a - \boldsymbol{v}_b$, coefficients of the Fokker–Planck collision term, $\langle \Delta\boldsymbol{v} \rangle / \Delta t$ and $\langle \Delta\boldsymbol{v} \Delta\boldsymbol{v} \rangle / 2\Delta t$ are given as follows due to the symmetry around $\boldsymbol{u}_{ab}$ axis,

$$\langle \Delta\boldsymbol{v} \rangle / \Delta t = \begin{bmatrix} \langle \Delta v_{\parallel} \rangle / \Delta t \\ 0 \\ 0 \end{bmatrix} , \tag{5.36}$$

$$\frac{\langle \Delta\boldsymbol{v} \Delta\boldsymbol{v} \rangle}{2\Delta t} = \begin{bmatrix} \langle \Delta v_{\parallel}^2 / \Delta t \rangle & 0 & 0 \\ 0 & \langle \Delta v_{\perp}^2 / 2\Delta t \rangle & 0 \\ 0 & 0 & \langle \Delta v_{\perp}^2 / 2\Delta t \rangle \end{bmatrix} .$$

Substitution of $\delta n_b = f_b(\boldsymbol{v}_b) \mathrm{d}\boldsymbol{v}_b$ into Equation 5.33 and integration by $\boldsymbol{v}_b$ gives,

$$\left\langle \frac{\Delta \boldsymbol{v}_a}{\Delta t} \right\rangle = -\sum_b \frac{e_a^2 e_b^2 \ln \Lambda}{4\pi m_a^2 \varepsilon_0^2} \left(1 + \frac{m_a}{m_b} \right) \int \frac{\boldsymbol{u}_{ab}}{u_{ab}^3} f_b(\boldsymbol{v}_b) \mathrm{d}\boldsymbol{v}_b \,. \tag{5.37}$$

Here, it should be remembered that the contribution at small b in the integral expression of Equation 5.33 is not small-angle scattering although it is small. Similarly, using the relation $\sin^2\theta = 2[1 - b_0^2/(b_0^2 + b^2)] b_0^2/(b_0^2 + b^2)$, we obtain,

$$\left\langle \frac{\Delta v_{\perp}^2}{\Delta t} \right\rangle = \sum_b \frac{e_a^2 e_b^2}{4\pi m_a^2 \varepsilon_0^2} \left(\frac{1}{2} + \ln \Lambda \right) \int \frac{1}{u_{ab}} f_b(\boldsymbol{v}_b) \mathrm{d}\boldsymbol{v}_b \,, \tag{5.38}$$

$$\left\langle \frac{\Delta v_\parallel^2}{\Delta t} \right\rangle = \sum_b \frac{e_a^2 e_b^2}{16\pi m_a^2 \varepsilon_0^2} \int \frac{1}{u} f_b(\boldsymbol{v}_b) \mathrm{d}\boldsymbol{v}_b \,. \tag{5.39}$$

Usually, terms that do not include the Coulomb integral (1/2 of Equations 5.38 and 5.39) are neglected. Then, we obtain

$$\left\langle \frac{\Delta \boldsymbol{v}_a}{\Delta t} \right\rangle = \sum_b \frac{e_a^2 e_b^2 \ln \Lambda}{4\pi m_a^2 \varepsilon_0^2} \frac{\partial h_{ab}(\boldsymbol{v}_a)}{\partial \boldsymbol{v}_a} \,, \tag{5.40}$$

$$\left\langle \frac{\Delta \boldsymbol{v}_a \Delta \boldsymbol{v}_a}{2\Delta t} \right\rangle = \sum_b \frac{e_a^2 e_b^2 \ln \Lambda}{8\pi m_a^2 \varepsilon_0^2} \int \frac{u_{ab}^2 \boldsymbol{I} - \boldsymbol{u}_{ab}\boldsymbol{u}_{ab}}{u_{ab}^3} f_b(\boldsymbol{v}_b) \mathrm{d}\boldsymbol{v}_b \tag{5.41}$$

$$= \sum_b \frac{e_a^2 e_b^2 \ln \Lambda}{8\pi m_a^2 \varepsilon_0^2} \frac{\partial^2 g_{ab}(\boldsymbol{v}_a)}{\partial \boldsymbol{v}_a \partial \boldsymbol{v}_a} \,,$$

$$h_{ab}(\boldsymbol{v}_a) = \left(1 + \frac{m_a}{m_b}\right) \int \frac{f_b(\boldsymbol{v}_b)}{u_{ab}} \mathrm{d}\boldsymbol{v}_b \,, \tag{5.42}$$

$$g_{ab}(\boldsymbol{v}_a) = \int u_{ab} f_b(\boldsymbol{v}_b) \mathrm{d}\boldsymbol{v}_b \,. \tag{5.43}$$

Here, $\partial u_{ab}^{-1}/\partial \boldsymbol{v}_a = -\boldsymbol{u}_{ab}/u_{ab}^3$, $\partial^2 u_{ab}/\partial \boldsymbol{v}_a \partial \boldsymbol{v}_a = u_{ab}^{-3}(u_{ab}^2 \boldsymbol{I} - \boldsymbol{u}_{ab}\boldsymbol{u}_{ab}) = \boldsymbol{U}_{ab}$ are used, where h_{ab}, g_{ab} is called the Rosenbluth potential [23]. Substituting these into Equation 5.35 gives,

$$C(f_a) = \sum_b \frac{e_a^2 e_b^2 \ln \Lambda}{4\pi m_a^2 \varepsilon_0^2} \left[-\frac{\partial}{\partial \boldsymbol{v}_a} \cdot \left(\frac{\partial h_{ab}}{\partial \boldsymbol{v}_a} f_a \right) + \frac{1}{2} \frac{\partial^2}{\partial \boldsymbol{v}_a \partial \boldsymbol{v}_a} : \left(\frac{\partial^2 g_{ab}}{\partial \boldsymbol{v}_a \partial \boldsymbol{v}_a} f_a \right) \right] . \tag{5.44}$$

Taking the partial integral for $\boldsymbol{v}_b$ integral and using the relation $\partial \boldsymbol{U}_{ab}/\partial \boldsymbol{v}_a = -2\boldsymbol{u}_{ab}/u_{ab}^3$, we obtain $2\partial h_{ab}/\partial \boldsymbol{v}_a = -(1 + m_a/m_b) \int (\partial \boldsymbol{U}_{ab}/\partial \boldsymbol{v}_b) f_b(\boldsymbol{v}_b) \mathrm{d}\boldsymbol{v}_b = (1+m_a/m_b) \int (\boldsymbol{U}_{ab} \partial f_b/\partial \boldsymbol{v}_b) \mathrm{d}\boldsymbol{v}_b$. Also, $\partial/\partial \boldsymbol{v}_a \cdot [(\partial^2 g_{ab}/\partial \boldsymbol{v}_a \partial \boldsymbol{v}_a) f_a(\boldsymbol{v}_a)] = \partial/\partial \boldsymbol{v}_a \cdot [\int \boldsymbol{U}_{ab} f_b(\boldsymbol{v}_b) \mathrm{d}\boldsymbol{v}_b f_a(\boldsymbol{v}_a)] = \int \boldsymbol{U}_{ab}[-\partial f_b/\partial \boldsymbol{v}_b f_a(\boldsymbol{v}_a) + f_b(\boldsymbol{v}_b) \partial f_a/\partial \boldsymbol{v}_a] \mathrm{d}\boldsymbol{v}_b$. Substituting these equations into Equation 5.44 gives,

$$C(f_a) = \sum_b \frac{e_a^2 e_b^2 \ln \Lambda}{8\pi \varepsilon_0^2 m_a} \frac{\partial}{\partial \boldsymbol{v}} \cdot \int \mathrm{d}\boldsymbol{v}_b \boldsymbol{U} \cdot \left[\frac{f_b(\boldsymbol{v}_b)}{m_a} \frac{\partial f_a(\boldsymbol{v})}{\partial \boldsymbol{v}} - \frac{f_a(\boldsymbol{v})}{m_b} \frac{\partial f_b(\boldsymbol{v}_b)}{\partial \boldsymbol{v}_b} \right] . \tag{5.45}$$

This form of collision term is given by Landau in 1936 [5–24]. He obtained this collision term from the Boltzmann collision term, which is not valid for plasma but he obtained a correct result. Indeed, the Boltzmann collision term can be applied outside the collision between molecules as discussed by Balescu in detail [22].

Let us return to the discussion on Debye shielding. As discussed in Section 5.2, the Coulomb collision term of particle species s, $C(f_s)$ is the ensemble average of the gradient in the velocity space of the fluctuating part of the discrete distribution function and acceleration by a microscopic Coulomb field.

$$C(f_a) = -\left\langle \tilde{\boldsymbol{a}} \cdot \frac{\partial \tilde{F}_a}{\partial \boldsymbol{v}} \right\rangle_{\text{ensemble}} \tag{5.46}$$

Substitution of $\tilde{\boldsymbol{a}} = -(e_s/m_s)\nabla\phi$ (ϕ: electrostatic potential) into Equation 5.5 gives $\partial\tilde{F}_s/\partial t + \boldsymbol{v}\cdot\partial\tilde{F}_s/\partial\boldsymbol{x} = (e_s/m_s)\nabla\phi\cdot\partial f_s/\partial\boldsymbol{v}$ and considering $-\nabla^2\phi = \sum(e_s/\varepsilon_0)\int\tilde{F}_s\mathrm{d}\boldsymbol{x}$, we obtain $\tilde{F}_s(\boldsymbol{x},\boldsymbol{v},t) = \tilde{F}_s(\boldsymbol{x}-\boldsymbol{v}t,\boldsymbol{v},0) + (e_s/m_s)\int_0^\infty \mathrm{d}\tau\nabla\phi(\boldsymbol{x}-\boldsymbol{v}\tau,t-\tau)\cdot\partial f_s/\partial\boldsymbol{v}$ from the convolution integral. Substitution into Equation 5.46 gives the Balescu–Lenard collision term in Landau form after some assumptions and manipulations [10,25].

$$C(f_a) = -\frac{e_a^2e_b^2}{8\pi\varepsilon_0^2 m_a}\frac{\partial}{\partial\boldsymbol{v}_a}\cdot\int \mathrm{d}\boldsymbol{v}_b\boldsymbol{K}_{ab}(\boldsymbol{v}_a,\boldsymbol{v}_b)\cdot\left[f_a\frac{1}{m_b}\frac{\partial f_b}{\partial\boldsymbol{v}_b} - f_b\frac{1}{m_a}\frac{\partial f_a}{\partial\boldsymbol{v}_a}\right], \tag{5.47}$$

$$\boldsymbol{K}_{ab}(\boldsymbol{v}_a,\boldsymbol{v}_b) = \int \mathrm{d}\boldsymbol{k}\delta(\boldsymbol{k}\cdot(\boldsymbol{v}_a-\boldsymbol{v}_b))\frac{\boldsymbol{k}\boldsymbol{k}}{k^4\left|\kappa(\boldsymbol{k},\boldsymbol{k}\cdot\boldsymbol{v}_a)\right|^2}, \tag{5.48}$$

$$\kappa(\boldsymbol{k},\omega) = 1 + \frac{e_b^2}{\varepsilon_0 m_b k^2}\int \mathrm{d}\boldsymbol{v}\frac{\boldsymbol{k}\cdot\partial f_b/\partial\boldsymbol{v}}{\omega-\boldsymbol{k}\cdot\boldsymbol{v}}. \tag{5.49}$$

Approximating $\kappa = (k^2+k_\mathrm{D}^2)/k^2 (k_\mathrm{D} = 1/\lambda_\mathrm{D})$ (corresponding to Debye potential), K_{ab} is given by,

$$K_{ab} \sim \int \mathrm{d}\boldsymbol{k}\frac{\boldsymbol{k}\boldsymbol{k}\delta(\boldsymbol{k}\cdot(\boldsymbol{v}_a-\boldsymbol{v}_b))}{\left|k^2+k_\mathrm{D}^2\right|^2} = \frac{1}{u^3}[u^2\boldsymbol{I}-\boldsymbol{u}\boldsymbol{u}]\int\frac{\mathrm{d}k}{k}\frac{k^4}{\left|k^2+k_\mathrm{D}^2\right|^2}. \tag{5.50}$$

Divergence in the short wave number (long wave length) regime is suppressed by the Debye shielding. If we take $k_\mathrm{max} = 1/b_0$ where b_0 is Landau parameter, integration in wave number gives the Coulomb logarithm $\ln(\lambda_\mathrm{D}/b_0)$. In this approximation the Balescu–Lenard collision term is consistent with the Landau collision term.

Note: M. N. Rosenbluth and B. B. Kadomtsev

M. N. Rosenbluth (1927–2003; Figure 5.10 (a)) and B. B. Kadomtsev (1928–1998 Figure 5.10 (b)) were great US and Russian theoreticians in plasma physics, respectively. They made significant contributions to the development of plasma physics for fusion research.

(a) (b)

Figure 5.10 (a) M.N. Rosenbluth (Courtesy of the University of Texas at Austin) and (b) B. B. Kadomtsev (with kind permission of Physics-Uspekhi journal)

5.7 Gyro-center Kinetic Theory: Drift and Gyro Kinetic Theory

The plasma kinetic equation including the collision is given by,

$$\frac{\partial f}{\partial t} + \boldsymbol{v} \cdot \frac{\partial f}{\partial \boldsymbol{x}} + (\boldsymbol{E} + \boldsymbol{v} \times \boldsymbol{B}) \cdot \frac{\partial f}{\partial \boldsymbol{v}} = C(f) \,. \tag{5.51}$$

Here, C is the collision term. This equation is not appropriate for studying collisional transport or turbulent transport due to drift wave turbulence with the wavelength near the ion gyro radius ρ_i since this equation includes both the slow drift motion and fast gyro motion. So, we derive the kinetic equation for the guiding center (called drift kinetic equation) using the guiding center equations, Equations 4.39 and 4.40 derived in Section 4.3. We define the guiding center Poisson brackets $\{,\}$ as follows,

$$\begin{aligned} \{X, Y\} = {} & \frac{e_a}{m_a}\left(\frac{\partial X}{\partial \theta}\frac{\partial Y}{\partial \mu} - \frac{\partial X}{\partial \mu}\frac{\partial Y}{\partial \theta}\right) - \frac{\boldsymbol{b}}{e_a\, B_\parallel^*} \cdot \nabla X \times \nabla Y \\ & + \frac{\boldsymbol{B}^*}{m_a\, B_\parallel^*}\left(\nabla X \frac{\partial Y}{\partial v_\parallel} - \frac{\partial X}{\partial v_\parallel}\nabla Y\right) . \end{aligned} \tag{5.52}$$

Here, X and Y are arbitrary functions of $\boldsymbol{z} = (\boldsymbol{r}, v_\parallel, \mu, \theta)$. Our target is to transform this without ε dependence into new coordinates $\bar{z} = \{\bar{z}^\mu\} = \{t, \bar{z}^i\}$ as $L\mathrm{d}t = \Gamma = \Gamma_i(\bar{z})\mathrm{d}\bar{z}^i - H(\bar{z})\mathrm{d}t + \mathrm{d}S(\bar{z})$ (see Equation 4.91). Then using the Hamiltonian Equation 4.30, the magnetic moment and gyro angle evolution equation and Equations 4.39 and 4.40 are given as follows,

$$\frac{\mathrm{d}\mu}{\mathrm{d}t} = \{\mu, H\} = 0 \,, \quad \frac{\mathrm{d}\theta}{\mathrm{d}t} = \{\theta, H\} \,, \tag{5.53}$$

$$\frac{\mathrm{d}v_\parallel}{\mathrm{d}t} = \{v_\parallel, H\} = -\frac{\boldsymbol{B}^*}{m_a\, B_\parallel^*}\nabla H \,, \tag{5.54}$$

$$\frac{\mathrm{d}\boldsymbol{r}}{\mathrm{d}t} = \{\boldsymbol{r}, H\} = \frac{\boldsymbol{b}}{e_a\, B_\parallel^*} \times \nabla H + \frac{\boldsymbol{B}^*}{m_a\, B_\parallel^*}\frac{\partial H}{\partial v_\parallel} \,. \tag{5.55}$$

The drift kinetic equation for guiding center velocity distribution function F to treat slow drift motion is given as follows by considering $\partial f/\partial\theta = 0$ and $\mathrm{d}\mu/\mathrm{d}t = 0$.

$$\frac{\partial F}{\partial t} + \dot{\boldsymbol{z}} \cdot \frac{\partial F}{\partial \boldsymbol{z}} = \frac{\partial F}{\partial t} + \{F, H\} = \frac{\partial F}{\partial t} + \dot{\boldsymbol{r}} \cdot \frac{\partial F}{\partial \boldsymbol{r}} + \dot{v}_\parallel \frac{\partial F}{\partial v_\parallel} = C(F) \,. \tag{5.56}$$

Turbulent fluctuation in plasma has been observed at wavelengths near the ion gyro radius. In order to treat the electromagnetic fluctuation of the order of the ion gyro radius, the gyro kinetic equation is developed by formulating the motion in time and varying electromagnetic field. In particular, the polarization drift (see Section 7.3) must be considered in the drift wave turbulence. For the set of electrostatic and vector potential fluctuation, $(\delta\varphi, \delta\boldsymbol{A})$, the perturbation Lagrangian δL is given by

$$\delta L\,\mathrm{d}t = e_a \delta_* \boldsymbol{A} \cdot (\mathrm{d}\boldsymbol{r} + \mathrm{d}\boldsymbol{\rho}) - e_a \delta_* \varphi\,\mathrm{d}t = -\delta H\,\mathrm{d}t\ , \qquad (5.57)$$
$$\delta_* \boldsymbol{A} = \delta \boldsymbol{A}(\boldsymbol{r} + \boldsymbol{\rho}), \delta_* \varphi = \delta\varphi(\boldsymbol{r} + \boldsymbol{\rho})\ .$$

The perturbed Hamiltonian δH is given by,

$$\delta H = e_a \delta_* \varphi - e_a \delta_* \boldsymbol{A} \cdot \boldsymbol{v}\ . \qquad (5.58)$$

Since Equation 5.58 has gyro radius and $\boldsymbol{v}_\perp$ dependences, perturbed Hamiltonian depends on gyro phase θ. Therefore, the magnetic moment μ is no longer a conserved quantity ($\{\mu, \delta H\}\ \neq 0$). So, we construct new coordinates $\boldsymbol{z} = (\boldsymbol{r}, v_\parallel, \mu, \theta) \Rightarrow \bar{\boldsymbol{z}} = (\bar{\boldsymbol{r}}, \bar{v}_\parallel, \bar{\mu}, \bar{\theta})$ where the newly defined magnetic moment becomes a conserved quantity, and a mathematical tool called Lie perturbation theory is used. An important point here is the gauge arbitrariness of the Lagrangian $(L + \mathrm{d}S/\mathrm{d}t)$ gives the same equation of motion as that of L, see Section 4.1). Using the perturbation expansion of Hamiltonian $\bar{H} = \bar{H}_0 + \bar{H}_1 + \bar{H}_2 + \ldots (H_0 =$ Equation 4.31 formula), we obtain following formula similar to Equations (4.100) and (4.105).

$$\bar{H}_1 = \delta H - \frac{\mathrm{d}S_1}{\mathrm{d}t}\ , \quad \bar{H}_2 = \frac{e_a^2}{2m_a}|\delta_* \boldsymbol{A}|^2 - \frac{1}{2}\{S_1, \delta H\} - \frac{\mathrm{d}S_2}{\mathrm{d}t} \qquad (5.59)$$

and are solved by evaluating S_1 and S_2. Then, the coordinates after transformation are given by,

$$\bar{z}_a = z_a + \{S_1, z_a\} + e_a \delta_* \boldsymbol{A} \cdot \{\boldsymbol{r} + \boldsymbol{\rho}, z_a\} + \ldots \qquad (5.60)$$

Using this new coordinate system with the Hamiltonian obtained in this way, the gyrokinetic equation is given as follows,

$$\frac{\partial \bar{F}}{\partial t} + \{\bar{F}, \bar{H}\} = \bar{C}(\bar{F}) \left(\text{or } \frac{\partial \bar{F}}{\partial t} + \dot{\bar{\boldsymbol{r}}} \cdot \frac{\partial \bar{F}}{\partial \bar{\boldsymbol{r}}} + \dot{\bar{v}}_\parallel \frac{\partial \bar{F}}{\partial \bar{v}_\parallel} = \bar{C}(\bar{F}) \right)\ . \qquad (5.61)$$

The Hamilton equation of motion for gyro-center is given by,

$$\frac{\mathrm{d}\bar{v}_\parallel}{\mathrm{d}t} = \{\bar{v}_\parallel, \bar{H}\} = -\frac{\boldsymbol{B}^*}{m_a\, B_\parallel^*}\bar{\nabla}\bar{H}\ , \qquad (5.62)$$

$$\frac{\mathrm{d}\bar{\boldsymbol{r}}}{\mathrm{d}t} = \{\bar{\boldsymbol{r}}, \bar{H}\} = \frac{\boldsymbol{b}}{e_a\, B_\parallel^*} \times \bar{\nabla}\bar{H} + \frac{\boldsymbol{B}^*}{m_a\, B_\parallel^*}\frac{\partial \bar{H}}{\partial \bar{v}_\parallel}\ . \qquad (5.63)$$

Details are given in Brizard–Hahm [26] and more plainly in Brizard [27].

References

1. Kubo R (1965) Statistical Mechanics. Elsevier.
2. Gibbs JW (1892) Elementary Principles in Statistical Mechanics. Yale University Press.
3. Goldstein H (1950) Classical Mechanics. Addison-Wesley.
4. Poincare H (1890) Acta Math., 13, 67.
5. Lindley D (2001) Boltzmann's Atom. Free Press.
6. Diacu F and Holmes P (1996) Celestial Encounters. Springer Math.Club.
7. Arnold VI (1978) Mathematical Methods of Classical Mechanics. Springer.
8. Maxwell JC (1860) Phylos. Mag. (4), Vol. 19, 19; Vol. 20, 21, 33.
9. Klimontovich YuL (1964) The Statistical Theory of Nonequilbrium Processes in a Plasma. Pergamon Press.
10. Hazeltine RD, Waelblock FL (2004) The Framework of Plasma Physics. Westview Press.
11. Tomonaga S (1979) What is Physics? (in Japanese). Iwanami Shinsho.
12. Toda M (1997) Maxwell's Demon (in Japanese). Iwanami Books.
13. Davies PC (1974) Physics of Time Asymmetry. Surrey University Press.
14. Vlasov AA (1945) J. Phys. USSR, 9, 25.
15. Krall NA, Trivelpiece AW (1973) Principles of Plasma Physics. McGraw-Hill.
16. Van Kampen NG, Felderhof BU (1967) Theoretical Methods in Plasma Physics. North Holland Pub.
17. Kadomtzev BB (1976) Collective Phenomena in Plasmas. Nauka Moscow.
18. Kolmogorov AN, Fomin SV (1999) Elements of the Theory of Functions and Functional Analysis. Dover Books.
19. Friedman B (1990) Principles and Techniques of Applied Mathematics. Dover Books.
20. Yoshida Z (1995) Mathematics of Collective Phenomena (in Japanese). Iwanami Books, Chapter 5.
21. Landau LD (1946) J. Phys. USSR 10, 25.
22. Balescu R (1997) Statistical Dynamics-Matter out of Equilibrium. Imperial College Press.
23. Rosenbluth M, MacDonald W, Judd D (1957) Phys. Rev., 107, 1–6.
24. Landau LD (1936) Phys. Z. Sov, 10, 154.
25. Balescu R (1960) Phys. Fluids 3, 52; also, A. Lenard (1960) Ann. Phys., 10, 390.
26. Brizard AJ, Hahm TS (2007) Rev. Mod. Phys., 79, 421–468.
27. Brizard AJ (1995) Phys. Plasma, 2, 459–471.

Chapter 6
Magnetohydrodynamic Stability: Energy Principle, Flow, and Dissipation

It is not easy to discuss general plasma stability since plasma is a nonlinear and dissipative medium. In this chapter, after a survey of the general stability, linear stability, in particular, ideal magnetohydrodynamic stability with an Hermitian (self-adjoint) linear operator is discussed. Then, nonlinear tearing forming a magnetic island by magnetic reconnection caused by the dissipation, and the stability of plasma flow with a non-Hermitian operator are outlined.

6.1 Stability: Introduction

To confine high-temperature plasma in a torus is topologically reasonable, but it actually requires careful consideration. Plasma is "soft" matter, and often becomes unstable when the internal energy is large. In this section, we introduce a general definition of stability, "stability in the sense of Lyapunov" for the general evolution equation of the system. The property of the linear operator of the evolution equation are described as the basis of stability.

The mathematical theory of stability was developed through the investigation of stability in stellar dynamics by the French mathematician S. D. Poisson (1781–1840). A complete general mathematical definition of stability was given by the Russian mathematician A. M. Lyapunov (1857–1918) [1]. Assume that the behavior of the plasma is given by the following evolution equation.

$$\frac{\mathrm{d}\boldsymbol{X}}{\mathrm{d}t} = N(\boldsymbol{X}) \,. \tag{6.1}$$

Here, it is important to note that time evolution is determined only by the present value of $\boldsymbol{X}$. Such a system is called a "dynamical system". The equilibrium point $\boldsymbol{X}_0(N(\boldsymbol{X}_0) = 0)$ is called "unstable in the sense of Lyapunov" if there is another solution that rapidly moves away from the first solution over time when a small change is applied to $\boldsymbol{X}$. Conversely, Lyapunov stability is given as follows.

Lyapunov stability: If there exists a neighborhood V for any neighborhood U of X so that orbit starting from inside of V stays within U, X is called stable in the sense of Lyapunov.

In other words, if the solution of $\mathrm{d}\Delta \boldsymbol{X}(t)/\mathrm{d}t = N(\boldsymbol{X}_0 + \boldsymbol{\xi})$ is always bounded, it is Lyapunov stable. Linearizing the evolution equation 6.1, we obtain the linearized equation $\mathrm{d}\boldsymbol{\xi}/\mathrm{d}t = L\boldsymbol{\xi}$. Here, $L = N'(\boldsymbol{X}_0)$ and $\boldsymbol{\xi} = \boldsymbol{X} - \boldsymbol{X}_0$. If the steady-state flow is zero, we have $\mathrm{d}\boldsymbol{\xi}/\mathrm{d}t = \partial\boldsymbol{\xi}/\partial t$, and the equation is led to an eigenvalue problem $L\boldsymbol{\xi} = \lambda\boldsymbol{\xi}$ by setting $\partial/\partial t = \lambda$. A linear operator L can be expressed by a finite dimensional matrix if the set of solutions of $L\boldsymbol{\xi} = \lambda\boldsymbol{\xi}$ is covered by a finite number of eigen functions. But, in general, an infinite number of eigen functions can exist and will form a "functional space" [2].

If the matrix L defined in the finite dimensional linear space is "regular" ($LL^* = L^*L$), a complete set of orthogonal eigen functions can be obtained. And the "unitary transformation" $U^{-1}LU$ diagonalizes the matrix L and eigenvalues appear as diagonal elements. If L is "self-adjoint" ($L^* = L$), the eigenvalues are all real. A negative eigenvalue means the system is unstable. The eigenvalue problem of linear operator L defined in the functional space (infinite dimensional linear space) is different in nature from that in the finite-dimensional linear space. An important difference is the existence of a "continuous spectrum." The solution of the eigenvalue problem in the functional space in general, consists of a discrete eigenvalue ("point spectrum") and continuous eigenvalue ("continuous spectrum") on the segment in the real axis. In quantum mechanics, the point spectrum appears in bound states, while the continuous spectrum appears in non-bound states. In plasma physics, the continuous spectrum appears in the Alfven waves and longitudinal waves in collisionless plasma (Section 5.3) [3].

Linear operators such as the linear Vlasov operator (see Section 5.3) and linear MHD (magnetohydrodynamic) operator F (see Section 6.2) appear in plasma physics. These linear operators are infinite dimensional linear operators and cause special behaviors such as Landau damping and Alfven continuum damping through the continuous spectrum.

The continuous spectrum has a singular eigen function (such as the Dirac δ function) not defined in the functional space for the linear operator ("Hilbert space"). Let us determine the operator to give the continuous spectrum. Consider position operator $Au(x) = xu(x)$, the eigenvalue problem for A is given by $xu = \lambda u$. From $(x - \lambda)u = 0$,we have $u = \delta(x - \lambda)$ (δ is the "Dirac delta function"). Since λ can take any real number, operator A gives a continuous spectrum.

Let A be the linear operator. The eigenvalue problem is to obtain eigenvalue $\lambda(\in C)$ and eigenvector u to satisfy $Au = \lambda u$. Rewriting this equation as $(\lambda I - A)u = 0$, the problem becomes finding null points for the linear operator $(\lambda I - A)$ or singular points of the operator $(\lambda I - A)^{-1}$. The theory requires a generalization of the concept of eigenvalue and eigen function for the operator in infinite dimensional linear space [2, 4].

Magnetohydrodynamic behavior of the plasma can be formulated in the form of a variational principle using the Lagrangian. If there is no dissipation, the total energy of the system (sum of potential energy and kinetic energy) is conserved. The system is unstable if a negative change occurs in the potential energy leading the kinetic energy to grow. Conversely, the system is stable if a positive change occurs in potential energy leading the kinetic energy to decrease. The method used to investigate the stability of the system through its potential energy in this way is called the "energy principle" [5].

The linear MHD equation can be expressed as $\rho\partial_t^2\boldsymbol{\xi} = F[\boldsymbol{\xi}]$. Here, F is the self-adjoint operator (Hermitian operator) as discussed in Section 6.2. The eigenvalue of the Hermitian operator (spectrum) is real. However, if the flow is included in the steady state, the linear MHD operator includes a non-Hermitian operator and is not easy to handle (see Section 6.7) [6].

The linear stability of the system is closely related to "bifurcation" in nonlinear phenomena. If the change of the system is described by a control parameter, bifurcation occurs when the eigenvalue of the linearized equation crosses the imaginary axis.

6.2 Ideal Magnetohydrodynamics: Action Principles and the Hermitian Operator

The conductivity of high temperature plasma is very similar to that of metals, and the motion of the magnetic field is strongly restricted. According to Alfven, the magnetic field is frozen into the plasma motion. Such plasma is treated in the continuum approximation, and is called "Ideal Magnetohydrodynamics" (Ideal MHD). As described in Goldstein [7], Lagrange mechanics in the continuum is reduced to the variational principle, whose action integral is given by the time and space integral of the Lagrangian density L. In ideal MHD [8], L is given by,

$$L = \frac{1}{2}\rho\boldsymbol{v}^2 - \frac{P}{\gamma - 1} - \frac{\boldsymbol{B}^2}{2\mu_0} . \tag{6.2}$$

Here, ρ, $\boldsymbol{v}$, P, $\boldsymbol{B}$ are the mass density, fluid velocity, plasma pressure, and the magnetic field, respectively. The first term of integral is the plasma kinetic energy, the second term is the plasma energy in the adiabatic approximation, and the third term is the magnetic energy. Using this Lagrangian, action S is represented by,

$$S = \int_{t_1}^{t_2} \mathrm{d}t \int L\mathrm{d}V . \tag{6.3}$$

Let $\boldsymbol{\xi}$ be the plasma displacement, variations of ρ, $\boldsymbol{v}$, P, and $\boldsymbol{B}$ are given by

$$\delta\boldsymbol{v} = \boldsymbol{v}\cdot\nabla\boldsymbol{\xi} - \boldsymbol{\xi}\cdot\nabla\boldsymbol{v} + \partial\boldsymbol{\xi}/\partial t , \tag{6.4}$$

$$\delta\rho = -\nabla\cdot(\rho\boldsymbol{\xi}) , \tag{6.5}$$

$$\delta P = -\gamma P\nabla\cdot\boldsymbol{\xi} - \boldsymbol{\xi}\cdot\nabla P , \tag{6.6}$$

$$\delta\boldsymbol{B} = \nabla\times(\boldsymbol{\xi}\times\boldsymbol{B}) . \tag{6.7}$$

Applying these relations, the action integral δS is given as follows,

$$\delta S = \int_{t_1}^{t_2} \mathrm{d}t \int \mathrm{d}V \left[\delta\rho \frac{\boldsymbol{v}^2}{2} + \rho \boldsymbol{v} \cdot \delta \boldsymbol{v} - \frac{\delta P}{\gamma - 1} - \frac{\boldsymbol{B} \cdot \delta \boldsymbol{B}}{\mu_0} \right]$$
$$= \int_{t_1}^{t_2} \mathrm{d}t \int \mathrm{d}V \Bigg[-\nabla \cdot (\rho \boldsymbol{\xi}) \frac{\boldsymbol{v}^2}{2} + \rho \boldsymbol{v} \cdot (\boldsymbol{v} \cdot \nabla \boldsymbol{\xi} - \boldsymbol{\xi} \cdot \nabla \boldsymbol{v} + \partial \boldsymbol{\xi} / \partial t)$$
$$+ \frac{\gamma P \nabla \cdot \boldsymbol{\xi} + \boldsymbol{\xi} \cdot \nabla P}{\gamma - 1} - \frac{\boldsymbol{B} \cdot \nabla \times (\boldsymbol{\xi} \times \boldsymbol{B})}{\mu_0} \Bigg] . \tag{6.8}$$

Partial integration for the displacement vector gives

$$\delta S = -\int_{t_1}^{t_2} \mathrm{d}t \int \mathrm{d}V \delta \boldsymbol{\xi} \cdot \left[\frac{\partial (\rho \boldsymbol{v})}{\partial t} + \nabla \cdot (\rho \boldsymbol{v} \boldsymbol{v}) + \nabla P - \boldsymbol{J} \times \boldsymbol{B} \right] . \tag{6.9}$$

Therefore, the variational principle $\delta S = 0$ is equivalent to the following equation:

$$\rho \frac{\partial \boldsymbol{v}}{\partial t} + \rho \boldsymbol{v} \cdot \nabla \boldsymbol{v} = \boldsymbol{J} \times \boldsymbol{B} - \nabla P . \tag{6.10}$$

Here, the continuity equation for mass density $\partial \rho / \partial t + \nabla \cdot (\rho \boldsymbol{v}) = 0$ is used. If there is no flow ($\boldsymbol{v} = 0$), plasma is in static force equilibrium. The variational principle in this case was given by Kruskal–Krusrud in 1958 [9].

$$S = \int L \, \mathrm{d}V = \int \left[\frac{B^2}{2\mu_0} + \frac{P}{\gamma - 1} \right] \mathrm{d}V . \tag{6.11}$$

For this variational principle, substitution of Equations 6.6 and 6.7 into the above equation gives

$$\delta S = -\int \boldsymbol{\xi} \cdot [\mu_0^{-1} (\nabla \times \boldsymbol{B}) \times \boldsymbol{B} - \nabla P] \mathrm{d}V . \tag{6.12}$$

Hence, the variational principle $\delta S = 0$ using Equation 6.11 is equivalent to the equilibrium condition $\boldsymbol{J} \times \boldsymbol{B} = \nabla P$. In the case of force equilibrium (the first order term of the action integral with respect to the displacement = 0), the variation δS is given by a quadratic form of the displacement. The stability of the equilibrium can be determined by its sign. Linearization of Equation 6.10 gives the following linear evolution equation considering $\rho \partial^2 \boldsymbol{\xi} / \partial t^2 = \delta \boldsymbol{J} \times \boldsymbol{B} + \boldsymbol{J} \times \delta \boldsymbol{B} - \nabla \delta P$, and (6.6), (6.7) and $\delta \boldsymbol{J} = \nabla \times \delta \boldsymbol{B}$.

$$\rho \frac{\partial^2 \boldsymbol{\xi}}{\partial t^2} = \boldsymbol{F}(\boldsymbol{\xi}) , \tag{6.13}$$
$$\boldsymbol{F}(\boldsymbol{\xi}) = \mu_0^{-1} \{ \nabla \times [\nabla \times (\boldsymbol{\xi} \times \boldsymbol{B})] \} \times \boldsymbol{B}$$
$$+ \mu_0^{-1} (\nabla \times \boldsymbol{B}) \times [\nabla \times (\boldsymbol{\xi} \times \boldsymbol{B})] + \nabla [\gamma P \nabla \cdot \boldsymbol{\xi} + \boldsymbol{\xi} \cdot \nabla P] .$$

The linear operator $\boldsymbol{F}$ is characterized by its important Hermitian property (F is a self-adjoint operator). This property can be proved by a fairly complicated modification of Equation 6.13 [5], but simple derivation is possible by using the fact that energy is conserved in an ideal MHD fluid [8]. In fact, the energy E of the ideal MHD fluid is given by the sum of kinetic and potential energies and is constant from the equation of motion, Equation 6.10

$$E = \int \left[\frac{1}{2}\rho \boldsymbol{v}^2 + \frac{P}{\gamma - 1} + \frac{\boldsymbol{B}^2}{2\mu_0} \right] \mathrm{d}V \, . \tag{6.14}$$

The total energy can be expressed as a function of ξ

$$E = \int \frac{1}{2}\rho \left(\frac{\partial \boldsymbol{\xi}}{\partial t} \right)^2 \mathrm{d}V + W(\boldsymbol{\xi}, \boldsymbol{\xi}) \, . \tag{6.15}$$

Here, W is the quadratic form of displacement $\boldsymbol{\xi}$ and is actually quite complex. We proceed without the detailed structure of W and consider up to the second order expansion of W with respect to $\boldsymbol{\xi}$.

$$W(\boldsymbol{\xi}, \boldsymbol{\xi}) = W_0 + W_1(\boldsymbol{\xi}) + W_2(\boldsymbol{\xi}, \boldsymbol{\xi}) \, . \tag{6.16}$$

Since energy is conserved, $\mathrm{d}E/\mathrm{d}t = 0$. In other words,

$$\frac{\mathrm{d}E}{\mathrm{d}t} = \int \rho \frac{\partial \boldsymbol{\xi}}{\partial t} \frac{\partial^2 \boldsymbol{\xi}}{\partial t^2} \mathrm{d}V + W_1 \left(\frac{\partial \boldsymbol{\xi}}{\partial t} \right) + W_2 \left(\frac{\partial \boldsymbol{\xi}}{\partial t}, \boldsymbol{\xi} \right) + W_2 \left(\boldsymbol{\xi}, \frac{\partial \boldsymbol{\xi}}{\partial t} \right) = 0 \, . \tag{6.17}$$

Defining $\boldsymbol{\eta} = \partial \boldsymbol{\xi}/\partial t$ and substituting $\rho \partial^2 \boldsymbol{\xi}/\partial t^2 = F(\boldsymbol{\xi})$ into Equation 6.17 we obtain

$$\int \boldsymbol{\eta} \cdot F(\boldsymbol{\xi}) \, \mathrm{d}V + W_1(\boldsymbol{\eta}) + W_2(\boldsymbol{\eta}, \boldsymbol{\xi}) + W_2(\boldsymbol{\xi}, \boldsymbol{\eta}) = 0 \, . \tag{6.18}$$

Since the system is in equilibrium, $W_1(\boldsymbol{\eta}) = 0$ for arbitrary $\boldsymbol{\eta}$. Also, taking into account that $W_2(\boldsymbol{\xi}, \boldsymbol{\eta}) + W_2(\boldsymbol{\xi}, \boldsymbol{\eta})$ in the left hand of Equation 6.18 is symmetric with respect to the exchange of $\boldsymbol{\xi}$ and $\boldsymbol{\eta}$, we obtain

$$\int \boldsymbol{\eta} \cdot F(\boldsymbol{\xi}) \, \mathrm{d}V = \int \boldsymbol{\xi} \cdot F(\boldsymbol{\eta}) \, \mathrm{d}V \, . \tag{6.19}$$

This property of the linear ideal MHD operator F is called the Hermitian (self-adjoint). The explicit expression of F as the Hermitian is given by Freidberg [10] as follows,

$$\begin{aligned} \int \boldsymbol{\eta} \cdot \boldsymbol{F}(\boldsymbol{\xi}) \, \mathrm{d}V = - \int \mathrm{d}V \bigg[& \frac{1}{\mu_0} (\boldsymbol{B} \cdot \nabla \boldsymbol{\xi}_\perp) \cdot (\boldsymbol{B} \cdot \nabla \boldsymbol{\eta}_\perp) + \gamma P (\nabla \cdot \boldsymbol{\xi})((\nabla \cdot \boldsymbol{\eta}) \\ & + \frac{B^2}{\mu_0} (\nabla \cdot \boldsymbol{\xi}_\perp + 2\boldsymbol{\xi}_\perp \cdot \boldsymbol{\kappa})(\nabla \cdot \boldsymbol{\eta}_\perp + 2\boldsymbol{\eta}_\perp \cdot \boldsymbol{\kappa}) \\ & - \frac{4B^2}{\mu_0} (\boldsymbol{\xi}_\perp \cdot \boldsymbol{\kappa})(\boldsymbol{\eta}_\perp \cdot \boldsymbol{\kappa}) + (\boldsymbol{\eta}_\perp \boldsymbol{\xi}_\perp : \nabla\nabla) \left(P + \frac{B^2}{2\mu_0} \right) \bigg] . \end{aligned} \tag{6.20}$$

6.3 Energy Principle: Potential Energy and Spectrum

The energy conservation law for a small displacement can be given by the integration of $(\partial\boldsymbol{\xi}/\partial t)\cdot$ (Equation 6.13) over time using the Hermitian property, as follows,

$$\frac{1}{2}\int \rho\left(\frac{\partial \boldsymbol{\xi}}{\partial t}\right)^2 \mathrm{d}V = \frac{1}{2}\int \boldsymbol{\xi}\cdot F(\boldsymbol{\xi})\,\mathrm{d}V\,. \tag{6.21}$$

Here, $\delta K = (1/2)\int \rho(\partial\boldsymbol{\xi}/\partial t)^2 \mathrm{d}V$ is the change of kinetic energy, $\delta W = -(1/2)\int \boldsymbol{\xi}\cdot F(\boldsymbol{\xi})\,\mathrm{d}V$ is the change of potential energy. Since total energy $E = K + W$ is conserved, a negative change in potential energy ($\delta W < 0$) gives an increase in the kinetic energy ($\delta K > 0$) and the system is unstable. Conversely, a positive change in potential energy ($\delta W > 0$) gives a reduction in kinetic energy ($\delta K < 0$) and the system is stable. In this way, the stability of the system can be examined by the potential energy and this method is called the "Energy Principle." Potential energy can be given by a quadratic form of $\boldsymbol{\xi}$ and Furth [11] gave such a form that is easy to understand as follows,

$$\delta W(\boldsymbol{\xi}) = \int \mathrm{d}V[\delta W_{\mathrm{SA}} + \delta W_{\mathrm{MS}} + \delta W_{\mathrm{SW}} + \delta W_{\mathrm{IC}} + \delta W_{\mathrm{KI}}]\,, \tag{6.22}$$

$$\delta W_{\mathrm{SA}} = B_1^2/2\mu_0\,,\quad \boldsymbol{B}_1 = \nabla\times(\boldsymbol{\xi}\times\boldsymbol{B})\,,$$
$$\delta W_{\mathrm{MS}} = \boldsymbol{B}^2(\nabla\cdot\boldsymbol{\xi}_\perp + 2\boldsymbol{\xi}_\perp\cdot\boldsymbol{\kappa})^2/2\mu_0\,,$$
$$\delta W_{\mathrm{SW}} = \gamma P(\nabla\cdot\boldsymbol{\xi})^2/2\,,\quad \delta W_{\mathrm{EX}} = (\boldsymbol{\xi}_\perp\cdot\nabla P)(\boldsymbol{\xi}_\perp\cdot\boldsymbol{\kappa})/2\,,$$
$$\delta W_{\mathrm{KI}} = -J_\parallel \boldsymbol{b}\cdot(\boldsymbol{B}_{1\perp}\times\boldsymbol{\xi}_\perp)/2\,.$$

Here, δW_{SA} is the "bending energy of the magnetic field" and is a source of "shear Alfven wave." δW_{MS} is the "compressing energy of the magnetic field" and is a source of "magnetosonic waves." δW_{SW} is the "compressing energy of the plasma" and a source of the "sound wave." All these terms are positive and stabilizing. Meanwhile, δW_{IC} is the "interchange energy" of plasma pressure in the curved magnetic field and can take positive or negative value. δW_{KI} is "kinking energy" of the current and can take a positive or negative value. Here, the curvature vector is given by $\boldsymbol{\kappa} = \boldsymbol{b}\cdot\nabla\boldsymbol{b}$. If $\boldsymbol{\kappa}\cdot\nabla P < 0$, the interchange energy is the source of instability.

Using F is Hermitian operator, we can show that the eigenvalue ω^2 is real. Setting $\boldsymbol{\xi} = \boldsymbol{\xi}\exp(i\omega t)$ in the linear MHD Equation 6.13 and taking the volume integral of $\boldsymbol{\xi}^*\cdot$ (Equation 6.13), we obtain

$$\omega^2\int \rho\,|\boldsymbol{\xi}|^2\,\mathrm{d}V = -\int \boldsymbol{\xi}^*\cdot\boldsymbol{F}(\boldsymbol{\xi})\,\mathrm{d}V\,. \tag{6.23}$$

Taking the difference with complex conjugate of Equation 6.23 and using the Hermitian relation $\int \boldsymbol{\xi}\cdot F(\boldsymbol{\xi}^*)\,\mathrm{d}V = \int \boldsymbol{\xi}^*\cdot F(\boldsymbol{\xi})\,\mathrm{d}V$, we obtain

$$\left(\omega^2 - \omega^{*2}\right)\int \rho\,|\boldsymbol{\xi}|^2\,\mathrm{d}V = 0\,. \tag{6.24}$$

Namely, eigenvalue ω^2 is real. The case of $\omega^2 > 0$ shows oscillation without damping and is stable, while the case of $\omega^2 < 0$ grows exponentially and is unstable. The transition from the stable to the unstable state occurs at $\omega^2 = 0$. The locus of the root moves on a real and imaginary axis in the complex plane.

Considering F is Hermitian, we can prove the orthogonality of the eigen functions weighted by ρ. Eigen functions $\boldsymbol{\xi}_m$ and $\boldsymbol{\xi}_n$ with different eigenvalues ω_m^2 and ω_n^2 satisfy $-\rho\omega_m^2\boldsymbol{\xi}_m = \boldsymbol{F}(\boldsymbol{\xi}_m)$ and $-\rho\omega_n^2\boldsymbol{\xi}_n = \boldsymbol{F}(\boldsymbol{\xi}_n)$. Taking the inner product with $\boldsymbol{\xi}_m$ and $\boldsymbol{\xi}_n$ and integrating over the volume, we obtain

$$(\omega_m^2 - \omega_n^2)\int \rho\boldsymbol{\xi}_m \cdot \boldsymbol{\xi}_n \mathrm{d}V = \int [\boldsymbol{\xi}_m \cdot \boldsymbol{F}(\boldsymbol{\xi}_n) - \boldsymbol{\xi}_n \cdot \boldsymbol{F}(\boldsymbol{\xi}_m)]\,\mathrm{d}V = 0\,. \qquad (6.25)$$

If there is only a discrete spectrum, this orthogonality leads to the energy integral $\delta W = \sum a_n^2\omega_n^2$ for $\boldsymbol{\xi} = \sum a_n\boldsymbol{\xi}_n$. Hence, we may judge the stability by the sign of the minimum eigenvalue ω_j^2 $(j = 1, \ldots, n)$. However, the existence of a continuous spectrum in the linear MHD operator causes this argument to break down.

To examine the continuous spectrum case, we describe the general properties of linear MHD Equation 6.13. Setting $\omega^2 = -\lambda$ Equation 6.13 can be expressed as follows,

$$[\lambda - \boldsymbol{F}/\rho]\boldsymbol{\xi} = \boldsymbol{a}\,. \qquad (6.26)$$

Here, $\boldsymbol{a}$ is either the initial value of the Laplace transform of Equation 6.13 or the external force which is not considered in Equation 6.13 (for example, the Alfven mode can be excited with external coils). Then,

$$\boldsymbol{\xi} = [\lambda - \boldsymbol{F}/\rho]^{-1}\boldsymbol{a}\,. \qquad (6.27)$$

The linear MHD operator has an infinite number of independent eigen functions and eigenvalues (often it they are not countable and termed a "spectrum"). The spectrum of $\boldsymbol{F}$ corresponds to the singular points of $(\lambda - \boldsymbol{F}/\rho)^{-1}$. If $(\lambda - \boldsymbol{F}/\rho)\boldsymbol{x} = 0$ has a nontrivial solution, a point spectrum appears. If $(\lambda - \boldsymbol{F}/\rho)^{-1}$ exists but is unbounded, a continuous spectrum will appear (see Note).

In non-uniform plasma, MHD waves such as Alfven waves and slow and fast magnetosonic waves can have a continuous spectrum. For example, $\lambda - \boldsymbol{F}/\rho = \lambda - k_\parallel^2 V_A^2$ in a cylindrical inhomogeneous plasma and the Alfven wave has a phase velocity $V_p = V_A$ in the direction of a magnetic field. If the density changes in the direction perpendicular to the magnetic field, the Alfven wave will propagate with a different phase velocity to its local Alfven velocity for each layer of different density. The oscillation phase difference between adjacent layers increases and the arbitrary initial perturbation will decay with time. In non-uniform plasma, damping of waves occurs due to phase mixing in the radial direction, while Landau damping occurs by phase mixing in velocity space [3]. Thus this damping is called "continuum damping."

Note: Hermitian (self-adjoint) Operator and Spectral Theory [2,4]

During the construction phase of quantum mechanics, it was necessary to establish the spectral theory to generalize the concept of the eigenvalue problem. The operator in quantum mechanics is self-adjoint and J. von Neumann (1903–1957) created spectral theory in the functional analysis. However, the theory is limited to the self-adjoint operator and the general properties of the non-self-adjoint operator are not well understood. Among the operators in the infinite-dimensional space, the spectral resolution is possible in general only for self-adjoint operators (or unitary operators). Among the various functional spaces, the most frequently used space (or set) is the Hilbert space dubbed H-space of square integrable functions ([Chapter VIII of 2]).

For the linear operator A, the eigenvalue problem is to obtain the eigenvalues $\lambda(\in C)$ and eigenvectors u to satisfy $Au = \lambda u$. This can be rewritten as $(\lambda I - A)u = 0$ and the problem is to find a set of null points of the linear operator $(\lambda I - A)$. In the operator in infinite dimensional linear space, spectrum analysis is used to investigate singularity of $(\lambda I - A)^{-1}$. For complex values of λ, the following three classes of the spectrum arise [4].

1. Point spectrum: In the case where $(\lambda I - A)^{-1}$ does not exist since $(\lambda I - A)u = 0$ has a non-trivial u, the corresponding set of λ is called a "point spectrum."
   Example: $A = -\partial_x^2$, solves the eigenvalue problem $(\lambda I - A)u = 0$ are $\lambda = \{(n\pi)^2; n = 1, 2, \ldots\}$.
2. Continuous spectrum: In this case, the unbounded inverse $(\lambda I - A)^{-1}$ exists, the corresponding set of λ is called a "continuous spectrum."
   Example: $A = x$, the solution for $(\lambda x)u = 0$ is $u = \delta(x - \lambda)$. This Dirac delta function is not square integrable and does not belong to Hilbert space.
3. Residual spectrum: In the case where inverse $(\lambda I - A)^{-1}$ exists and is bounded, the corresponding set of λ is called a "residual spectrum." It is important to note that if λ is in the residual spectrum of A, λ is in the point spectrum of the adjoint operator A^*. So, there is no residual spectrum in Hermitian operator.

Here, a linear operator A is said to be "bounded" if there exists a constant N for all $u \in H$ such that

$$\|Au\| \leq N \|u\| \ . \tag{6.28}$$

6.4 Newcomb Equation: Euler–Lagrange Equation of Ideal MHD

Minimization of the energy integral of the linear ideal MHD equation in cylindrical plasma and axisymmetric plasma can be reduced to the Euler–Lagrange equation of the radial coordinate. This is called the "Newcomb equation." Newcomb [12] derived the equation for cylindrical plasma and Tokuda [13] derived the equation for axisymmetric plasma.

Cylindrical plasma

In the case of cylindrical symmetry, ξ_r, $\mathrm{i}\xi_\theta$, $\mathrm{i}\xi_z$ can be expressed as the real normal mode $\exp(\mathrm{i}m\theta + \mathrm{i}kz)$ without loss of generality considering the symmetry in the cylindrical coordinates (r, θ, z). The stability condition can be given for a pair (m, k). Minimization of energy integral Equation 6.22 for $\mathrm{i}\xi_\theta$ and $\mathrm{i}\xi_z$ gives incompressibility of displacement $\nabla \cdot \boldsymbol{\xi} = 0$ and $v = \mathrm{i}(\xi_\theta B_z - \xi_z B_\theta) = \zeta_0(\xi_r, \mathrm{d}\xi_r, /\mathrm{d}r)$, and the energy integral W for unit length along z direction is given using $\xi = \xi_r$ as follows,

$$W = \frac{\pi}{2\mu_0} \int_0^a \left[f \left| \frac{\mathrm{d}\xi}{\mathrm{d}r} \right|^2 + g\,|\xi|^2 \right] \mathrm{d}r + W_a + W_v\,, \tag{6.29}$$

$$g = \frac{1}{r} \frac{(kB_z - (m/r)B_\theta)^2}{k^2 + (m/r)^2} + r(kB_z + (m/r)B_\theta)^2 - \frac{2B_\theta}{r} \frac{\mathrm{d}(rB_\theta)}{\mathrm{d}r}$$
$$- \frac{\mathrm{d}}{\mathrm{d}r} \left(\frac{k^2 B_z^2 - (m/r)^2 B_\theta^2}{k^2 + (m/r)^2} \right),$$

$$f = \frac{r(kB_z + (m/r)B_\theta)^2}{k^2 + (m/r)^2}\,,$$

$$\zeta_0 \left(\xi, \frac{\mathrm{d}\xi}{\mathrm{d}r} \right) = \frac{r}{k^2 r^2 + m^2} \left[(krB_\theta - mB_z) \frac{\mathrm{d}\xi}{\mathrm{d}r} - (krB_\theta + mB_z) \frac{\xi}{r} \right].$$

Here, W_a and W_v are the surface terms from the partial integration and the energy integral in the vacuum, respectively. The Euler–Lagrange equation to minimize Equation 6.29 is given by the following equation:

$$\frac{\mathrm{d}}{\mathrm{d}r} \left(f \frac{\mathrm{d}\xi}{\mathrm{d}r} \right) - g\xi = 0\,. \tag{6.30}$$

This equation is known as the "Newcomb equation." A significant feature of the Newcomb equation is that it becomes singular at the rational surface given by $f(r) = 0$. Since $f \geq 0$, the $f(\mathrm{d}\xi/\mathrm{d}r)^2$ term in (6.29) is stabilizing. At the rational surface, the condition of the local solution to be non-oscillatory (oscillatory solution is unstable) gives the "Suydam condition" for local mode stability $(q'(r)/q(r))^2 + 8\mu_0 P'(r)/rB_z^2 > 0$ with $q = rB_z/RB_q$ (the stability condition in the torus is given by $r(\mathrm{d}\ln q/\mathrm{d}r)^2/4 + 2\mu_0(\mathrm{d}P/\mathrm{d}r)(1-q^2)/B_z^2 > 0$ and is usually satisfied since $\mathrm{d}P/\mathrm{d}r(1-q^2) > 0$ in the $q > 1$ regime, even if $\mathrm{d}P/\mathrm{d}r < 0$ is a large negative value, the "Mercier stability criteria" [14]). The $q'(r)/q(r)$ term is stabilized by the magnetic shear. Considering the case with multiple singularities in the plasma $(r_1, r_2, \ldots)$, the Euler–Lagrange solution is separated at the singular point and the energy integral between adjacent singular points can be minimized independently. In this case, the energy integral of the Euler–Lagrange solution between the singular points r_1 and r_2 is given by $W = (\pi/2\mu_0)[f\xi\mathrm{d}\xi/\mathrm{d}r]_{r_1}^{r_2}$. For $x = r - r_s$, the solution near the singular points is given by two eigen solutions $x \sim x_1^{-n}$ and

x_2^{-n}, where n_1 and n_2 are solutions of $n^2 - n + \gamma = 0$ ($\gamma = -rP'(r)/(B_z^2/2\mu_0)s^2$: $s = r(\mathrm{d}q/\mathrm{d}r)$ is the magnetic shear). Assuming $n_1 < n_2$, $x \sim x_1^{-n}$ is called the "small solution" and $x \sim x_2^{-n}$ is called the "large solution." Newcomb derived 14 theorems of the Euler–Lagrange solution of Equation 6.30 [12]. Theorem 10 is particularly important.

Newcomb's theorem 10: For specific values of m and k, cylindrical plasma is stable in an independent interval I if and only if (1) Suydam's condition is fulfilled at the left endpoint if the point is singular, and (2) the Euler–Lagrange solutions that are small at the left endpoint never vanish in the interior of I. In marginal cases, the solution is also small at the right endpoint.

If the numerical integration of this Euler–Lagrange equation using, for example, the Runge-Kutta method with a boundary condition $\xi = 0$, $\mathrm{d}\xi/\mathrm{d}r = 1$ at the left edge gives a crossing $\xi = 0$ within the interval, the plasma is unstable according to this theorem.

Axisymmetric plasma

In the case of an axisymmetric torus, the energy integral is minimized under the incompressibility condition $\nabla \cdot \boldsymbol{\xi} = 0$ as in the case of cylindrical symmetry. The magnetic field is expressed by Equation 3.59 in an axisymmetric torus, and the Grad–Shafranov equation is given in the flux coordinates (r, θ, ζ) with $r = [2R_0 \int_0^{\psi} (q/F)\mathrm{d}\psi]^{1/2}$ and Jacobian $J = g^{1/2} = R^2 r/R_0$ as follows,

$$\frac{\partial}{\partial r}\left[r\frac{\mathrm{d}\psi}{\mathrm{d}r}|\nabla r|^2\right] + \frac{\partial(\nabla r \cdot \nabla\theta)}{\partial\theta}\frac{\mathrm{d}\psi}{\mathrm{d}r} = -\mu_0 R^2 \frac{\mathrm{d}P}{\mathrm{d}\psi} - F\frac{\mathrm{d}F}{\mathrm{d}\psi}\,. \tag{6.31}$$

By using $X = \boldsymbol{\xi} \cdot \nabla r$ and $V = r\boldsymbol{\xi} \cdot \nabla(\theta - \zeta/q)$ in the flux coordinates (r, θ, ζ), the energy integral W under $\nabla \cdot \boldsymbol{\xi} = 0$ can be expressed in a following form,

$$W_p = \frac{\pi}{2\mu_0}\int_0^a \mathrm{d}r \int_0^{2\pi} \mathrm{d}\theta\, L\left(X, \frac{\partial X}{\partial\theta}, \frac{\partial X}{\partial r}, V, \frac{\partial V}{\partial\theta}\right) \tag{6.32}$$

where $r = a$ is the plasma surface. Minimization of the energy integral with respect to V is easy in the cylindrical plasma. In the axisymmetric case, minimization with respect to V is a bit more complicated since the energy integral contains the $\partial V/\partial\theta$ term but the absence of the $\partial V/\partial r$ term in the energy integral leads to following Euler equation [13],

$$\frac{\partial}{\partial\theta}\left[\frac{\partial L}{\partial(\partial V/\partial\theta)}\right] - \frac{\partial L}{\partial V} = 0\,. \tag{6.33}$$

The solvability of Equation 6.33 imposes a condition for L, called the "solvable condition." By integrating of Equation 6.33, $\theta = 0 - 2\pi$, $\partial L/\partial(\partial V/\partial\theta)$ must have the same value at $\theta = 0$ and 2π (periodic boundary conditions). The solvable condition becomes,

$$\int_0^{2\pi} \frac{\partial L}{\partial V} \mathrm{d}\theta = 0 \,. \tag{6.34}$$

Fourier expansion of V and X for θ is defined as follows,

$$X(r,\theta) = \sum_{m=-\infty}^{m=\infty} X_m(r)\exp(im\theta) \,, \quad V(r,\theta) = -\mathrm{i} \sum_{m=-\infty}^{m=\infty} V_m(r)\exp(im\theta) \,. \tag{6.35}$$

Substitution of these equations into Equation 6.34 gives linear equations for V_m and the solution is substituted into Equation 6.32. The integrant L is now given by $\boldsymbol{X} = (\ldots, X_{-2}, X_{-1}, X, X_1, X_2, \ldots)^t$ (t is transposed) and $\mathrm{d}\boldsymbol{X}/\mathrm{d}r$ and the Euler–Lagrange equation is obtained [6–13].

$$W_p = \frac{\pi^2}{\mu_0} \int_0^a L\left(\boldsymbol{X}, \frac{\mathrm{d}\boldsymbol{X}}{\mathrm{d}r}\right) \mathrm{d}r$$
$$\frac{\mathrm{d}}{\mathrm{d}r} \frac{\partial L}{\partial(\mathrm{d}\boldsymbol{X}/\mathrm{d}r)} - \frac{\partial L}{\partial \boldsymbol{X}} = 0 \,. \tag{6.36}$$

Since L is given by a quadratic form of $\boldsymbol{X}, \mathrm{d}\boldsymbol{X}/\mathrm{d}r$, the Euler–Lagrange equation follows the form of the second order ordinary differential equation,

$$\frac{\mathrm{d}}{\mathrm{d}r} \boldsymbol{f} \frac{\mathrm{d}\boldsymbol{X}}{\mathrm{d}r} + \boldsymbol{g} \frac{\mathrm{d}\boldsymbol{X}}{\mathrm{d}r} + \boldsymbol{h}\boldsymbol{X} = 0 \,. \tag{6.37}$$

where $\boldsymbol{f}, \boldsymbol{g}$, and $\boldsymbol{h}$ are matrices. This is called the "two-dimensional Newcomb equation." Diagonal elements of $\boldsymbol{f}$ have $(n/m - 1/q)^2$ dependence similar to the one-dimensional Newcomb equation and the radius of $q = m/n$ is the singular point. Small and large solutions exist near the singular point and the Mercier condition is derived as the local stability condition. Once the Mercier condition is met, a similar method can be applied as Newcomb's theorem 10 to determine the stability. Also, "kink" and "peeling modes*" can be studied using the two-dimensional Newcomb equation.

6.5 Tension of Magnetic Field: Kink and Tearing

As described in Chapter 3, the magnetic field is bent helically and densely covers the torus to confine high temperature plasma. As Maxwell's equations teach us, the

* Peeling mode: Finite edge current can drive external modes localized near the plasma edge. This mode is called the peeling mode. The peeling mode becomes most unstable when a rational surface is located just outside the plasma surface. This mode can be coupled to the pressure driven ballooning mode and is thought to be a cause of ELM (Edge Localized Modes) in tokamak.

tension of the magnetic field works in the direction of the magnetic field and works to make the field lines straight. When the field becomes straight, plasma is deformed helically. This is the generation mechanism of instabilities called the "kink" mode and "tearing" mode. Kink is the deformation in the limit of zero plasma resistivity (ideal MHD plasma), while tearing is the deformation allowed by the magnetic reconnection with the change in magnetic field topology. This reconnection occurs at the rational surface, which is a singular point of the Newcomb equation of ideal MHD. There is an "external kink mode" and "internal kink mode" in the kink mode. The energy integral $W = W_p + W_v$ in the cylindrical plasma approximation (low beta (beta is the ratio of volume average plasma pressure $\langle P \rangle$ to magnetic pressure $B^2/2\mu_0$), large aspect ratio, circular cross section tokamak approximation) is obtained from Equation 6.29 as follows,

$$W_p = \frac{\pi^2 B_\xi^2}{\mu_0 R_0} \left\{ \int_0^a \left[\left(r\frac{\mathrm{d}\xi}{\mathrm{d}r} \right)^2 + (m^2-1)\xi^2 \right] \left(\frac{n}{m} - \frac{1}{q} \right)^2 r\mathrm{d}r \right\}, \tag{6.38}$$

$$W_v = \frac{\pi^2 B_\xi^2}{\mu_0 R_0} \left[\frac{2}{q_a} \left(\frac{n}{m} - \frac{1}{q_a} \right) + (1+m\lambda) \left(\frac{n}{m} - \frac{1}{q_a} \right)^2 \right] a^2 \xi_a^2 . \tag{6.39}$$

Here, $\lambda = (1+(a/b)^{2m})/(1-(a/b)^{2m})$, a and b are the plasma minor radius and the radius of ideally conducting wall, respectively. Although the resistance of the wall is finite, the wall can be regarded as an ideal wall for timescales shorter than the wall time constant τ_{wall}. The energy integral inside the plasma is non-negative ($W_p \geq 0$), but the energy integral of the vacuum W_v can be negative when $(m/n)(1-2/(m\lambda+1)) < q_a < m/n$. The external kink is unstable for $q_a < m/n$ if the energy integral inside the plasma is small.

An unstable plasma mode with only internal displacement is possible, even if the surface displacement is zero, $\xi_a = 0$. This is called the internal kink mode. If $\xi_a = 0$, vacuum energy is zero, $W_v = 0$. Also, if a $q = 1$ surface exists in the plasma ($q(0) < 1$), internal energy can be zero ($W_p = 0$) for the non-trivial solution for $m = 1$ and $\mathrm{d}\xi/\mathrm{d}r = 0$, that means that the plasma is in neutral stability. This mode becomes weakly unstable if the poloidal beta value is above ~ 0.3 if we take into account the destabilizing effect of pressure by the toroidal effect.

The instability of practical importance is the tearing mode associated with the reconnection of the magnetic field at the resonant rational surface. This mode is destabilized by changing the topology of the magnetic field, while it is stable within the ideal MHD context. The linear growth rate of this mode is given as $\gamma \sim \eta^{3/5}$ but it soon goes into the nonlinear region. The nonlinear regime is the "Rutherford regime," derived by P. H. Rutherford [15]. Substituting Ohm's law $\boldsymbol{E} + \boldsymbol{v} \times \boldsymbol{B} = \eta \boldsymbol{J}$ into $\partial \boldsymbol{B}/\partial t = \nabla \times \boldsymbol{E}$, we can write down the major terms in r direction as follows,

$$\gamma B_r - \frac{B_\theta}{r}(m-nq)\mathrm{i}v_r = \frac{\eta}{\mu_0}\frac{\mathrm{d}^2 B_r}{\mathrm{d}r^2} . \tag{6.40}$$

In Equation 6.40, the second term on the left-hand side originates from $\boldsymbol{v} \times \boldsymbol{B}$ and becomes zero at the resonant surface. Then, the resistive diffusion term becomes important. Conversely, the effect of resistivity is not important except near the resonant surface. Defining $\psi = irB_r/m$, the following magnetic diffusion equation governs the dynamics near the rational surface,

$$\frac{\partial \psi}{\partial t} = \frac{\eta}{\mu_0}\frac{\partial^2 \psi}{\partial r^2} . \tag{6.41}$$

Integration of this equation within the "magnetic island" width w gives,

$$w\frac{\partial \psi}{\partial t} = \frac{\eta}{\mu_0}\left[\frac{\partial \psi}{\partial r}\left(r_\mathrm{s} + \frac{w}{2}\right) - \frac{\partial \psi}{\partial r}\left(r_\mathrm{s} - \frac{w}{2}\right)\right] . \tag{6.42}$$

where, r_s is a singular radius. Since w and ψ are related as $w = 4(q\psi/q'B_\theta)^{1/2}$,

$$\frac{\mathrm{d}w}{\mathrm{d}t} = \frac{\eta}{2\mu_0}\Delta'(w) . \tag{6.43}$$

Here, $\Delta'(w) = [\mathrm{d}\psi/\mathrm{d}r(r_\mathrm{s} + w/2) - \mathrm{d}\psi/\mathrm{d}r(r_\mathrm{s} - w/2)]/\psi(r_\mathrm{s})$. According to the White's detailed calculation [16], the time evolution of the island width is given by,

$$\frac{\mathrm{d}w}{\mathrm{d}t} = 1.66\frac{\eta}{\mu_0}(\Delta'(w) - \alpha w) . \tag{6.44}$$

Here, $\Delta'(w)$ is the solution ignoring resistive diffusion (external solution) and α is a constant. The external ψ can be obtained from the helically perturbed equilibrium equation from the cylindrical one as shown below. Except for the case $\psi = 0$ at the resonant surface for $m = 1$, ψ can be assumed to be constant near the resonant surface. This is essentially the same as the Newcomb equation. Near the resonant surface, the derivative diverges logarithmically and this term must be separated for the accurate evaluation of $\Delta'(w)$.

$$\frac{1}{r}\frac{\mathrm{d}}{\mathrm{d}r}\left(r\frac{\mathrm{d}\psi}{\mathrm{d}r}\right) - \frac{m^2}{r^2}\psi - \frac{\mu_0 \mathrm{d}J/\mathrm{d}r}{B_\theta(1 - nq/m)}\psi = 0 \tag{6.45}$$

As seen from Figure 6.1, the perturbed current inside the magnetic island is anti-parallel to the equilibrium plasma current forming counter-clockwise field lines around the island for the case of positive "magnetic shear" $s > 0$ ($s = (r/q)\mathrm{d}q/\mathrm{d}r$). The formation of magnetic islands reduces the pressure gradient and the reduction of the "bootstrap current" (see Section 8.5) occurs and accelerates the growth of magnetic islands. This mode is called the "neoclassical tearing mode" (NTM). On the other hand, the perturbed current is parallel to the equilibrium plasma current and reduction of the bootstrap current reduces the magnetic island for $s < 0$.

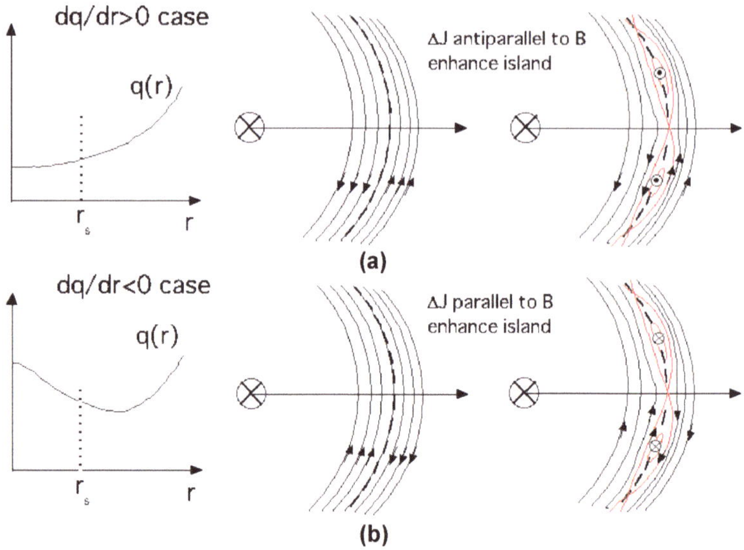

Figure 6.1 Relative magnetic field line flow inside and outside of the resonant surface in equilibrium (*left*) and the formation of magnetic island by the magnetic reconnection shown by *red line* (*right*) for (a) positive magnetic shear $s = r\,\mathrm{d}q/\mathrm{d}r/q > 0$ case and (b) negative magnetic shear $s = r\,\mathrm{d}q/\mathrm{d}r/q < 0$ case. For the positive shear case, perturbed current, which is antiparallel to the equilibrium current enhances the formation of a magnetic island and the perturbed field line encircles the magnetic island in a counter-clockwise direction. For the negative magnetic shear ($s < 0$) case, the perturbed current parallel to the equilibrium current enhances the formation of a magnetic island and the perturbed field line encircles the magnetic island in a clockwise direction

Salon: Harold Furth

Professor Harold Furth (1930–2002; Figure 6.2) was a US fusion physicist and was Director of Princeton Plasma Physics Laboratory (PPPL) between 1981 and 1990. Before coming to PPPL, he wrote a pioneering paper on resistive instabilities using matched asymptotic expansion [17].

Figure 6.2 H. P. Furth (1930–2002) (Courtesy of Princeton Plasma Physics Laboratory)

6.6 Curvature of Magnetic Field: Ballooning and Quasi-mode Expansion

The local mode without the amplitude variation along $\boldsymbol{B}$ is stabilized by the average minimum B effect ($(q^2 - 1)$ term in the Mercier condition). But the ballooning mode can be unstable when the amplitude along $\boldsymbol{B}$ is larger in the weak magnetic field regime (outside of the torus). This mode has a long wavelength along $\boldsymbol{B}$ ($\lambda_{\|} \sim Rq$), and short wavelength perpendicular to $\boldsymbol{B}$. It must satisfy the periodic boundary conditions in both poloidal and toroidal directions. The magnetic field in Clebsch coordinates (r, θ, α) is given by $\boldsymbol{B} = \nabla\alpha \times \nabla\psi$ ($\alpha = \zeta - q\theta$). The displacement perpendicular to the magnetic field $\boldsymbol{\xi}_{\perp}$ is given by using the stream function Φ as follows,

$$\boldsymbol{\xi}_{\perp} = \frac{\mathrm{i}\boldsymbol{B} \times \nabla\Phi}{B^2} , \tag{6.46}$$

$$\Phi = F(r, \theta) \exp(\mathrm{i}S(r, \alpha)) . \tag{6.47}$$

We use eikonal S depending on r and α that are perpendicular to $\boldsymbol{B}$, since ballooning has a short wavelength perpendicular to $\boldsymbol{B}$. Here, $r \equiv a(\phi/\phi_a)^{1/2}$ is the radius defined using the toroidal flux. Φ is a slowly varying function of r and θ. Toroidal symmetry allows Fourier expansion in the toroidal direction, as $\mathrm{i}S \sim -\mathrm{i}n\zeta$. Since $\alpha = \zeta - q\theta$, a possible form of S is $S(r, \alpha) = -n(\alpha + \alpha_0(r))$. Considering the relation $S(r, \theta + 2\pi, \zeta) = S(r, \theta, \zeta) + 2\pi q$, S will not satisfy the periodic condition on θ. This is expected from the nature of the magnetic field lines on the magnetic surface mentioned in Section 3.7. Wave number perpendicular to the magnetic field is given from the following expression,

$$\boldsymbol{k}_{\perp} = \nabla S = n\left[\nabla\alpha + \alpha_0'(r)\nabla r\right] . \tag{6.48}$$

With this wave number, the phase along $\boldsymbol{B}$ becomes uniform but the wave cannot be completed within $[0, 2\pi]$ for θ and should be extended to $\pm\infty$ since $\boldsymbol{B}$ is wound endlessly around the torus as discussed in Section 3.7. In other words, as in "Riemann surfaces," the solution must be obtained to infinity by inserting the cut for every poloidal circulation. Here, in relation to $\theta \in [0, 2\pi]$, $y \in [-\infty, \infty]$ is called the "covering space." Using the arbitrariness of $\alpha_0(r)$, we can construct a solution Φ satisfying the periodic boundary condition of θ from the solution in the covering space. Let $\Phi(y, r) = \varphi(y, r) \exp[-in(\alpha + \alpha_0(r))]$ as the function defined in the covering space, where $\varphi(y, r)$ is the non-periodic square integrable function defined in $[-\infty, \infty]$ ($\varphi \in L_2$). It can be seen that the sum of $\Phi(y, r)$ shifted $2\pi j$ ($j = -\infty, +\infty$) will satisfy the periodic condition $\Phi(\theta + 2\pi, r) = \Phi(\theta, r)$ (this is called a "quasi-mode expansion"),

$$\begin{aligned}\Phi(\theta, r) &= \sum_{j=-\infty}^{\infty} \Phi_0(\theta + 2\pi j, r) = \sum_{j=-\infty}^{\infty} \varphi(\theta + 2\pi j, r)\mathrm{e}^{inq(\theta - \theta_0 + 2\pi j)}\mathrm{e}^{-in\zeta} \\ &= F(\theta, r)\mathrm{e}^{-in\alpha} \\ \theta_0 &= \frac{\alpha_0(r)}{q} .\end{aligned} \tag{6.49}$$

When the mode is expressed by Equation 6.49, the major terms of energy integral δW (Equation 6.22) are given by,

$$\delta W_{\rm SA} = \frac{B_1^2}{2\mu_0} \sim \frac{(\nabla\alpha)^2}{2\mu_0 B^2} |\boldsymbol{B}\cdot\nabla F|^2 \tag{6.50}$$

$$\delta W_{\rm EX} = (\boldsymbol{\xi}_\perp\cdot\nabla P)(\boldsymbol{\xi}_\perp\cdot\boldsymbol{\kappa})/2 \sim P'(\psi)[(\boldsymbol{B}\times\nabla\alpha)\cdot\boldsymbol{\kappa}/B^2]|F|^2 \,. \tag{6.51}$$

The other terms are $O(1/n)$ and can be ignored in large n approximation [18]. Physically, ballooning mode stability is determined by the balance between the bending energy of the magnetic field and the interchange energy of plasma. Then,

$$W_p = \frac{1}{2\mu_0}\int\left[\frac{|\nabla\alpha|^2}{B^2}(\boldsymbol{B}\cdot\nabla F)^2 - 2\mu_0 P'(\psi)\kappa_w F^2\right]\mathrm{d}V \,. \tag{6.52}$$

Here $\kappa_w = (\boldsymbol{B}\times\nabla\alpha)\cdot\boldsymbol{\kappa}/B^2$ is negative with bad magnetic curvature and the term $\kappa_w P' F^2$ is destabilizing. The Euler–Lagrange equation to minimize the energy integral is obtained considering $\boldsymbol{B}\cdot\nabla = \boldsymbol{B}\cdot((\nabla\psi)\partial/\partial\psi + (\nabla\theta)\partial/\partial\theta + (\nabla\alpha)\partial/\partial\alpha) = J^{-1}\partial/\partial\theta$ as follows,

$$J^{-1}\frac{\partial}{\partial\theta}\left[\frac{|\nabla\alpha|^2}{JB^2}\frac{\partial F}{\partial\theta}\right] + \mu_0 P'(\psi)\kappa_w F = 0 \,. \tag{6.53}$$

This equation written in Clebsch coordinates (ψ, θ, α) is the same as Equation 9 of Connor and Taylor [19], who first derived a correct ballooning equation with orthogonal coordinates (ψ, χ, ζ) (ψ: poloidal flux RA_ζ, χ: poloidal angle, ζ: toroidal angle) [20]. Since Equation 6.53 does not include ψ derivative, we can solve it without considering the radial structure as given by $\Phi = F(\theta)\mathrm{e}^{-in\alpha}$. However, its meaning is given in the note. Considering Equation 6.53 is a linear equation and the quasi-mode expansion $F = \sum_j \varphi(\theta + 2\pi j)\mathrm{e}^{inq(2\pi j - \theta_0)} = \sum_j F_1(\theta + 2\pi j)$, $F_1(y)$ satisfies the same Euler–Lagrange equation for F but with its domain $(-\infty, \infty)$.

$$J^{-1}\frac{\partial}{\partial y}\left[\frac{|\nabla\alpha|^2}{JB^2}\frac{\partial F_1(y)}{\partial y}\right] + \mu_0 P'(\psi)\kappa_w F_1(y) = 0 \,. \tag{6.54}$$

The stability condition is $F_1(y)$ should not cross zero, as in Newcomb's theorem 10 and $F_1(y) \to 0$ at $y \to \pm\infty$ for marginal stability.

Note: Radial Structure of Quasi-modes [21]

Zakharov [21] gave a physical explanation showing that the quasi-mode is a superposition of infinite radially (perpendicular to flux surface) overlapping modes (see Figure 6.3). We start from the following Fourier expansion of F, since any periodic function of θ can be expanded in the Fourier series,

$$\Phi = \sum_{k=-\infty}^{\infty} \Phi_k(q) e^{i(m+k)\theta} e^{-in\zeta} \tag{6.55}$$

where $q = m/n$ (since we consider the case of $n \to \infty$ limit, we can use the fact that any irrational number can be given as the limit of a rational number).

It should be noted that the Fourier spectrum of Equation 6.55 resonates at a different safety factor (or radial position) $q(r) = (m+k)/n$. Namely, $\Phi_k(q)$ is a resonant mode at the rational surface $q + \Delta q = (m+k)/n$. Since radial variation of equilibrium quantities is weak, we can assume the translational symmetry for $\Phi_k(q)$ with the amplitude envelope $a(\Delta q)$. Namely, $\Phi_k(q) = a(\Delta q)\Phi_0(q - \Delta q)(\Delta q = k/n)$, $\Phi_0(q)$ is an eigen function for $k = 0$).

For the $n \to \infty$ ballooning mode, we can set $a(\Delta q) = 1$, since $\Delta q = k/n \to 0$. If we consider the expression of $\Phi_0(q)$ by the Fourier transform $\Phi_0(q) = (2\pi)^{-1} \int F_0(s) \exp(isnq)\, ds$ in the infinite domain of $nq \in (-\infty, \infty)$, we obtain following form of Φ,

$$\begin{aligned}
\Phi &= \frac{1}{2\pi} \sum_{k=-\infty}^{\infty} e^{ik\theta} e^{-in\alpha} \int_{-\infty}^{\infty} F_0(s) e^{is(nq-k)}\, ds \\
&= e^{-in\alpha} \int_{-\infty}^{\infty} F_0(s) e^{isnq} \sum_{j=-\infty}^{\infty} \delta(\theta - s + 2\pi j)\, ds \\
&= -e^{-in\alpha} \sum_{j=-\infty}^{\infty} F_0(\theta + 2\pi j) e^{inq(\theta+2\pi j)} = F(\theta, r) e^{-in\alpha} \, .
\end{aligned} \tag{6.56}$$

Here, we defined $F = -\sum_j F(\theta + 2\pi j) e^{2\pi i n j} = \sum_j F_1(\theta + 2\pi j)$ and used the following delta function formula,

$$\frac{1}{2\pi} \sum_{k=-\infty}^{\infty} e^{ik(\theta - s)} = \sum_{j=-\infty}^{\infty} \delta(\theta - s + 2\pi j) \, .$$

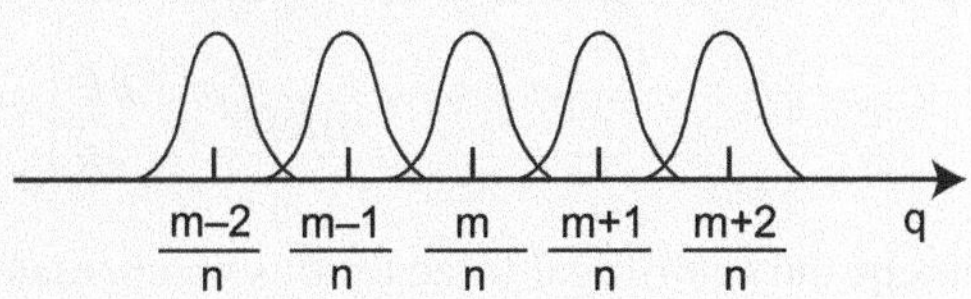

Figure 6.3 Radial mode overlap of ballooning modes in Equation 6.55

6.7 Flow: Non-Hermitian Frieman–Rotenberg Equation

In an axisymmetric system such as tokamak, the neoclassical viscosity in the toroidal direction is small and the toroidal rotation at a fraction of the speed of sound can be induced. In this case, we need to consider the flow in the force equilibrium as follows [22],

$$\rho(\boldsymbol{u}\cdot\nabla)\boldsymbol{u} + \nabla P - \boldsymbol{J}\times\boldsymbol{B} = 0 \tag{6.57}$$

$$\nabla\times(\boldsymbol{u}\times\boldsymbol{B}) = 0 \tag{6.58}$$

$$\boldsymbol{B} = \nabla\zeta\times\nabla\psi + F\nabla\zeta\ . \tag{6.59}$$

From Equation 6.58, we obtain $\boldsymbol{u}\times\boldsymbol{B} = -\nabla\Phi$ and considering $\boldsymbol{B}\cdot\nabla\Phi = 0$,

$$\nabla\Phi = \Omega(\psi)\nabla\psi\ . \tag{6.60}$$

Flow $\boldsymbol{u}$ on the flux surface can be expressed as

$$\boldsymbol{u} = \frac{\Phi_M}{\rho}\boldsymbol{B} + R^2\Omega\nabla\zeta\ . \tag{6.61}$$

In a tokamak, the poloidal rotation is small for neoclassical viscosity, so we consider the case of pure toroidal rotation $\Phi_M = 0$. In this case, we obtain $\rho(\boldsymbol{u}\cdot\nabla)\boldsymbol{u} = -\rho R\Omega^2\nabla R$ (the centrifugal force term) from $\boldsymbol{u} = R^2\Omega\nabla\zeta$. Substituting this equation into Equation 6.57 and taking ζ component, we get $(\boldsymbol{J}\times\boldsymbol{B})\cdot\nabla\zeta = 0$ because of axisymmetry. The following relation can be obtained by taking $\nabla\times$ Equation 6.59.

$$\boldsymbol{J} = \mu_0^{-1}[\nabla F\times\nabla\zeta + \Delta^*\psi\nabla\zeta]\ . \tag{6.62}$$

From $(\boldsymbol{J}\times\boldsymbol{B})\cdot\nabla\zeta = 0$, $(\nabla\psi\times\nabla F)\cdot\nabla\zeta = 0$ is obtained by using the vector formula and we obtain $F = F(\psi)$:

$$\mu_0\boldsymbol{J}\times\boldsymbol{B} = -\frac{FF'(\psi) + \Delta^*\psi}{R^2}\nabla\psi\ . \tag{6.63}$$

Therefore, the following relation is obtained from Equation 6.57:

$$-\rho R\Omega^2\nabla R = -\nabla P - \frac{FF'(\psi) + \Delta^*\psi}{\mu_0 R^2}\nabla\psi\ . \tag{6.64}$$

From the centrifugal force term in the left-hand side, the pressure is no longer a flux function. Taking Equation 6.64 $\cdot\partial\boldsymbol{x}/\partial R$, and considering the orthogonality relation (Equation 3.5), we get the following relation:

$$\rho R\Omega^2 = \left.\frac{\partial P}{\partial R}\right|_\psi\ . \tag{6.65}$$

Namely, the centrifugal force term is compensated by the radial pressure gradient. Furthermore, taking Equation 6.64 $\cdot\partial\boldsymbol{x}/\partial\psi$ and considering Equation 3.5, we obtain the following Grad–Shafranov equation with toroidal flow:

$$\Delta^*\psi = -\mu_0 R^2 \partial P(\psi, R)/\partial\psi - FF'(\psi) \,. \tag{6.66}$$

Assuming $T = T(\psi)$ and defining the ion mass $M = \rho/n$, R integration of Equation 6.6) gives the following formula:

$$P(\psi, R) = P_0(\psi)\exp\left[\frac{M}{2T}R^2\Omega^2\right] \,. \tag{6.67}$$

The action principle of magnetic fluid flow in the plasma, Frieman–Rotenberg [6] is given as follows:

$$S = \int L \,\mathrm{d}V \mathrm{d}t \tag{6.68}$$

$$L = \frac{1}{4}\rho\dot{\boldsymbol{\xi}}^2 - \rho\boldsymbol{\xi}\cdot(\boldsymbol{u}\cdot\nabla)\dot{\boldsymbol{\xi}} + \frac{1}{2}\rho\boldsymbol{\xi}\cdot\boldsymbol{F}(\boldsymbol{\xi}) \,. \tag{6.69}$$

From the Lagrangian L, the generalized momentum is given by $\boldsymbol{p} \equiv \partial L/\partial\boldsymbol{\xi} = \rho(\partial\boldsymbol{\xi}/\partial t) + \rho\boldsymbol{u}\cdot\nabla\boldsymbol{\xi}$ and the Hamiltonian $H(= \boldsymbol{p}\cdot\boldsymbol{\xi} - L)$ is given below:

$$H = \frac{1}{2\rho}[\boldsymbol{p} - \rho\boldsymbol{u}\cdot\nabla\boldsymbol{\xi}]^2 - \frac{1}{2}\rho\boldsymbol{\xi}\cdot\boldsymbol{F}(\boldsymbol{\xi}) \,. \tag{6.70}$$

The Hamilton equation $\mathrm{d}\boldsymbol{p}/\mathrm{d}t = -\partial H/\partial\boldsymbol{\xi}$ gives $\mathrm{d}\boldsymbol{p}/\mathrm{d}t = \boldsymbol{F}(\boldsymbol{\xi}) - \rho\boldsymbol{u}\cdot\nabla[(\boldsymbol{p}/\rho) - \boldsymbol{u}\cdot\nabla\boldsymbol{\xi}]$. From this equation, the following Frieman–Rotenberg equation is obtained as the linearized equation of motion with the magnetic fluid flow [6]:

$$\begin{aligned}
\rho\frac{\partial^2\boldsymbol{\xi}}{\partial t^2} + 2\rho(\boldsymbol{u}\cdot\nabla)\frac{\partial\boldsymbol{\xi}}{\partial t} &= \boldsymbol{F}(\boldsymbol{\xi}) \\
\boldsymbol{F}(\boldsymbol{\xi}) &= \boldsymbol{F}_s(\boldsymbol{\xi}) + \boldsymbol{F}_d(\boldsymbol{\xi}) \\
\boldsymbol{F}_s(\boldsymbol{\xi}) &= \nabla[\boldsymbol{\xi}\cdot\nabla P + \gamma P\nabla\cdot\boldsymbol{\xi}] + (\nabla\times\boldsymbol{B}_1)\times\boldsymbol{B} + \boldsymbol{J}\times\boldsymbol{B}_1 \\
\boldsymbol{F}_d(\boldsymbol{\xi}) &= \nabla\cdot[\rho\boldsymbol{\xi}(\boldsymbol{u}\cdot\nabla)\boldsymbol{u} - \rho\boldsymbol{u}(\boldsymbol{u}\cdot\nabla)\boldsymbol{\xi}] \\
\boldsymbol{B}_1 &= \nabla\times(\boldsymbol{\xi}\times\boldsymbol{B}) \,.
\end{aligned} \tag{6.71}$$

$\boldsymbol{F}_s(\boldsymbol{\xi})$ and $\boldsymbol{F}_d(\boldsymbol{\xi})$ are the static and dynamic operators, respectively, and both are Hermitian operators [22]. This Hermitian property of $\boldsymbol{F}$ is consistent with the energy conservation equation as given by

$$H = \frac{1}{2}\int\left[\rho\left(\frac{\partial\boldsymbol{\xi}}{\partial t}\right)^2 - \boldsymbol{\xi}\cdot\boldsymbol{F}\cdot\boldsymbol{\xi}\right]\mathrm{d}V = \text{const.} \tag{6.72}$$

On the other hand, the convective term $L = 2\rho(\boldsymbol{u}\cdot\nabla)\partial_t\boldsymbol{\xi}$ is an anti-Hermitian operator ($L(\boldsymbol{\zeta},\boldsymbol{\xi}) = -L^*(\boldsymbol{\xi},\boldsymbol{\zeta})$), and the system is not self-adjoint as a whole. It is difficult to solve the equation as an eigenvalue problem. Therefore, the Frieman–Rotenberg equation is solved as the initial value problem [23] or by the Laplace transform technique [24]. For example, the Laplace transform, $\boldsymbol{\xi}(t) \to \boldsymbol{\xi}(\omega)$ $(t \in R, \omega \in C)$ gives

$$L\boldsymbol{\xi}(\omega) = \boldsymbol{m}_0(\omega) \,. \tag{6.73}$$

Here, $L = \omega^2\rho + 2\mathrm{i}\omega\rho(\boldsymbol{u}\cdot\nabla) + \boldsymbol{F}$ and $\boldsymbol{m}_0(\omega) = \mathrm{i}\omega\rho\boldsymbol{\xi}_0 + \rho\boldsymbol{u}\times(\nabla\times\boldsymbol{\xi}_0) + \boldsymbol{B}\times(\nabla\times\boldsymbol{\eta}_0) + \rho\nabla\alpha - \beta\nabla s$. Let the eigenvalue of this equation (the spectrum) ω_j $(j = 1, \ldots)$, and the continuous eigenvalue (continuous spectrum) $\omega \in \sigma_c$, the eigen mode decomposition of the solution is given by

$$\xi(t) = \sum_j \xi(\omega_j)\exp(-\mathrm{i}\omega_j t) + \int_{\sigma_c} \xi(\omega)\exp(-\mathrm{i}\omega t)\,\mathrm{d}\omega \,. \tag{6.74}$$

Here, the eigen function $\boldsymbol{\xi}(\omega_j)$ for the point spectrum ω_j $(j = 1, \ldots)$, and the singular eigen function $\boldsymbol{\xi}(\omega)$ corresponding to the continuous spectrum are given as follows,

$$\boldsymbol{\xi}(\omega_j) = -(1/2\pi)\int_{\Gamma(\omega_j)} \boldsymbol{\xi}_\omega \mathrm{d}\omega \,, \tag{6.75}$$

$$\boldsymbol{\xi}(\omega) = (1/2\pi)[\boldsymbol{\xi}(\omega + \mathrm{i}0) - \boldsymbol{\xi}(\omega - \mathrm{i}0)] \,. \tag{6.76}$$

For cylindrical plasma, Newcomb equation 6.30 has to be modified to include Doppler shift $\Delta\omega = \boldsymbol{k}\cdot\boldsymbol{u}$ due to plasma flow. This Doppler shift splits the singular point of the Newcomb equation from a rational surface $(\boldsymbol{k}\cdot\boldsymbol{B} = 0)$ to two singular points $(\boldsymbol{k}\cdot\boldsymbol{v}_A = \pm\boldsymbol{k}\cdot\boldsymbol{u}$ $(\boldsymbol{v}_A = \boldsymbol{B}/(\mu_0\rho)^{1/2})$ for Alfven and slow magnetosonic resonances [25].

References

1. Diacu F, Holmes P (1996) Celestial Encounters: The Origin of Chaos and Stability. Princeton University Press.
2. Kolmogorov AN, Fomin SV (1999) Elements of the Theory of Functions and Functional Analysis. Dover Books.
3. Kadomtzev BB (1976) Collective Phenomena in Plasmas. Nauka Moscow.
4. Friedman B (1990) Principles and Techniques of Applied Mathematics. Dover Books.
5. Bernstein IB, Frieman EA, Kruskal MD, Kulsrud RM (1958) Proc. Roy. Soc., A244, 17.
6. Frieman E, Rotenberg M (1960) Rev. Mod. Phys., 32, 898.
7. Goldstein H (1950) Classical Mechanics. Addison-Wesley,
8. Kulsrud RM (2005) Plasma Physics for Astrophysics. Princeton University Press.
9. Kruskal MD, Krusrud RM (1958) Phys. Fluids, 1, 265.
10. Freidberg JP (1987) Ideal Magnetohydrodynamics. Plenum.
11. Furth HP et al. (1966) in Proc. Int. Conf. Plasma Phys. and Contr. Nucl. Fusion Research (Culham, 1965). IAEA, Vol. 1, p. 103.
12. Newcomb A (1960) Ann. Phys., 10, 232.
13. Tokuda S, Watanabe T (1999) Phys. of Plasmas, 6, 3012.

14. Mercier C (1960) Nuclear Fusion, 1, 47.
15. Rutherford PH (1973) Phys. Fluids, 16, 1903; see also Wesson J (1997) Tokamaks. Clarendon Press.
16. White RB et al. (1977) Phys. Fluids 20, 800–805.
17. Furth HP, Killeen J, Rosenbluth MN (1963) Phys. Fluids, 6, 459–484.
18. White RB (2006) The Theory of Toroidally Confined Plasmas. Imperial College Press, Appendix B.3.
19. Connor JW, Hastie RJ, Taylor JB (1979) Proc. Roy. Soc., A365, 1.
20. Connor JW, Hastie RJ, Taylor JB (1978) Phys. Rev. Lett., 40, 396.
21. Zakharov LE (1979) Proc. Int. Conf. Plasma Physics and Contr. Nucl. Fusion Res. (Innsbruck 1978) IAEA, Vol. 1, 689.
22. Tokuda S (1998) Plasma Fusion Res., 74, No. 5, 503.
23. Aiba N, Tokuda S, et al. (2009) Comput. Phys. Commum., 180, 1282.
24. Hirota M, Fukumoto Y (2008) Phys. Plasmas, 15, 122101.
25. Shiraishi J, Tokuda S, Aiba N (2010) Phys. Plasma, 17, 012504.

Chapter 7
Wave Dynamics: Propagation and Resonance in Inhomogeneous Plasma

Plasma is a dispersive medium and wave propagation can be described by the Eikonal equation derived by Landau. Wave energy is also defined. If there is no dissipation in the plasma, Lagrange–Hamilton formulation is applicable and the conservation law is obtained. The zero approximation as the dispersive medium is the dissipationless cold plasma approximation ignoring the thermal effect. Cutoff and resonance occurs in this approximation.

The wave propagation in non-uniform plasma is important in confined plasma and energy absorption occurs in the resonance layer. The drift wave appears universally in the confined plasma and is unstable above the critical temperature gradient, producing turbulence through wave–wave interactions.

7.1 Eikonal Equation: Dynamics of Wave Propagation

Wave propagation in a dispersive media with a local dispersion relation $\omega = \Omega(\boldsymbol{k}, \boldsymbol{x}, t)$ cannot be described by a simple plane–wave approximation $e^{i(k\cdot x - \Omega t)}$ but is described by the Eikonal form $\boldsymbol{x} = \boldsymbol{A} e^{i\zeta(x,t)} + cc$ with its phase (or eikonal) ζ and amplitude A [1]. The angular frequency ω and wave number $\boldsymbol{k}$ are related to eikonal ζ by the following equation,

$$\omega = -\frac{\partial \zeta}{\partial t}, \tag{7.1}$$

$$\boldsymbol{k} = \frac{\partial \zeta}{\partial \boldsymbol{x}}. \tag{7.2}$$

From the analytical condition of the second-order derivative $\partial^2\zeta/\partial t\,\partial \boldsymbol{x}$, ω and $\boldsymbol{k}$ should satisfy following equation:

$$\frac{\partial \boldsymbol{k}}{\partial t} = -\frac{\partial \omega}{\partial \boldsymbol{x}}. \tag{7.3}$$

From the expression of local dispersion relation $\omega = \Omega(\boldsymbol{k}, \boldsymbol{x}, t)$, we obtain

$$\frac{\partial \omega}{\partial \boldsymbol{x}} = \boldsymbol{v}_g \cdot \frac{\partial \boldsymbol{k}}{\partial \boldsymbol{x}} + \frac{\partial \Omega}{\partial \boldsymbol{x}} |_{\boldsymbol{k}} \ . \tag{7.4}$$

Here the group velocity of the wave packet $\boldsymbol{v}_g = \partial\omega/\partial\boldsymbol{k} = \partial\Omega/\partial\boldsymbol{k}|_x$ is used. From Equations 7.3 and 7.4, we can see that the velocity of the wave packet center $\mathrm{d}\boldsymbol{x}/\mathrm{d}t$ and the time variation of the wave number in the moving frame with the group velocity $\mathrm{d}\boldsymbol{k}/\mathrm{d}t = \partial\boldsymbol{k}/\partial t + \boldsymbol{v}_g \partial\boldsymbol{k}/\partial\boldsymbol{x}$ satisfy the following Hamilton equation:

$$\frac{\mathrm{d}\boldsymbol{x}}{\mathrm{d}t} = \left(\frac{\partial \Omega}{\partial \boldsymbol{k}}\right)_{\boldsymbol{x}} , \tag{7.5}$$

$$\frac{\mathrm{d}\boldsymbol{k}}{\mathrm{d}t} = -\left(\frac{\partial \Omega}{\partial \boldsymbol{x}}\right)_{\boldsymbol{k}} . \tag{7.6}$$

Here, Ω plays the role of the Hamiltonian and $\boldsymbol{k}$ the role of canonical momentum. They are also called Eikonal equations. The variational principle to give this eikonal equation is given by $\delta \int \mathcal{L}\,\mathrm{d}t = 0$ with $\mathcal{L} = \boldsymbol{k} \cdot \boldsymbol{x} - \Omega$. Substituting the eikonal expression into the Maxwell equation, $\boldsymbol{E}, \boldsymbol{B} \sim \hat{\boldsymbol{E}}, \hat{\boldsymbol{B}} \mathrm{e}^{\mathrm{i}\xi(x,t)} + cc$, we obtain

$$\boldsymbol{k} \times \hat{\boldsymbol{B}} = -i\mu_0 \hat{\boldsymbol{J}} - \frac{\omega}{c^2} \hat{\boldsymbol{E}} , \tag{7.7}$$

$$\boldsymbol{k} \times \hat{\boldsymbol{E}} = \omega \hat{\boldsymbol{B}} . \tag{7.8}$$

We assume that the following linear relation holds using the electrical conductivity tensor σ (see Section 7.3 for details):

$$\hat{\boldsymbol{J}} = \boldsymbol{\sigma} \hat{\boldsymbol{E}} . \tag{7.9}$$

This is Ohm's law and Equation 7.7 leads to

$$\boldsymbol{k} \times \hat{\boldsymbol{B}} = -\frac{\omega}{c^2} \boldsymbol{K}(\omega, \boldsymbol{k}) \cdot \hat{\boldsymbol{E}} , \tag{7.10}$$

$$\boldsymbol{K}(\omega, \boldsymbol{k}) = \boldsymbol{I} + \frac{\mathrm{i}\boldsymbol{\sigma}}{\varepsilon_0 \omega} . \tag{7.11}$$

where $\boldsymbol{K}$ is the dielectric tensor. From Equations 7.8 and 7.10, vectors $\boldsymbol{k}$, $\boldsymbol{K}(\omega, \boldsymbol{k}) \cdot \hat{\boldsymbol{E}}$, and $\boldsymbol{B}$ are orthogonal within the Eikonal approximation. Eliminating $\boldsymbol{B}$ from these equations, we obtain

$$\boldsymbol{M} \cdot \hat{\boldsymbol{E}} = 0 , \tag{7.12}$$

$$\boldsymbol{M} = (\boldsymbol{k}\boldsymbol{k} - k^2 \boldsymbol{I})/k_0^2 + \boldsymbol{K} \tag{7.13}$$

where $k_0 = \omega/c$. As is well known in electromagnetism, the following Poynting theorem holds among the electromagnetic energy $\boldsymbol{B}^2/2\mu_0 + \varepsilon_0 \boldsymbol{E}^2/2$, Joule losses $\boldsymbol{J} \cdot \boldsymbol{E}$, and Poynting vector $\boldsymbol{S} = \boldsymbol{E} \times \boldsymbol{B}/\mu_0$:

$$\frac{\partial}{\partial t}\left(\frac{\boldsymbol{B}^2}{2\mu_0} + \frac{\varepsilon_0 \boldsymbol{E}^2}{2}\right) = -\boldsymbol{J} \cdot \boldsymbol{E} - \nabla \cdot \boldsymbol{S} . \tag{7.14}$$

This equation shows that the Poynting vector $\boldsymbol{S}$ is the energy flux density. If we take time and space derivatives of the amplitude in slowly changing medium up to the first order $(\omega + \mathrm{i}\partial/\partial t, \mathrm{i}\boldsymbol{k} + \nabla)$, $\boldsymbol{J}$ and $\boldsymbol{B}$ are given as follows,

$$\hat{\boldsymbol{J}} + \delta\hat{\boldsymbol{J}} = \boldsymbol{\sigma}\left(\omega + \mathrm{i}\frac{\partial}{\partial t}\right)(\hat{\boldsymbol{E}} + \delta\hat{\boldsymbol{E}}) = \boldsymbol{\sigma}(\omega)\hat{\boldsymbol{E}} + \boldsymbol{\sigma}(\omega)\delta\hat{\boldsymbol{E}} + \frac{\partial\boldsymbol{\sigma}}{\partial\omega}\mathrm{i}\frac{\partial\hat{\boldsymbol{E}}}{\partial t}, \quad (7.15)$$

$$\mathrm{i}\omega\delta\hat{\boldsymbol{B}} = \mathrm{i}\boldsymbol{k} \times \delta\hat{\boldsymbol{E}} + \frac{\partial\hat{\boldsymbol{B}}}{\partial t} + \nabla \times \hat{\boldsymbol{E}}\,. \quad (7.16)$$

Here $\hat{\boldsymbol{B}} = \boldsymbol{k} \times \hat{\boldsymbol{E}}/\omega$. Substituting these equations into Equation 7.14, we can obtain the following equations:

$$\frac{\partial\mathcal{E}}{\partial t} + \nabla\cdot\hat{\boldsymbol{S}} = Q\,, \quad (7.17)$$

$$\mathcal{E} = \frac{1}{2}\left(\varepsilon_0\hat{\boldsymbol{E}}^* \cdot \frac{\partial(\omega\boldsymbol{K}_{\mathrm{h}})}{\partial\omega}\cdot\hat{\boldsymbol{E}} + \frac{1}{\mu_0}\hat{\boldsymbol{B}}^*\cdot\hat{\boldsymbol{B}}\right), \quad (7.18)$$

$$\hat{\boldsymbol{S}} = Re(\hat{\boldsymbol{E}}^* \times \hat{\boldsymbol{B}})/\mu_0 \quad (7.19)$$

$$Q = \hat{\boldsymbol{E}}^*\cdot\boldsymbol{\sigma}_{\mathrm{h}}\cdot\hat{\boldsymbol{E}}\,, \quad (7.20)$$

$$\boldsymbol{K}_{\mathrm{h}} = (\boldsymbol{K} + \boldsymbol{K}^+)/2\,, \quad (7.21)$$

$$\boldsymbol{\sigma}_{\mathrm{h}} = (\boldsymbol{\sigma} + \boldsymbol{\sigma}^+)/2\,. \quad (7.22)$$

Here, $\boldsymbol{K}_{\mathrm{h}}$, $\boldsymbol{\sigma}_{\mathrm{h}}$ are the Hermitian part in each component, $\hat{\boldsymbol{E}}^*$ and $\hat{\boldsymbol{B}}^*$ are complex conjugate vectors, $\boldsymbol{K}^+$ and $\boldsymbol{\sigma}^+$ are complex conjugate tensors. Using Equations 7.12 and 7.13, wave energy $\mathcal{E}$ is given as follows,

$$\mathcal{E} = \frac{\varepsilon_0}{2}\hat{\boldsymbol{E}}^*\cdot\frac{\partial(\omega\boldsymbol{M}_{\mathrm{h}})}{\partial\omega}\cdot\hat{\boldsymbol{E}}\,, \quad (7.23)$$

$$\mathcal{E} = \omega\mathcal{J}\,, \quad (7.24)$$

$$\mathcal{J} = \frac{\varepsilon_0}{2}\hat{\boldsymbol{E}}^*\cdot\frac{\partial\boldsymbol{M}_{\mathrm{h}}}{\partial\omega}\cdot\hat{\boldsymbol{E}}\,. \quad (7.25)$$

Here, J is a "wave action" and is the adiabatic invariant if there is no dissipation. The wave energy form seen in Equation 7.18 was derived by M. von Laue in 1905 for a dispersive medium [2]. In non-thermodynamic equilibrium plasma, this wave energy can take a negative value and is called "negative energy wave" [3]. If we regard $(\boldsymbol{x}, \boldsymbol{k})$ as independent variables, the wave action J is conserved along the trajectory in the phase space $(\boldsymbol{x}, \boldsymbol{k})$, in case there is no dissipation. So, J follows

$$\frac{\partial J}{\partial t} + \dot{\boldsymbol{k}}\cdot\frac{\partial J}{\partial\boldsymbol{k}} + \dot{\boldsymbol{x}}\cdot\frac{\partial J}{\partial\boldsymbol{x}} = 0\,. \quad (7.26)$$

Substituting Equations 7.5 and 7.6 into Equation 7.26, we obtain the following equation:

$$\frac{\partial J}{\partial t} + [J, \Omega] = 0\,, \quad (7.27)$$

where $[J, \Omega]$ is the Poisson bracket given by

$$[J, \Omega] = \frac{\partial \Omega}{\partial \boldsymbol{k}} \cdot \frac{\partial J}{\partial \boldsymbol{x}} - \frac{\partial \Omega}{\partial \boldsymbol{x}} \cdot \frac{\partial J}{\partial \boldsymbol{k}} . \tag{7.28}$$

Equation 7.27 is called the "wave kinetic equation."

7.2 Lagrange Wave Dynamics: Ideal and Dissipative Systems

As described in Goldstein [4], Lagrange mechanics in a continuum are reduced to the variational principle with the action integral of time and space integration of the Lagrangian density L. Whitham formulated the Lagrange mechanics for dissipationless ideal plasma in 1965 [5,6], while the dynamics of the dissipative plasma wave have not been formulated [7]. The action integral S is given by

$$S = \int \mathrm{d}t \int L \mathrm{d}V \tag{7.29}$$

$$L = L_M(\boldsymbol{A}, \Phi) + \sum_a L_a(\boldsymbol{\xi}_a, \boldsymbol{A}, \Phi) , \tag{7.30}$$

$$L_M(\boldsymbol{A}, \Phi) = \varepsilon_0 \left[\frac{\partial \boldsymbol{A}}{\partial t} + \nabla \Phi \right]^2 - \frac{1}{\mu_0} (\nabla \times \boldsymbol{A})^2 , \tag{7.31}$$

$$L_a(\boldsymbol{\xi}_a, \boldsymbol{A}, \Phi) = n_a \left[\frac{m_a}{2} \dot{\boldsymbol{\xi}}_a^2 + e_a \left(\dot{\boldsymbol{\xi}}_a \cdot \boldsymbol{A}(\boldsymbol{x} + \boldsymbol{\xi}_a, t) - \Phi(\boldsymbol{x} + \boldsymbol{\xi}_a, t) \right) \right] . \tag{7.32}$$

Here, L_M (Equation 4.25) and L_a are Lagrangians for fields and particles, respectively. The Lagrangian of particle a can be expanded as a quadratic form as follows,

$$[L_a(\boldsymbol{\xi}_a, \boldsymbol{A}, \Phi)]_{\text{lin}} = n_a \left[\frac{m_a}{2} \dot{\boldsymbol{\xi}}_a^2 + e_a \, \boldsymbol{\xi}_a \cdot (\dot{\boldsymbol{\xi}}_a \times \boldsymbol{B}_0) + e_a \dot{\boldsymbol{\xi}}_a \cdot \tilde{\boldsymbol{A}} - e_a \boldsymbol{\xi}_a \cdot \nabla \tilde{\Phi} \right] . \tag{7.33}$$

Lagrangian density can be rewritten in the following form by using the wave eikonal form $\boldsymbol{\xi} = \boldsymbol{\xi} \mathrm{e}^{\mathrm{i}\zeta(x,t)} + \text{c.c.}$ and $\boldsymbol{E} = -\partial \boldsymbol{A}/\partial t - \nabla \Phi$ as follows,

$$[L]_{\text{lin}} = \varepsilon_0 \hat{\boldsymbol{E}}^* \cdot \boldsymbol{M} \cdot \hat{\boldsymbol{E}} . \tag{7.34}$$

Here, cold plasma is a typical example of non-dissipative plasma and its $\boldsymbol{M}$ is given in next section. Variation of S with respect to $\hat{\mathbf{E}}^*$ gives the local dispersion relation,

$$\boldsymbol{M} \cdot \hat{\boldsymbol{E}} = 0 . \tag{7.35}$$

In addition, the minimization with respect to the eikonal ζ gives the following Euler–Lagrange equation.

$$\frac{\partial}{\partial t} \left[\frac{\partial L}{\partial \omega} \right] + \frac{\partial}{\partial \boldsymbol{x}} \cdot \left[\frac{\partial L}{\partial \boldsymbol{k}} \right] = 0 . \tag{7.36}$$

Here,

$$J = \frac{\partial L}{\partial \omega} = \frac{\partial L}{\partial \dot{\zeta}} = \varepsilon_0 \hat{\boldsymbol{E}}^* \cdot \frac{\partial \boldsymbol{M}}{\partial \omega} \cdot \hat{\boldsymbol{E}} \tag{7.37}$$

is the momentum conjugate to eikonal ζ and is the adiabatic invariant. Using $\partial L/\partial \boldsymbol{k} = (\partial L/\partial \omega)(\partial \omega/\partial \boldsymbol{k}) = \boldsymbol{v}_g(\partial L/\partial \omega) = \boldsymbol{v}_g J$, J corresponds to the number of photons in the wave packet and Equation 7.36 gives the conservation law for the number of photons.

$$\frac{\partial J}{\partial t} + \frac{\partial}{\partial \boldsymbol{x}} \cdot (\boldsymbol{v}_g J) = 0 \,. \tag{7.38}$$

Hamilton mechanics is a powerful technique in solving mathematical problems in plasma dynamics, but it cannot be applied to a dissipative system as it stands. Hamilton mechanics of dissipative systems can be formulated by the adjoint-variable method [8]. Consider ordinary differential equations in general with n-dimensional variable $\boldsymbol{x}$:

$$\frac{\mathrm{d}\boldsymbol{x}}{\mathrm{d}t} = \boldsymbol{f}(\boldsymbol{x}, t) \,. \tag{7.39}$$

We introduce a new n dimensional variable $\boldsymbol{p}$, and define L as

$$L = \boldsymbol{p} \cdot \left(\frac{\mathrm{d}\boldsymbol{x}}{\mathrm{d}t} - \boldsymbol{f} \right) = \boldsymbol{p} \cdot \frac{\mathrm{d}\boldsymbol{x}}{\mathrm{d}t} - H \,. \tag{7.40}$$

Here $H = \boldsymbol{p} \cdot \boldsymbol{f}(\boldsymbol{x}, t)$ plays the role of the Hamiltonian. Also, it is easy to see that

$$\frac{\partial L}{\partial (\mathrm{d}\boldsymbol{x}/\mathrm{d}t)} = \boldsymbol{p} \,. \tag{7.41}$$

This means that $\boldsymbol{p}$ is momentum conjugate to $\boldsymbol{x}$. Then, we obtain the following Hamilton equation:

$$\frac{\mathrm{d}\boldsymbol{x}}{\mathrm{d}t} = \frac{\partial H}{\partial \boldsymbol{p}} = \boldsymbol{f}(\boldsymbol{x}, t) \,, \tag{7.42}$$

$$\frac{\mathrm{d}\boldsymbol{p}}{\mathrm{d}t} = -\frac{\partial H}{\partial \boldsymbol{x}} = -\frac{\partial \boldsymbol{f}(\boldsymbol{x}, t)}{\partial \boldsymbol{x}} \cdot \boldsymbol{p} \,. \tag{7.43}$$

In other words, any system of ordinary differential equations including dissipation can be attributed to the Hamilton system by doubling the variables. Application of this formulation to dissipative plasma dynamics is left for future study.

7.3 Plasma as a Dielectric Medium: Cold and Hot Plasmas

Plasma is a dielectric media in which various waves can propagate. We assume that perturbed electromagnetic fields are given by a plane wave, $\boldsymbol{E}_1 = \boldsymbol{E} \exp(\mathrm{i}\boldsymbol{k} \cdot \boldsymbol{x} - \mathrm{i}\omega t)$, $\boldsymbol{B}_1 = \boldsymbol{b} \exp(\mathrm{i}\boldsymbol{k} \cdot \boldsymbol{x} - \mathrm{i}\omega t)$ and $\boldsymbol{J}_1 = \boldsymbol{j} \exp(\mathrm{i}\boldsymbol{k} \cdot \boldsymbol{x} - \mathrm{i}\omega t)$. The Maxwell equation gives $\boldsymbol{k} \times \boldsymbol{B}_1 = -\mathrm{i}\omega\mu_0 \boldsymbol{J}_1 - \omega \boldsymbol{E}_1$. Combining this with Ohm's law $\boldsymbol{J}_1 = \boldsymbol{\sigma} \cdot \boldsymbol{E}_1$ leads to the relation $\boldsymbol{k} \times \boldsymbol{B}_1 = -(\omega/c^2)\boldsymbol{K} \cdot \boldsymbol{E}_1$, where $\boldsymbol{K} = \boldsymbol{I} + \mathrm{i}\boldsymbol{\sigma}/\varepsilon_0\omega$ is the dielectric tensor. Substitution of $\boldsymbol{k} \times \boldsymbol{B}_1 = -(\omega/c^2)\boldsymbol{K} \cdot \boldsymbol{E}_1$ into $\boldsymbol{k} \times \boldsymbol{E} = \omega \boldsymbol{B}_1$ gives,

$$\boldsymbol{M} \cdot \boldsymbol{E} = 0\,, \ \boldsymbol{M} = (\boldsymbol{kk} - \boldsymbol{I})\left(\frac{kc}{\omega}\right)^2 + \boldsymbol{K}\,. \tag{7.44}$$

The solvable condition is given as $M = \det(\boldsymbol{M}) = 0$ and is called the dispersion relation. The group velocity of the wave $\boldsymbol{v}_{\mathrm{g}}$ can be obtained from the $\boldsymbol{k}$ derivative of the dispersion relation $M(\omega, \boldsymbol{k}, \boldsymbol{x}, t) = \det(\boldsymbol{M}) = 0$ as $\boldsymbol{v}_{\mathrm{g}} = \partial\Omega/\partial\boldsymbol{k} = (\partial M/\partial\boldsymbol{k})/(\partial M/\partial\omega)$.

When the plasma temperature is low or the wave phase speed (ω/k) is much higher than the thermal speed ($v_{\mathrm{th}} \ll \omega/k$), the effect of thermal motion can be neglected and is called "cold plasma." When the plasma temperature is not too low, wave propagation is influenced by the sound wave and the pressure effect must be taken into account for the dielectric tensor. This is "warm plasma." When the resonant wave-particle interaction such as Landau damping plays some role, the Vlasov equation must be solved to obtain the dielectric tensor, and this is called "hot plasma."

In the cold plasma case, an important characteristic is that dielectric tensor $\boldsymbol{K}$ does not have any k dependence and is given as follows [9],

$$\boldsymbol{K} = \begin{bmatrix} S & -\mathrm{i}D & 0 \\ \mathrm{i}D & S & 0 \\ 0 & 0 & P \end{bmatrix}. \tag{7.45}$$

$$S = 1 - \sum_a \frac{\omega_{pa}^2}{\omega^2 - \Omega_a^2}\,, \quad D = \sum_a \frac{\Omega_a}{\omega}\frac{\omega_{pa}^2}{\omega^2 - \Omega_a^2}\,, \quad P = 1 - \sum_a \frac{\omega_{pa}^2}{\omega^2}\,,$$

$$\omega_{pa}^2 = \frac{n_a e_a^2}{\varepsilon_0 m_a}\,, \quad \Omega_a = \frac{e_a B}{m_a}\,.$$

Let θ be the angle between $\boldsymbol{B}$ and $\boldsymbol{k}$ and the refractive index $n = kc/\omega$, the following dispersion relation is obtained:

$$[S \sin^2\theta + P\cos^2\theta]n^4 - [RL\sin^2\theta + PS(1+\cos^2\theta)]n^2 + PRL = 0 \tag{7.46}$$

where $R = (S + D)/2$ and $L = (S - D)/2$. The condition of the refractive index $n = 0$ (phase velocity $= \omega/k = \infty$) is called the "cut-off," and $n = \infty$ (phase velocity $= \omega/k = 0$) is called "resonance." From Equation 7.46

$$\text{Cut-off condition } (n = 0): \qquad PRL = 0\,, \tag{7.47}$$

$$\text{Resonance condition } (n = \infty): \qquad \tan^2\theta = -\frac{P}{S}\,. \tag{7.48}$$

If we have a cut-off layer in the plasma, the plasma wave can only propagate an evanescent wave. On the other hand, the resonance is important for plasma heating and as a damping mechanism for instability. Considering propagation parallel to the magnetic field ($\theta = 0$), S becomes ∞ at $\omega = \Omega_a$ and gives resonance. This is the cyclotron resonance. In the case of propagation perpendicular to magnetic field ($\theta = \pi/2$), the resonance condition is $S = 0$. Finally, it is worth noting that the cold plasma dielectric tensor has various symmetries [5].

Time symmetry: $\boldsymbol{K}(-\omega) = \boldsymbol{K}^*(\omega)$ by the time symmetry of the equation of motion.
Onsager Symmetry: $\boldsymbol{K}(-\boldsymbol{B}) = \boldsymbol{K}^t(\boldsymbol{B})$ corresponding to the Onsager theorem.
Hermitian: $\boldsymbol{K} = \boldsymbol{K}^+$ (+: complex conjugate) corresponding to energy conservation. If energy is not conserved, the system is not Hermitian.

In the hot plasma case, the plasma response as a dielectric medium can be expressed by the dielectric tensor $\boldsymbol{K}$, as for the cold plasma wave. But the structure of $\boldsymbol{K}$ is more complicated compared to that in cold plasma. Since $\boldsymbol{K} = \boldsymbol{I} + \mathrm{i}\boldsymbol{\sigma}/\varepsilon_0\omega$ and the conductivity tensor is related to the perturbed velocity distribution function $f_{a1k}(\boldsymbol{v})$ as $\boldsymbol{J} = \boldsymbol{\sigma}\boldsymbol{E}$ and $\boldsymbol{J} = \sum e_a \int \boldsymbol{v} f_{a1k}(\boldsymbol{v})\mathrm{d}\boldsymbol{v}$, we have to solve the following linearized Vlasov equation:

$$\frac{\partial f_{a1k}}{\partial t} + \boldsymbol{v}\cdot\frac{\partial f_{a1k}}{\partial \boldsymbol{x}} + \frac{e_a}{m_a}(\boldsymbol{v}\times\boldsymbol{B})\cdot\frac{\partial f_{a1k}}{\partial \boldsymbol{v}} = -\frac{e_a}{m_a}(\boldsymbol{E}_1 + \boldsymbol{v}\times\boldsymbol{B}_1)\cdot\frac{\partial f_{a0}}{\partial \boldsymbol{v}} . \tag{7.49}$$

Here, the suffix for equilibrium value for $\boldsymbol{B}$ is suppressed for simplicity. Since the left-hand side of Equation 7.49 is the Lagrange derivative Df_{a1k}/Dt along the unperturbed charged particle orbit, $f_{a1k}(\boldsymbol{v})$ is readily obtained as follows [9],

$$f_{a1k}(\boldsymbol{x},\boldsymbol{v},t) = -\frac{e_a}{m_a}\int_{-\infty}^{t}\left(\boldsymbol{E}_1(\boldsymbol{x}(t'),t') + \frac{1}{\omega}\boldsymbol{v}(t')\times(\boldsymbol{k}\times\boldsymbol{E}_1(\boldsymbol{x}(t'),t'))\right) \cdot\frac{\partial f_0(\boldsymbol{v}(t'))}{\partial \boldsymbol{v}}\mathrm{d}t' . \tag{7.50}$$

Here, $\boldsymbol{B}_1 = (\boldsymbol{k}\times\boldsymbol{E}_1)/\omega$ is used. The particle position at t', $\boldsymbol{x}(t')$ is given by a combination of cyclotron excursion and the original position as follows,

$$\begin{aligned} x(t') &= x(t) + \frac{v_\perp}{\Omega}\left(\sin(\theta + \Omega(t'-t)) - \sin\theta\right) , \\ y(t') &= y(t) - \frac{v_\perp}{\Omega}\left(\cos(\theta + \Omega(t'-t)) - \cos\theta\right) , \\ z(t') &= z(t) + v_z(t'-t) . \end{aligned} \tag{7.51}$$

Here, the direction of $\boldsymbol{B}$ is chosen as z. Substitution of Equation 7.51 into Equation 7.50 gives the following form for $f_{a1k}(\boldsymbol{v})$:

$$\begin{aligned} f_{a1k}(\boldsymbol{x},\boldsymbol{v},t) &= -\frac{e_a}{m_a}\mathrm{e}^{\mathrm{i}(k_x x + k_z z - \omega t)} \\ &\times \int_{-\infty}^{t}\left(\left(1 - \frac{\boldsymbol{k}\cdot\boldsymbol{v}(t')}{\omega}\right)\boldsymbol{E} + (\boldsymbol{v}(t')\cdot\boldsymbol{E})\frac{\boldsymbol{k}}{\omega}\right)\cdot\frac{\partial f_0(\boldsymbol{v}(t'))}{\partial \boldsymbol{v}} \\ &\times \exp\left(\mathrm{i}\frac{k_x v_\perp}{\Omega}\left(\sin(\theta + \Omega(t'-t)) - \sin\theta\right) + i(k_z v_z - \omega)(t'-t)\right)\mathrm{d}t' . \end{aligned} \tag{7.52}$$

Here, the direction of the perpendicular wave vector is chosen as x. Using the Bessel function formula $\exp(\mathrm{i}a\sin\theta) = \sum J_m(a)\exp(\mathrm{i}m\theta)$ and $J_{-m}(a) = (-1)^m J_m(a)$, final form of the perturbed distribution function $f_{a1k}(\boldsymbol{x}, \boldsymbol{v}, t)$ is obtained and the dielectric tensor $\boldsymbol{K}$ is obtained by using $\boldsymbol{J} = \sum e_a \int \boldsymbol{v} f_{a1k}(\boldsymbol{v}) \mathrm{d}\boldsymbol{v}$, $\boldsymbol{J} = \boldsymbol{\sigma E}$ and $\boldsymbol{K} = \boldsymbol{I} + \mathrm{i}\boldsymbol{\sigma}/\varepsilon_0\omega$ as follows for Maxwell distribution with $T_\parallel$ and $T_\perp$ [10] (note: the definition of v_{Ta} is different in [10]).

$$\boldsymbol{K} = \boldsymbol{I} + \sum_{a=i,e} \frac{\omega_{pa}^2}{\omega^2} \Bigg[2\eta_{0a}^2 \lambda_{Ta} \boldsymbol{L} + \sum_n \left(\zeta_{0a} Z(\zeta_{na}) - \left(1 - \frac{1}{\lambda_{Ta}}\right)(1 + \zeta_{na} Z(\zeta_{na})) \right) \times \exp(-b_a) \boldsymbol{X}_{na} \Bigg], \tag{7.53}$$

$$\boldsymbol{X}_{na} = \begin{bmatrix} (n^2/b_a) I_n & \mathrm{i}n(I_n' - I_n) & -2nI_n\lambda_{Ta}^{1/2}\eta_{na}/\alpha_a \\ -\mathrm{i}n(I_n' - I_n) & (n^2/b_a + 2b_a)I_n - 2b_a I_n' & \mathrm{i}\lambda_{Ta}^{1/2}\eta_{na}\alpha_a(I_n' - I_n) \\ -2nI_n\lambda_{Ta}^{1/2}\eta_{na}/\alpha_a & -\mathrm{i}\lambda_{Ta}^{1/2}\eta_{na}\alpha_a(I_n' - I_n) & 2\lambda_{Ta}\eta_{na}^2 I_n \end{bmatrix}, \tag{7.54}$$

$$\eta_{na} = \frac{\omega + n\Omega_a}{k_z v_{T\parallel a}}, \quad \lambda_{Ta} = \frac{T_{\parallel a}}{T_{\perp a}}, \quad b_a = \frac{1}{2}\left(\frac{k_\perp v_{T\perp a}}{\Omega_a}\right)^2, \quad v_{Ta} = \left(\frac{2T_a}{m_a}\right)^{1/2},$$

$$\zeta_{na} = \frac{\omega - k_\parallel u_{\parallel a} + n\Omega_a}{k_z v_{T\parallel a}}, \quad \alpha_a = \frac{k_x v_{T\perp a}}{\Omega_a}, \quad Z(\zeta) = \frac{1}{\sqrt{\pi}} \int_{-\infty}^{\infty} \frac{\exp(-\beta^2)}{\beta - \zeta},$$

Here, Z is called the plasma dispersion function and $u_{\parallel a}$ is the parallel flow velocity, the argument of the modified Bessel function is b_a. $\boldsymbol{L}$ is a tensor with $L_{zz} = 1$ and other components are 0.

Salon: Professor Thomas Stix

Professor Thomas Stix (Figure 7.1) was a plasma physicist. He wrote a pioneering text on plasma waves, *The Theory of Plasma Waves* in 1962 [9].

Figure 7.1 Prof. T. H. Stix (1924–2001) (Courtesy of Princeton Plasma Physics Laboratory)

Note: Causality and translational symmetry in plasma waves

Plasma exhibits characteristics of a dielectric medium and excites the "wave" as a collective motion [11]. Let $\boldsymbol{E}(\boldsymbol{x}', t')$ be an excited perturbation at $\boldsymbol{x}'$ in the time t'. In response to this excitation, the current $\boldsymbol{J}(\boldsymbol{x}, t)$ is created at another location in the plasma $\boldsymbol{x}$ at time t

$$\boldsymbol{J}(\boldsymbol{x}, t) = \boldsymbol{\sigma}(\boldsymbol{x}, \boldsymbol{x}', t, t')\boldsymbol{E}(\boldsymbol{x}', t') . \tag{7.55}$$

Since $\boldsymbol{J}$ will not be created before the excitation of $\boldsymbol{E}$, $\boldsymbol{\sigma}(\boldsymbol{x}, \boldsymbol{x}'; t, t') = 0$ for $t < t'$ (causality requirement). In the usual Ohm's law, $\boldsymbol{\sigma}(\boldsymbol{x}, \boldsymbol{x}'; t, t') = \delta(\boldsymbol{x} - \boldsymbol{x}')\delta(t - t')$. Considering the linear response, the superposition principle is valid and the induced current is given as follows,

$$\boldsymbol{J}(\boldsymbol{x}, t) = \frac{1}{2\pi}\iint \mathrm{d}\boldsymbol{x}'\mathrm{d}t'\boldsymbol{\sigma}(\boldsymbol{x}, \boldsymbol{x}', t, t')\boldsymbol{E}(\boldsymbol{x}', t') . \tag{7.56}$$

Moreover, if we assume "translational symmetry" in space and time (medium is uniform and stationary), $\boldsymbol{\sigma}$ becomes a function of $\boldsymbol{x} - \boldsymbol{x}'$ and $t - t'$.

$$\boldsymbol{J}(\boldsymbol{x}, t) = \frac{1}{2\pi}\iint \mathrm{d}\boldsymbol{x}'\mathrm{d}t'\boldsymbol{\sigma}(\boldsymbol{x} - \boldsymbol{x}', t - t')\boldsymbol{E}(\boldsymbol{x}', t') . \tag{7.57}$$

It is important to show that Ohm's law $\boldsymbol{J}(\omega, \boldsymbol{k}) = \boldsymbol{\sigma}(\omega, \boldsymbol{k})\boldsymbol{E}(\omega, \boldsymbol{k})$ is equivalent to the electrical conductivity depending only on $\boldsymbol{x} - \boldsymbol{x}'$ and $t - t'$. The Fourier transform and its inverse Fourier transform of the electric field perturbation $\boldsymbol{E}(\omega, \boldsymbol{k})$, $\boldsymbol{E}(\boldsymbol{x}, t)$ and $\boldsymbol{J}(\boldsymbol{x}, t)$ are given by the following equations:

$$\boldsymbol{E}(\omega, \boldsymbol{k}) = \frac{1}{2\pi}\iint \mathrm{d}\boldsymbol{x}\mathrm{d}t\mathrm{e}^{-\mathrm{i}(\boldsymbol{k}\cdot\boldsymbol{x} - \omega t)}\boldsymbol{E}(\boldsymbol{x}, t) , \tag{7.58}$$

$$\boldsymbol{E}(\boldsymbol{x}, t) = \frac{1}{2\pi}\iint \mathrm{d}\boldsymbol{k}\mathrm{d}\omega\mathrm{e}^{\mathrm{i}(\boldsymbol{k}\cdot\boldsymbol{x} - \omega t)}\boldsymbol{E}(\omega, \boldsymbol{k}) . \tag{7.59}$$

$$\begin{aligned}
\boldsymbol{J}(\boldsymbol{x}, t) &= \frac{1}{2\pi}\iint \mathrm{d}\omega\mathrm{d}\boldsymbol{k}\mathrm{e}^{\mathrm{i}(\boldsymbol{k}\cdot\boldsymbol{x} - \omega t)}\boldsymbol{\sigma}(\omega, \boldsymbol{k})\boldsymbol{E}(\omega, \boldsymbol{k}) \\
&= \frac{1}{2\pi}\iint \mathrm{d}\omega\mathrm{d}\boldsymbol{k}\mathrm{e}^{\mathrm{i}(\boldsymbol{k}\cdot\boldsymbol{x} - \omega t)}\left[\frac{1}{2\pi}\iint \mathrm{d}\boldsymbol{x}''\mathrm{d}t''\mathrm{e}^{-\mathrm{i}(\boldsymbol{k}\cdot\boldsymbol{x}'' - \omega t'')}\boldsymbol{\sigma}(\boldsymbol{x}'', t'')\right] \\
&\quad \times \left[\frac{1}{2\pi}\iint \mathrm{d}\boldsymbol{x}'\mathrm{d}t'e^{-i(\boldsymbol{k}\cdot\boldsymbol{x}' - \omega t')}\boldsymbol{E}(\boldsymbol{x}', t')\right] \\
&= \frac{1}{(2\pi)^3}\iint \mathrm{d}\omega\mathrm{d}\boldsymbol{k}\mathrm{e}^{\mathrm{i}(\boldsymbol{k}\cdot(\boldsymbol{x} - \boldsymbol{x}' - \boldsymbol{x}'') - \omega(t - t' - t''))}\iint \mathrm{d}\boldsymbol{x}''\mathrm{d}t''\boldsymbol{\sigma}(\boldsymbol{x}'', t'') \\
&\quad \times \iint \mathrm{d}\boldsymbol{x}'\mathrm{d}t'\boldsymbol{E}(\boldsymbol{x}', t') \\
&= \frac{1}{2\pi}\iint \mathrm{d}\boldsymbol{x}'\mathrm{d}t'\boldsymbol{\sigma}(\boldsymbol{x} - \boldsymbol{x}', t - t')\boldsymbol{E}(\boldsymbol{x}', t') .
\end{aligned} \tag{7.60}$$

Here, the properties of the Dirac delta function $\delta(\boldsymbol{x} - \boldsymbol{x}' - \boldsymbol{x}'') = (2\pi)^{-1} \int \mathrm{d}\boldsymbol{k} e^{\mathrm{i}\boldsymbol{k}\cdot(\boldsymbol{x}-\boldsymbol{x}'-\boldsymbol{x}'')}$, $\delta(t - t' - t'') = (2\pi)^{-1} \int \mathrm{d}\omega e^{-\mathrm{i}\omega(t-t'-t'')}$ are used. The causality saying "the response in the stable medium always appears after its excitation" requires that the response should follow the excitation and not come before.

7.4 Non-uniform Plasma: Alfven Wave Resonance and Continuous Spectrum

We describe here the Alfven resonance in non-uniform plasma since it plays an important role in the damping of the Alfven Eigen (AE) mode or resistive wall mode in toroidal confinement plasmas. Defining $n_\parallel = n\cos\theta$ and $n_\perp = n\sin\theta$, we can rewrite Equation 7.46 as follows,

$$n_\perp^2 = \frac{(R - n_\parallel^2)(L - n_\parallel^2)}{S - n_\parallel^2} . \tag{7.61}$$

In this case, the resonance condition is $S = n_\parallel^2$. Since $S < 0$ at $\Omega_i < \omega < \omega_{LH}$ ($\omega_{LH} = 1/[(\Omega_i^2 + \omega_{pe}^2)^{-1} + 1/|\Omega_i \Omega_e|]^{1/2}$ is the lower hybrid frequency), the resonance occurs ω below the ion cyclotron frequency. Since $S \sim 1 + (c^2/V_A^2)[\Omega_i^2/(\Omega_i^2 - \omega^2)] \sim c^2/V_A^2$ at $\omega \ll \Omega_i$, the Alfven resonance condition is given as follows,

$$\omega = k_\parallel V_A . \tag{7.62}$$

Figure 7.2 shows the schematics of Alfven resonance in high-density confined plasma. In the Alfven resonance, triple layer of cut-off–resonance–cut-off appears spatially close to each other. From the outward side this is the cut-off of the shear Alfven wave ($L - n_\parallel = 0$: $n_\perp = 0$ at $r = r_s$), Alfven resonance ($S - n_\parallel = 0$: $n_\perp = \infty$ at $r = r_A$), the cut-off of compressional Alfven wave ($R - n_\parallel = 0$: $n_\perp = 0$ at $r = r_s$). The wave equation in the vicinity of the resonance can be obtained by replacing $n_\perp$ of Equation 7.61 to $-\mathrm{i}(c/\omega)\mathrm{d}/\mathrm{d}x (x = r - r_A)$ as follows,

$$\frac{c^2}{\omega^2}\frac{\mathrm{d}^2 E}{\mathrm{d}x^2} + \frac{(R - n_\parallel^2)(L - n_\parallel^2)}{S - n_\parallel^2} E = 0 . \tag{7.63}$$

Assuming that density near the resonance points is proportional to x including the cut-off point, we define $y = xS'(0)/D(0)$ to convert Equation 7.63 to the following singular turning-point equation [9]:

$$\frac{\mathrm{d}^2 E}{\mathrm{d}y^2} + \frac{\lambda^2(y^2 - 1)}{y + \mathrm{i}\varepsilon} E = 0 , \tag{7.64}$$

$$\lambda^2 = \left| \frac{D^3 \omega^2}{c^2(\mathrm{d}S(0)/\mathrm{d}x)^2} \right| \tag{7.65}$$

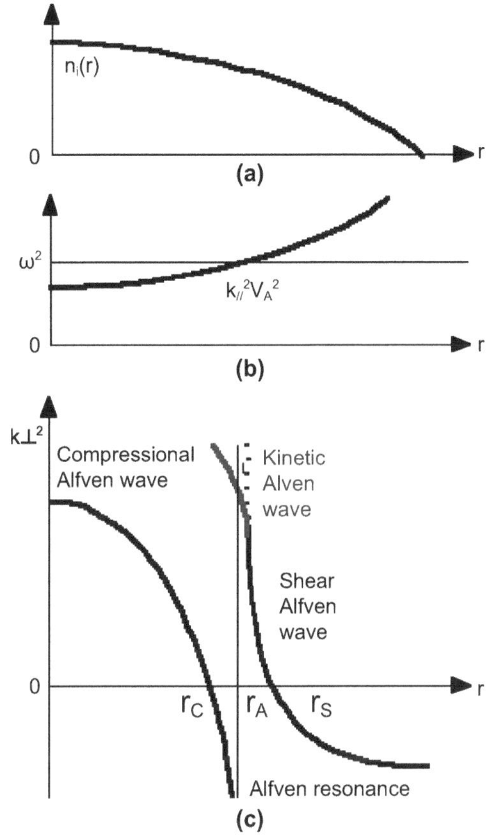

Figure 7.2 Schematics of Alfven resonance in tokamak plasma (cold plasma approximation) taking into account the effect of kinetic Alfven waves [13]. (a) Density profile; (b) Alfven frequency; (c) perpendicular wave number

where ε is a negative infinitesimal constant. As Budden showed in [12], complete absorption of the plasma wave occurs at the singular turning point. Singularity at $y = 0$ produces phase mixing discussed in Section 6.2 and Alfven wave energy is absorbed by the particles.

Near the Alfven resonance, kinetic effects become significant and the mode is converted to the Kinetic Alfven wave (KAW) [13]. Derivation of the kinetic dielectric tensor is lengthy. The dispersion relation of the KAW is given as follows at $(k_\perp \rho_i)^2 \ll 1$:

$$\omega^2 = k_\parallel^2 V_A^2 \left[1 + k_\perp^2 \rho_i^2 \left(\frac{3}{4} + \frac{T_e}{T_i} \right) \right] . \tag{7.66}$$

After conversion to the KAW mode, the Alfven wave damps via electron Landau damping. This wave damping mechanism is called "continuum damping" since it is caused by the existence of the continuous spectrum of the Alfven wave. This mechanism is important as a stabilizing mechanism of the resistive wall mode (RWM) destabilized by the finite electrical resistivity of the wall around the plasma.

If energetic particles exist in the plasma with velocities close to the Alfven velocity, the Alfven wave tends to be destabilized by getting energy from the energetic particles, but is stabilized if there is Alfven resonance in the plasma due to its strong continuum damping. On the other hand, if the Alfven resonance does not exist for some frequency band due to the toroidal coupling of the modes, etc., the Alfven wave becomes unstable due to a lack of continuum damping and is called the Alfven Eigen mode [14].

7.5 Drift Waves: Universal Waves in Confined Plasma

In confined plasma, there is a radial temperature gradient (∇T), density gradient (∇n), and a electrostatic potential gradient ($\nabla \Phi$) (see also Chapter 8). These gradients create first order flows on the flux surface and the coupling of these flows (drift) with the ion sound wave produces a drift wave. We only consider the density gradient (namely $\nabla T_i = \nabla T_e = 0$, in this case). Assuming collisionless ($\eta = 0$) and uniform T_e along $\boldsymbol{B}(\nabla_\parallel T_e = 0)$, particle conservation ($\partial n/\partial t + \nabla \cdot (n\boldsymbol{V}) = 0$), Ohm's law parallel to $\boldsymbol{B}(en_e E_\parallel + \nabla_\parallel P_e = 0)$, and the momentum balance equation parallel to $\boldsymbol{B}(m_i n \partial V_\parallel/\partial t = -\nabla_\parallel P)$ are given as follows,

$$\text{Particle balance:} \quad \mathrm{i}\omega \tilde{n}_e + \mathrm{i}k_\parallel n_e \tilde{V}_\parallel + \mathrm{i}\omega_* n_e (e\tilde{\Phi}/T_e) = 0\,, \tag{7.67}$$

$$\text{Ohm's Law:} \quad -\mathrm{i}k_\parallel e n_e \tilde{\Phi} + \mathrm{i}k_\parallel T_e \tilde{n}_e = 0\,, \tag{7.68}$$

$$\text{Momentum balance:} \quad -\mathrm{i}\omega m_i n_i \tilde{V}_\parallel + \mathrm{i}k_\parallel (\tilde{p}_e + \tilde{p}_i) = 0\,. \tag{7.69}$$

Here, $\omega_* = -(k_\perp T/eB)(\mathrm{d}\ln n_e/\mathrm{d}r)$ is the electron drift wave frequency. Equation 7.68 ($\tilde{n}_e/n_e = e\tilde{\Phi}/T_e$) is the "Boltzmann condition." If ions satisfy the adiabatic law $\tilde{p}_i = \gamma_i T_i \tilde{n}_i$, a combination of these equations leads to the following simple dispersion relation using $C_s = ((Z_i T_e + \gamma_i T_i)/m_i)^{1/2}$:

$$\omega(\omega - \omega_*) = k_\parallel^2 C_s^2\,. \tag{7.70}$$

This drift wave is stable (ω is real) but is destabilized by the inclusion of the ion temperature gradient ($\nabla T_i \neq 0$). The ion energy equation $((3/2)(\partial p_i/\partial t + V_E \cdot \nabla p_i + (5/2)p_i \nabla_\parallel V_\parallel) = 0)$ is given by the following equation with $\gamma_i = 5/2$,

$$-\frac{3}{2}\mathrm{i}\omega \tilde{P}_i + \frac{3}{2}\mathrm{i}\omega_* \tau (1 + \eta_i) e n_e \tilde{\Phi} + \mathrm{i}\gamma_i n_i T_i k_\parallel \tilde{V}_\parallel = 0\,. \tag{7.71}$$

Here $\tau = T_i/T_e, \eta_i = \mathrm{d}\ln T_i/\mathrm{d}\ln n_e$. Combining this with Equations 7.67, 7.68, and 7.69 including the effect of the ion temperature gradient, we obtain the dispersion relation of the ion temperature gradient (ITG) mode as follows,

$$\omega(\omega - \omega_*) = k_\parallel^2 C_s^2 \left[1 + \frac{\omega_*}{\omega}\frac{Z_i}{\gamma_i \tau + Z_i}\left(\eta_i - \frac{\gamma_i - Z_i}{Z_i}\right)\right]. \tag{7.72}$$

For the case of $Z_i = 1, \gamma_i = 5/2, \omega \sim k_\parallel C_s \ll \omega_*$, we obtain

$$\omega^2 \sim -\frac{k_\parallel^2 C_s^2}{2.5\tau + 1}\left(\eta_i - \frac{2}{3}\right) . \tag{7.73}$$

So, this drift wave becomes unstable when the ion temperature gradient exceeds a critical value, $\mathrm{d}T_i^{\mathrm{Crit}}/\mathrm{d}r$. In the real torus, the mode structure becomes more complex and is not so simple, but the threshold in the ion temperature gradient for destabilization exists [15].

The most basic equation describing the drift wave turbulence is the Hasegawa–Mima equation [16]. We derive this equation to see the fundamental process of drift wave turbulence. The guiding center motion is used for simplicity. Since the electrostatic approximation ($\boldsymbol{E} = -\nabla\Phi$) is applicable for the drift wave, we have

$$\boldsymbol{v}_{i\perp} = -\frac{\nabla\tilde{\Phi} \times \boldsymbol{B}}{B^2} - \frac{m_i}{eB^2}\frac{\mathrm{d}}{\mathrm{d}t}\nabla_\perp\tilde{\Phi} . \tag{7.74}$$

Here, the first term is $\boldsymbol{E} \times \boldsymbol{B}$ drift and the second term is the polarization drift. The origin of the polarization drift is as follows. Consider the particle motion when the electric field is suddenly applied. The particle moves in the direction of the electric field $\boldsymbol{E}$ first. When $\boldsymbol{E}$ becomes stationary, the particle moves with $\boldsymbol{E} \times \boldsymbol{B}$ drift. This transient drift by $\mathrm{d}\boldsymbol{E}/\mathrm{d}t$ is called the "polarization drift" and given as follows,

$$\boldsymbol{v}_{pa} = -\frac{m_a}{e_a B^2}\frac{\mathrm{d}\boldsymbol{E}}{\mathrm{d}t} . \tag{7.75}$$

The direction of this drift is opposite for ions and electrons and causes charge separation unlike the $\boldsymbol{E} \times \boldsymbol{B}$ drift. The ion polarization drift is important while the electron one is negligible since this drift is proportional to the mass. The electron and ion motion along the magnetic field compensates the charge separation by the polarization drift. Since the electron has a small mass, we can assume the Boltzmann relation $\tilde{n}_e/n_e = e\tilde{\Phi}/T_e$ for the electron. The ion equation of motion and particle conservation law are given as,

$$m_i n_i \frac{\mathrm{d}\boldsymbol{v}_{i\parallel}}{\mathrm{d}t} = -\nabla_\parallel(en\tilde{\Phi} + \gamma n_i T_i) , \tag{7.76}$$

$$\frac{\partial n_i}{\partial t} + \nabla\cdot(n\boldsymbol{v}_i) = 0 . \tag{7.77}$$

Here, we give the electrostatic potential as follows,

$$\tilde{\Phi} = \Phi_{\boldsymbol{k},\omega}\exp[\mathrm{i}k_\perp y + k_\parallel z - \mathrm{i}\omega t] . \tag{7.78}$$

From Equations 7.74 and 7.76, we can obtain an expression for $v_{\parallel i}$ and $v_{\perp i}$ and substituting them into the particle conservation law taking the quasi-neutrality condition $\tilde{n}_e = \tilde{n}_i$ into account, the drift wave dispersion relation is obtained from the linear term:

$$\omega^2(1 + \tau k_\perp^2 \rho_i^2/2) - \omega\omega_* = k_\parallel^2 C_s^2 \,. \tag{7.79}$$

Here, $C_s = [(T_e + \gamma T_i)/m_i]^{1/2}$ and ρ_i is the Larmor radius. The difference from Equation 7.71 comes from the polarization drift. In any case, the drift wave is a kind of ion-acoustic wave in non-uniform plasma. Let the spatial Fourier expansion of the electrostatic potential be $\tilde{\Phi} = \Phi_{\boldsymbol{k}}(t)\exp[\mathrm{i}k_\perp y + k_\parallel z]$. Selecting the wave number k from the nonlinear wave–wave coupling term in the $n_i \boldsymbol{v}_i$ term of Equation 7.77, we reach the Hasegawa-Mima equation:

$$\frac{\partial \Phi_{\boldsymbol{k}}(t)}{\partial t} + \mathrm{i}\omega_{\boldsymbol{k}*}\Phi_{\boldsymbol{k}}(t) = \sum_{\boldsymbol{k}=\boldsymbol{k}_1+\boldsymbol{k}_2} V_{\boldsymbol{k}_1,\boldsymbol{k}_2}\Phi_{\boldsymbol{k}1}(t)\Phi_{\boldsymbol{k}2}(t) \,, \tag{7.80}$$

$$V_{\boldsymbol{k}_1,\boldsymbol{k}_2} = \frac{\rho_s^2}{(1+\tau k^2\rho_s^2)B}(\boldsymbol{k}_1 \times \boldsymbol{k}_2)\cdot \boldsymbol{e}_z[k_2^2 - k_1^2] \,. \tag{7.81}$$

Here, $\omega_{\boldsymbol{k}}{}^* = \omega_*/(1 + \tau k_\perp^2\rho_i^2/2)$, and $\rho_s = (T_e/m_i)^{1/2}/\Omega_i$.

The nonlinear term of Hasegawa–Mima Equation 7.80 comes from the polarization drift and is used as the most basic equation for plasma turbulence.

Salon: Hasegawa–Mima Equation

Professor A. Hasegawa (Figure 7.3 (a)) and Professor K. Mima (Figure 7.3 (b)) who derived the Hasegawa-Mima equation. The Hasegawa–Mima equation produces zonal flow through the inverse cascade, which is essentially the same as the zonal flow in Jovian atmosphere [17]. At the time, Professor Hasegawa was at the Bell Laboratories and Professor Mima was a visitor from ILE-Osaka. Professor Mima became a director of ILE-Osaka, which promotes laser fusion.

Figure 7.3 (a) Professor A. Hasegawa (with kind permission of Prof. Hasegawa) and (b) Professor K. Mima (with kind permission of Prof. Mima)

References

1. Landau LD, and Lifschitz EM (1975) Classical Theory of Fields. 4th edn. Pergamon Press, p. 130.
2. von Laue M (1905) Ann. Phys., 18, 523.
3. Hasegawa A (1975) Plasma Instabilities and Nonlinear Effects. Springer-Verlag Berlin.
4. Goldstein H (1950) Classical Mechanics. Addison-Wesley.
5. Whitham GB (1974) Linear and Nonlinear Waves. John Wiley & Sons.
6. Whitham GB (1965) J. Fluid Mechanics, 22, 273.
7. Hazeltine RD, Waelbroeck FL (2004) The Framework of Plasma Physics: Frontiers in Physics. Westview Press.
8. Tokuda S (2008) Plasma and Fusion Research, 3, 57
9. Stix TH (1962) Theory of Plasma Waves. McGraw-Hill.
10. Miyamoto K (1980) Plasma Physics for Nuclear Fusion. MIT Press.
11. Ichimaru S (1973) Basics Principles of Plasma Physics: A Statistical Approach. Frontiers in Physics.
12. Budden KG (1955) Physics of the Ionosphere. Report of Phys. Soc. Conf. Cavendish Lab., p. 320.
13. Hasegawa A, Chen L (1975) Phys. Rev. Lett., 35, 370; also (1976) Phys. Fluids, 19, 1924.
14. Cheng CZ, Chen L, Chance MS (1985) Ann. Phys., 161, 21.
15. Itoh K, Itoh S-I, Fukuyama A (1999) Transport and Structure Formation in Plasmas. IOP.
16. Hasegawa A, Mima T (1977) Phys. Rev. Lett., 39, 205–208.
17. Hasegawa A (1985) Adv. Phys., 34, 1–42.

Chapter 8
Collisional Transport: Neoclassical Transport in a Closed Magnetic Configuration

In high temperature confined plasma, the mean free path of the Coulomb collision becomes much longer than the geometric (torus) dimensions, and this is called a "collisionless regime." In this case, the particles are categorized into particles trapped by a magnetic mirror and the particles not trapped by the magnetic mirror. The difference in these particle orbits produces distortion of the velocity distribution function and deviates from the Maxwell distribution.

This will affect the physical properties of the plasma. The thermal force produced by the trapped particle operates on untrapped particles to reduce the electrical conductivity or to produce a plasma current (bootstrap current), or to enhance the thermal diffusion across the magnetic field. They are collectively called neoclassical transport.

8.1 Collisionless Plasma: Moment Equation and Neoclassical Viscosity

When the collision mean free path becomes longer than the geometric dimension (*e.g.*, connection length along the magnetic field Rq) (called the collisionless regime), the local Maxwellian approximation is not valid and we have to use the drift kinetic equation introduced in Chapter 5 to evaluate collisional transport of the high-temperature plasma in a closed magnetic configuration. As Hirshman and Sigmar [1] showed, however, the use of the kinetic equation can be minimized by using the moment method. Taking $\boldsymbol{v}$ and $v^2\boldsymbol{v}$ moments of the Vlasov-Fokker Planck equation of species a of the plasma and subtracting convective heat flux from the $v^2\boldsymbol{v}$ moment, we obtain following moment equations for momentum and heat flux:

$$m_a n_a \frac{\mathrm{d}\boldsymbol{u}_a}{\mathrm{d}t} = e_a n_a(\boldsymbol{E} + \boldsymbol{u}_a \times \boldsymbol{B}) - \nabla P_a - \nabla \cdot \boldsymbol{\Pi}_a + \boldsymbol{F}_{a1} + \boldsymbol{M}_a\,, \quad (8.1)$$

$$m_a \frac{\partial}{\partial t}\left(\frac{\boldsymbol{q}_a}{T_a}\right) = \frac{e_a}{T_a}\boldsymbol{q}_a \times \boldsymbol{B} - \frac{5}{2}n_a \nabla T_a - \nabla \cdot \boldsymbol{\Theta}_a + \boldsymbol{F}_{a2} + \boldsymbol{Q}_a\,. \quad (8.2)$$

Here, n_a, $\boldsymbol{u}_a$, $\boldsymbol{q}_a$, P_a, $\boldsymbol{\Pi}_a$, $\boldsymbol{\Theta}_a$, $\boldsymbol{F}_{a1}$, $\boldsymbol{F}_{a2}$, $\boldsymbol{M}_a$, and $\boldsymbol{Q}_a$ are density, velocity, conduction heat flux, the average plasma pressure, viscosity tensor (anisotropic component of the pressure) and viscous heat tensor, friction force, heat friction force, momentum source, and heat momentum source, respectively. Definition of tensor and its algebra including its divergence are given in appendix "Tensor algebra". The velocity distribution function in strong magnetic fields shows anisotropy parallel and perpendicular to the magnetic field as shown by G. F. Chew, M. L. Goldberg, and F. E. Low [2], and $\boldsymbol{\Pi}_a$ and $\boldsymbol{\Theta}_a$ can be expressed as

$$\boldsymbol{\Pi}_a = (P_{\|a} - P_{\perp a})\left(\boldsymbol{bb} - \frac{1}{3}\boldsymbol{I}\right) + O(\delta^2) , \tag{8.3}$$

$$\boldsymbol{\Theta}_a = (\Theta_{\|a} - \Theta_{\perp a})\left(\boldsymbol{bb} - \frac{1}{3}\boldsymbol{I}\right) + O(\delta^2) . \tag{8.4}$$

Here, $\boldsymbol{b} = \boldsymbol{B}/B$ is the unit vector parallel to $\boldsymbol{B}$ and $\delta = \rho_a/L$ is a smallness parameter that is the ratio of the Larmor radius and the macroscopic length scale L of the plasma, $\rho_a = v_{Ta}/\Omega_a$ is Larmor radius, $v_{Ta} = (2T_a/m_a)^{1/2}$ is thermal velocity, and $\Omega_a = e_a B/m_a$ is the cyclotron angular frequency. Taking the cross product $\boldsymbol{B}$ with Equations 8.1 and 8.2, and neglecting the time derivative for the drift timescale $O((\delta^2\Omega)^{-1})$, which is much longer than the Alfven timescale $O((\delta\Omega)^{-1})$, and neglecting other $O(\delta^2)$ terms smaller than the ∇P, $\nabla\Phi$, and ∇T terms, the major components of particle and heat flows of particle species a perpendicular to $\boldsymbol{B}$ are give as

$$\boldsymbol{u}_{\perp a}^{(1)} = \frac{\boldsymbol{E}\times\boldsymbol{B}}{B^2} + \frac{\boldsymbol{b}\times\nabla P_a}{m_a n_a \Omega_a} , \tag{8.5}$$

$$\boldsymbol{q}_{\perp a}^{(1)} = \frac{5}{2} P_a \frac{\boldsymbol{b}\times\nabla T_a}{m_a \Omega_a} . \tag{8.6}$$

Here, the first term on the right-hand side of Equation 8.5 is the $E\times B$ drift flow, the second term is the diamagnetic drift flow caused by the pressure gradient. Equation 8.6 is the diamagnetic heat flux caused by the temperature gradient. In the drift timescale, the $-\partial\boldsymbol{A}/\partial t$ term is negligible and we can write $\boldsymbol{E} = -\nabla\Phi$. Taking the dot product of the magnetic field $\boldsymbol{B}$ with Equations 8.1 and 8.2 and taking the flux surface average (Section 3.8), the following flux surface averaged momentum and heat flux balance equation are obtained:

$$\langle \boldsymbol{B}\cdot\nabla\cdot\boldsymbol{\Pi}_a\rangle = \langle \boldsymbol{B}\cdot\boldsymbol{F}_{a1}\rangle + e_a n_a \langle \boldsymbol{B}\cdot\boldsymbol{E}\rangle + \langle \boldsymbol{B}\cdot\boldsymbol{M}_a\rangle , \tag{8.7}$$

$$\langle \boldsymbol{B}\cdot\nabla\cdot\boldsymbol{\Theta}_a\rangle = \langle \boldsymbol{B}\cdot\boldsymbol{F}_{a2}\rangle + \langle \boldsymbol{B}\cdot\boldsymbol{Q}_a\rangle . \tag{8.8}$$

Here $\langle \boldsymbol{B}\cdot\boldsymbol{F}_{a1}\rangle$ and $\langle \boldsymbol{B}\cdot\boldsymbol{F}_{a2}\rangle$ are the frictional forces on species a of the parallel flow on the magnetic surface, and $\langle \boldsymbol{B}\cdot\nabla\cdot\boldsymbol{\Pi}_a\rangle$ and $\langle \boldsymbol{B}\cdot\nabla\cdot\boldsymbol{\Theta}_a\rangle$ are viscous forces parallel to $\boldsymbol{B}$, which originate from the relaxation of velocity space anisotropy parallel and perpendicular to $\boldsymbol{B}$. The origin of this velocity space anisotropy is explained for the case of electrons in Figure 8.1 [3]. The magnetic moment μ is conserved in high temperature plasma when the electron moves along the magnetic field (see Section 4.5). So, the orbit of the electrons satisfying $B_{\max} \geq E/\mu$ is trapped in the weak magnetic field regime reflected by the magnetic mirror (trapped particle or banana orbit). For the case where density is lower towards the outside ($\mathrm{d}n/\mathrm{d}r < 0$),

Figure 8.1 Sketch of the trapped electron orbits and distortion of 1- and 2-dimensional velocity distribution functions in a tokamak

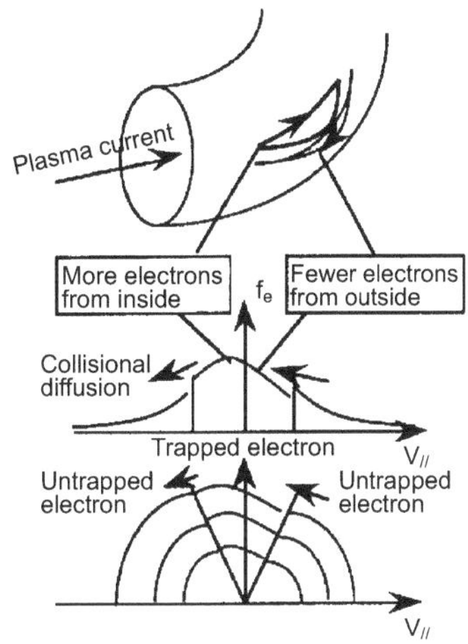

consider the velocity distribution function on a magnetic surface. There are fewer electrons for the trapped electrons with $v_\parallel > 0$ since it originates radially outside, while there are more electrons for the trapped electrons with $v_\parallel < 0$ since it originates from radially inside. The orbit of untrapped electrons stays much closer to the magnetic surface and the number of electrons for $v_\parallel > 0$ is roughly equal to that for $v_\parallel < 0$. Then, there appears a discontinuity in the trapped–untrapped boundary of the velocity distribution function. Small Coulomb collision smoothes this gap and causes particle diffusion in the velocity space. This collisional diffusion in velocity space acts as a viscous force in the magnetic field direction. Consider the drift and elliptic distortions from the Maxwell distribution, the velocity distribution function $f_a(\boldsymbol{v})$ can be expanded as follows:

$$f_a(\boldsymbol{v}) = f_{aM}(\boldsymbol{v}) + f_{a1}(\boldsymbol{v}) + f_{a2}(\boldsymbol{v}) , \tag{8.9}$$

$$f_{aM}(\boldsymbol{v}) = \frac{n_a(\psi)}{\pi^{3/2} v_{Ta}^3} \exp\left(-v^2/v_{Ta}^2\right) , \tag{8.10}$$

$$f_{a1}(\boldsymbol{v}) = \frac{2\boldsymbol{v}}{v_{Ta}^2} \cdot \left[\boldsymbol{u}_a - \left(1 - \frac{2}{5} x_a^2\right) \frac{\boldsymbol{q}_a}{P_a}\right] f_{aM}(\boldsymbol{v}) , \tag{8.11}$$

$$f_{a2}(\boldsymbol{v}) = 2 \frac{\boldsymbol{v}\boldsymbol{v} - \frac{v^2}{3}\boldsymbol{I}}{m_a n_a v_{Ta}^4} : \left[\boldsymbol{\Pi}_a + (\boldsymbol{\Theta}_a + \boldsymbol{\Pi}_a)\left(1 - \frac{2x_a^2}{7}\right)\right] f_{aM}(\boldsymbol{v}) . \tag{8.12}$$

Here, $x_a^2 = m_a v^2 / 2T_a$ and $v_{Ta}^2 = 2T_a/m_a$.

Note: Definition of friction and viscous forces

$$\boldsymbol{F}_{a1} \equiv \int m_a \boldsymbol{v} C(f_a)\, \mathrm{d}\boldsymbol{v} , \tag{8.13}$$

$$\boldsymbol{F}_{a2} \equiv \int m_a \boldsymbol{v} \left(\frac{m_a v^2}{2T_a} - \frac{5}{2}\right) C(f_a)\, \mathrm{d}\boldsymbol{v} , \tag{8.14}$$

$$\boldsymbol{\Pi}_a \equiv \int m_a \left(\boldsymbol{v}\boldsymbol{v} - \frac{1}{3} v^2 \boldsymbol{I} \right) f_a(\boldsymbol{x}, \boldsymbol{v}, t)\, \mathrm{d}\boldsymbol{v} \,, \tag{8.15}$$

$$P_{\|a} \equiv \int m_a (v_\| - u_{\|a})^2 f_a(\boldsymbol{x}, \boldsymbol{v}, t)\, \mathrm{d}\boldsymbol{v} \,, \tag{8.16}$$

$$P_{\perp a} \equiv \int \frac{m_a}{2} (\boldsymbol{v}_\perp - \boldsymbol{u}_{\perp a})^2 f_a(\boldsymbol{x}, \boldsymbol{v}, t)\, \mathrm{d}\boldsymbol{v} \,, \tag{8.17}$$

$$\boldsymbol{\Theta}_a \equiv \int m_a \left(\boldsymbol{v}\boldsymbol{v} - \frac{1}{3} v^2 \mathbf{I} \right) \left(\frac{m_a v^2}{2T_a} - \frac{5}{2} \right) f_a(\boldsymbol{x}, \boldsymbol{v}, t)\, \mathrm{d}\boldsymbol{v} \,, \tag{8.18}$$

$$\Theta_{\|a} \equiv \int m_a (v_\| - u_{\|a})^2 \left(\frac{m_a v^2}{2T_a} - \frac{5}{2} \right) f_a(\boldsymbol{x}, \boldsymbol{v}, t)\, \mathrm{d}\boldsymbol{v} \,, \tag{8.19}$$

$$\Theta_{\perp a} \equiv \int \frac{m_a}{2} (\boldsymbol{v}_\perp - \boldsymbol{u}_{\perp a})^2 \left(\frac{m_a v^2}{2T_a} - \frac{5}{2} \right) f_a(\boldsymbol{x}, \boldsymbol{v}, t)\, \mathrm{d}\boldsymbol{v} \tag{8.20}$$

8.2 Incompressible Flow: First Order Flow on a Magnetic Surface

If a magnetic surface is formed by the magnetic field, many physical quantities become the surface quantities (*e.g.*, function of ψ only). Pressure P is the surface quantity derived from the condition of force equilibrium. The temperature (also density) becomes an approximate surface quantity since the temperature equilibrates in the direction of the magnetic field lines due to high parallel thermal conductivity. In addition, the electrostatic potential Φ is an approximate surface quantity due to the high parallel electrical conductivity. Therefore, Equations 8.5 and 8.6 are given as follows:

$$\boldsymbol{u}_{\perp a}^{(1)} = -\frac{\nabla \Phi \times \boldsymbol{B}}{B^2} + \frac{\boldsymbol{b} \times \nabla P_a}{m_a n_a \Omega_a} = \frac{1}{B} \left[\frac{\mathrm{d}\Phi}{\mathrm{d}\psi} + \frac{1}{e_a n_a} \frac{\mathrm{d}P_a}{\mathrm{d}\psi} \right] \boldsymbol{b} \times \nabla \psi \,, \tag{8.21}$$

$$\boldsymbol{q}_{\perp a}^{(1)} = \frac{5}{2} P_a \frac{\boldsymbol{b} \times \nabla T_a}{m_a \Omega_a} = \frac{5 P_a}{2 e_a B} \frac{\mathrm{d}T_a}{\mathrm{d}\psi} \mathbf{b} \times \nabla \psi \,. \tag{8.22}$$

From these expressions, the following relations hold to show first order particle and heat flows perpendicular to $\boldsymbol{B}$ on the magnetic surface:

$$\boldsymbol{u}_{\perp a}^{(1)} \cdot \nabla \psi = 0 \,, \quad \boldsymbol{q}_{\perp a}^{(1)} \cdot \nabla \psi = 0 \,. \tag{8.23}$$

These are sometimes called “magnetization flows.” The first order flow on the magnetic surface is the sum of the flow along the magnetic field and the magnetization flow as follows,

$$\boldsymbol{u}_a^{(1)} = \boldsymbol{u}_{\perp a}^{(1)} + u_{\|a} \boldsymbol{b} \,, \tag{8.24}$$

$$\boldsymbol{q}_a^{(1)} = \boldsymbol{q}_{\perp a}^{(1)} + q_{\|a} \boldsymbol{b} \tag{8.25}$$

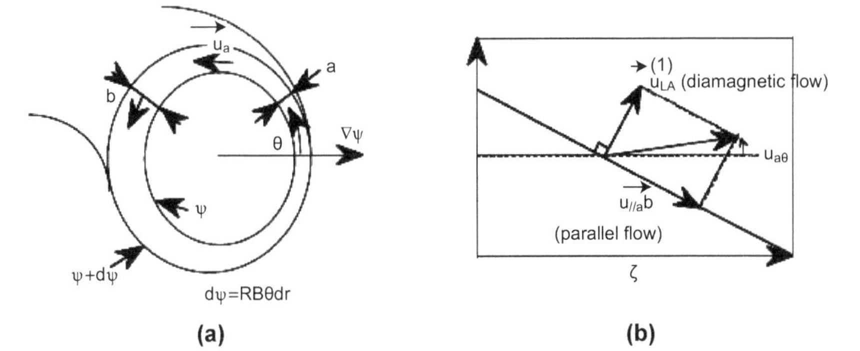

Figure 8.2 Geometry of (a) flow between magnetic surface ψ and $\psi+\mathrm{d}\psi$ and (b) first order flow as combination of diamagnetic flow and parallel flow

Continuity of poloidal flows in the flux tube between ψ and $\psi+\mathrm{d}\psi$ ($\mathrm{d}U = \boldsymbol{u}_a^{(1)}\cdot(\nabla\theta/|\nabla\theta|)2\pi R\mathrm{d}r = \boldsymbol{u}_a^{(1)}\cdot(\nabla\theta/|\nabla\theta|)\,\mathrm{d}\psi/\boldsymbol{B}\cdot(\nabla\theta/|\nabla\theta|) = \boldsymbol{u}_a^{(1)}\cdot\nabla\theta/\boldsymbol{B}\cdot\nabla\theta\mathrm{d}\psi$ and $\mathrm{d}U/\mathrm{d}\psi = u^*_{aq}(\psi)$ gives the following relations (see Figure 8.2)

$$\frac{\boldsymbol{u}_a^{(1)}\cdot\nabla\theta}{\boldsymbol{B}\cdot\nabla\theta} = u^*_{a\theta}(\psi)\,, \tag{8.26}$$

$$\frac{\boldsymbol{q}_a^{(1)}\cdot\nabla\theta}{\boldsymbol{B}\cdot\nabla\theta} = q^*_{a\theta}(\psi)\,. \tag{8.27}$$

Here, it should be noted that $u^*_{a\theta}(\psi)$ and $q^*_{a\theta}(\psi)$ do not have the dimensions of the particle and heat fluxes. Taking the dot product of Equations 8.24 and 8.25 with $\nabla\theta$ and considering the axisymmetric equilibrium relation (Equation 3.59) $\boldsymbol{B} = \nabla\zeta\times\nabla\psi + F\nabla\zeta$ (from this, we obtain $\boldsymbol{B}\times\nabla\psi\cdot\nabla\theta = F(\psi)\boldsymbol{B}\cdot\nabla\theta$), we obtain the following relations,

$$Bu^*_{a\theta}(\psi) = u_{\|a} - V_{1a}\,, \tag{8.28}$$

$$Bq^*_{a\theta}(\psi) = q_{\|a} - \frac{5}{2}P_aV_{2a} \tag{8.29}$$

where

$$V_{1a} = -\frac{\boldsymbol{u}_{\perp a}^{(1)}\cdot\nabla\theta}{\mathbf{b}\cdot\nabla\theta} = -\frac{F(\psi)}{B}\left(\frac{\mathrm{d}\Phi}{\mathrm{d}\psi} + \frac{1}{e_an_a}\frac{\mathrm{d}P_a}{\mathrm{d}\psi}\right), \tag{8.30}$$

$$V_{2a} = -\frac{2}{5P_a}\frac{\boldsymbol{q}_{\perp a}^{(1)}\cdot\nabla\theta}{\boldsymbol{b}\cdot\nabla\theta} = -\frac{F(\psi)}{e_aB}\frac{\mathrm{d}T_a}{\mathrm{d}\psi}\,. \tag{8.31}$$

Equations 8.28 and 8.29 show that the poloidal flow is the sum of the poloidal components of the parallel flow and diamagnetic flow. V_{1a} and V_{2a} are the "thermodynamic force." As shown in Section 8.1, $\boldsymbol{E}\times\boldsymbol{B}$ flow, diamagnetic drift flow, and diamagnetic heat flow produces thermodynamic forces. Substitution of Equations 8.21 and 8.22 into Equations 8.24 and 8.25 gives:

$$\boldsymbol{u}_a^{(1)} = u_{\parallel a}\boldsymbol{b} + \boldsymbol{u}_{\perp a}^{(1)} = u_{\parallel a}\boldsymbol{b} + \frac{BV_{1a}}{F(\psi)}\frac{\nabla\psi\times\boldsymbol{b}}{B}\,, \tag{8.32}$$

$$\boldsymbol{q}_a^{(1)} = q_{\parallel a}\boldsymbol{b} + \boldsymbol{q}_{\perp a}^{(1)} = q_{\parallel a}\boldsymbol{b} + \frac{5P_a}{2}\frac{BV_{2a}}{F(\psi)}\frac{\nabla\psi\times\boldsymbol{b}}{B}\,. \tag{8.33}$$

Combining these with the first order flow relations Equations 8.28 and 8.29, we obtain the following:

$$\boldsymbol{u}_a^{(1)} = u_{a\theta}^*(\psi)\boldsymbol{B} + \frac{BV_{1a}}{F(\psi)}R^2\nabla\zeta\,, \tag{8.34}$$

$$\boldsymbol{q}_a^{(1)} = q_{a\theta}^*(\psi)\boldsymbol{B} + \frac{5P_a}{2}\frac{BV_{2a}}{F(\psi)}R^2\nabla\zeta\,. \tag{8.35}$$

Here, we use the following axisymmetric relation (using Equation 3.59 and taking $\nabla\psi\times\boldsymbol{B}$ and $|\nabla\psi|^2 = R^2(B^2 - B_\zeta^2)$ for derivation):

$$\boldsymbol{b}\times\nabla\psi = F(\psi)\boldsymbol{b} - R^2B\nabla\zeta\,. \tag{8.36}$$

Taking ζ component of Equations 8.34 and 8.35 gives

$$u_{a\zeta}^{(1)} = u_{a\theta}^*(\psi)B_\zeta + \frac{BV_{1a}}{F(\psi)}R\,, \tag{8.37}$$

$$q_{a\zeta}^{(1)} = q_{a\theta}^*(\psi)B_\zeta + \frac{5P_a}{2}\frac{BV_{2a}}{F(\psi)}R\,. \tag{8.38}$$

From Equations 8.28 and 8.29, we obtain

$$\langle B^2\rangle u_{a\theta}^*(\psi) = \langle Bu_{\parallel a}\rangle - \langle BV_{1a}\rangle \tag{8.39}$$

$$\langle B^2\rangle q_{a\theta}^*(\psi) = \langle Bq_{\parallel a}\rangle - \frac{5P_a}{2}\langle BV_{2a}\rangle\,. \tag{8.40}$$

Here BV_{1a} and BV_{2a} are actually surface quantities and $\langle\,\rangle$ is not necessary. Substituting the expressions of $u_{a\theta}^*(\psi)$ and $q_{a\theta}^*(\psi)$ into Equations 8.37 and 8.38, we obtain the following results:

$$u_{a\zeta}^{(1)} = \frac{B_\zeta}{\langle B^2\rangle}\langle Bu_{\parallel a}\rangle + \left[1 - \frac{B_\zeta^2}{\langle B^2\rangle}\right]\frac{BV_{1a}}{B_\zeta}\,, \tag{8.41}$$

$$q_{a\zeta}^{(1)} = \frac{B_\zeta}{\langle B^2\rangle}\langle Bq_{\parallel a}\rangle + \frac{5P_a}{2}\left[1 - \frac{B_\zeta^2}{\langle B^2\rangle}\right]\frac{BV_{2a}}{B_\zeta}\,. \tag{8.42}$$

The second terms of the right-hand side of Equations 8.41 and 8.42 are called the Pfirsch–Schlüter terms.

8.3 Friction and Viscous Forces: Momentum and Heat Flow Balance

The distribution function (Equations 8.1–8.9) includes the anisotropic term produced in the collisionless regime. Substituting this expression into the definitions of friction force $\boldsymbol{F}_{a1}$ (Equation 8.13) and heat friction force $\boldsymbol{F}_{a2}$ (Equation 8.14) while the collision operator $C(f_a)$ given by Equation 5.45 is linearized around the Maxwellian. We obtain following formula:

$$\begin{bmatrix} \boldsymbol{F}_{a1} \\ \boldsymbol{F}_{a2} \end{bmatrix} = \sum_b \begin{pmatrix} l_{11}^{ab} & -l_{12}^{ab} \\ -l_{21}^{ab} & l_{22}^{ab} \end{pmatrix} \begin{bmatrix} \boldsymbol{u}_b^{(1)} \\ 2\boldsymbol{q}_b^{(1)}/5P_b \end{bmatrix}. \tag{8.43}$$

Here, l_{ij}^{ab} is called the friction coefficient, which has the symmetry $l_{ij}^{ab} = l_{ji}^{ba}$ due to self-adjoint property of the Coulomb collision term. A useful expression is given by Hirshman–Sigmar [1] as shown in the note. Since a viscous force operates when the particle moves poloidally in response to variation in the toroidal field, the viscous force is proportional to the poloidal flow:

$$\begin{bmatrix} \langle \boldsymbol{B} \cdot \nabla \cdot \boldsymbol{\Pi}_a \rangle \\ \langle \boldsymbol{B} \cdot \nabla \cdot \boldsymbol{\Theta}_a \rangle \end{bmatrix} = \langle B^2 \rangle \begin{bmatrix} \mu_{a1} & \mu_{a2} \\ \mu_{a2} & \mu_{a3} \end{bmatrix} \begin{bmatrix} u_{a\theta}^*(\psi) \\ 2q_{a\theta}^*(\psi)/5P_a \end{bmatrix}. \tag{8.44}$$

Here, μ_{a1}, μ_{a2}, μ_{a3} are called parallel viscosity coefficients and are obtained by substituting Equations 8.11 and 8.12 into the drift kinetic equation, and solving the equation for the anisotropic component $\boldsymbol{\Pi}_a$ and $\boldsymbol{\Theta}_a$, approximately [1]. There are three types of collisional transport regimes in tokamak: (1) banana regime where the collision time is longer than the bounce time of the trapped particle orbit ($\nu_c < \Omega_b$; ν_c: collision frequency, Ω_b: bounce frequency); (2) Pfirsch–Schlüter regime where collision time is shorter than the transit time of the untrapped particle ($\nu_c > \Omega_t$; ν_c: collision frequency, Ω_t: transit frequency $\sim v_{Ta}/Rq$); (3) plateau region, which is between the other two. The expression for the viscosity coefficient is derived for each regime and the velocity partitioned approximate viscosity coefficient valid for all velocity region is derived. The viscosity coefficient is obtained by integration in the velocity space as follows,

$$\mu_{a1} = K_{11}^a\,, \tag{8.45}$$

$$\mu_{a2} = K_{12}^a - \frac{5}{2}K_{11}^a\,, \tag{8.46}$$

$$\mu_{a3} = K_{22}^a - 5K_{12}^a + \frac{25}{4}K_{11}^a\,. \tag{8.47}$$

Here,

$$K_{ij}^a = \frac{m_a n_a}{\tau_{aa}} \frac{f_t}{f_c} \left\{ x_a^{2(i+j-2)} \nu_{\text{tot}}^a(v) \tau_{aa} \right\}, \tag{8.48}$$

$$\{A(v)\} = \frac{8}{3\pi^{1/2}} \int_0^\infty \exp(-x_a^2) x_a^4 A(x_a v_a)\,\mathrm{d}x_a \,, \tag{8.49}$$

$$\nu_{\mathrm{tot}}^a(v) = \frac{\nu_D^a(v)}{\left[1 + 2.48\nu_a^* \nu_D^a(v)\tau_{aa}/x_a\right]\left[1 + 1.96\nu_T^a(v)/x_a\omega_{Ta}\right]}, \tag{8.50}$$
$$\nu_T^a(v) = 3\nu_D^a(v) + \nu_E^a(v), x_a = v/v_{Ta}, \omega_{Ta} = v_{Ta}/L_c \,.$$

The first term of the denominator of Equation 8.50 is the correction term to connect the banana and plateau regimes. The second term is the Pfirsch–Schlüter correction term. $x_a = v/v_{Ta}$, $\nu_D^a(v)$ is the 90-degree deflection frequency, $\nu_E^a(v)$ energy exchange frequency (see note), ν_a^* is the collisionality defined by the ratio of collision frequency $1/\tau_{aa}$, the transit frequency ω_{Ta}, and $\varepsilon(\psi) \equiv (B_{\max} - B_{\min})/(B_{\max} + B_{\min})$. Using connection length $L_c \sim Rq$ and $\varepsilon \sim r/R$,

$$\nu_a^* \equiv \frac{1}{\varepsilon^{1.5}\omega_{Ta}\tau_{aa}} \sim \left(\frac{R}{r}\right)^{1.5} \frac{Rq}{v_{Ta}\tau_{aa}} \,. \tag{8.51}$$

Here, f_t is a trapped particle fraction and is related to the untrapped particle fraction f_c through $f_t + f_c = 1$. f_c is given by the following equation (see also Equation 8.69).

$$f_c = \frac{3\langle B^2\rangle}{4} \int_0^{1/B_{\max}} \frac{\lambda \mathrm{d}\lambda}{\left\langle\sqrt{1-\lambda B}\right\rangle} \,. \tag{8.52}$$

Substituting these formulas for viscosity and friction coefficients into Equation 8.7 and 8.8, the following balance equation of friction and viscous forces are obtained.

$$\begin{bmatrix} \mu_{a1} & \mu_{a2} \\ \mu_{a2} & \mu_{a3} \end{bmatrix} \begin{bmatrix} \langle Bu_{\|a}\rangle - BV_{1a} \\ \langle 2Bq_{\|a}/5P_a\rangle - BV_{2a} \end{bmatrix} = \sum_b \begin{bmatrix} l_{11}^{ab} & -l_{12}^{ab} \\ -l_{21}^{ab} & l_{22}^{ab} \end{bmatrix} \begin{bmatrix} \langle Bu_{\|b}\rangle \\ \langle 2Bq_{\|b}/5P_b\rangle \end{bmatrix}$$
$$+ \begin{bmatrix} e_a n_a \langle BE_\|\rangle \\ 0 \end{bmatrix} + \begin{bmatrix} \langle BM_{a\|}\rangle \\ \langle BQ_{a\|}\rangle \end{bmatrix} \,. \tag{8.53}$$

Note: Expression and Definitions on Friction and Viscous Coefficients

$$l_{ij}^{ab} = \frac{m_a n_a}{\tau_{aa}} \left[\left(\sum_k \frac{\tau_{aa}}{\tau_{ak}} M_{ak}^{i-1,j-1}\right)\delta_{ab} + \frac{\tau_{aa}}{\tau_{ab}} N_{ab}^{i-1,j-1}\right], \tag{8.54}$$

$$M_{ab}^{00} = -\left(1 + \frac{m_a}{m_b}\right)(1 + x_{ab}^2)^{-3/2} \,,$$

$$
\begin{aligned}
M_{ab}^{01} &= M_{ab}^{10} = -\frac{3}{2}\left(1+\frac{m_a}{m_b}\right)(1+x_{ab}^2)^{-5/2}\,,\\
M_{ab}^{11} &= -\left(\frac{13}{4}+4x_{ab}^2+\frac{15}{2}x_{ab}^4\right)(1+x_{ab}^2)^{-5/2}\,,\\
N_{ab}^{00} &= \left(1+\frac{m_a}{m_b}\right)(1+x_{ab}^2)^{-3/2}\,,\\
N_{ab}^{10} &= \frac{3}{2}\left(1+\frac{m_a}{m_b}\right)(1+x_{ab}^2)^{-5/2}\,,\\
N_{ab}^{01} &= \frac{3}{2}\frac{T_a}{T_b}x_{ab}^{-1}\left(1+\frac{m_b}{m_a}\right)(1+x_{ba}^2)^{-5/2}\,,\\
N_{ab}^{11} &= \frac{27}{4}\frac{T_a}{T_b}x_{ab}^2(1+x_{ab}^2)^{-5/2}\,,\\
\tau_{ab} &= \frac{3\pi^{3/2}\varepsilon_0^2 m_a^2 v_{Ta}^3}{n_b e_a^2 e_b^2 \ln\Lambda}\,,\\
x_{ab}^2 &= \frac{m_a T_b}{m_b T_a}\,,\qquad v_{Ta} = \sqrt{\frac{2T_a}{m_a}}
\end{aligned}
\tag{8.55}
$$

The friction coefficient has the following symmetry due to the self-adjointness of the Coulomb collision operator:

$$
l_{ij}^{ab} = l_{ji}^{ba}\,,\quad M_{ab}^{ij} = M_{ab}^{ji}\,,\quad N_{ab}^{j0} = -M_{ab}^{j0}\,,\quad N_{ab}^{ij} = \frac{T_a v_{Ta}}{T_b v_{Tb}} N_{ba}^{ji}\,. \tag{8.56}
$$

Collision frequencies $\nu_D^a(v)$ and $\nu_E^a(v)$ in the expression of the viscosity coefficient for the Maxwellian are expressed as follows:

$$
\nu_D^a(v) = \frac{3\sqrt{\pi}}{4}\frac{1}{\tau_{aa}}\sum_b \frac{n_b Z_b^2}{n_a Z_a^2}\frac{\Phi(x_b)-G(x_b)}{x_a^3}\,, \tag{8.57}
$$

$$
\nu_E^a(v) = \frac{3\sqrt{\pi}}{4}\frac{1}{\tau_{aa}}\sum_b \frac{n_b Z_b^2}{n_a Z_a^2}\left[4\left(\frac{T_a}{T_b}+x_{ab}^{-2}\right)\frac{G(x_b)}{x_a}-\frac{2\Phi(x_b)}{x_a^3}\right], \tag{8.58}
$$

$$
\Phi(x) = \frac{2}{\sqrt{\pi}}\int_0^x \exp(-u^2)\,\mathrm{d}u,\; G(x) = \left[\Phi(x) - x\Phi'(x)\right]/2x^2\,. \tag{8.59}
$$

8.4 Parallel Current: Generalized Ohm's Law

The following system of linear equations is obtained by writing down Equation 8.53 for electrons, ions, and impurities [4],

$$
\boldsymbol{M}(\boldsymbol{U}_\parallel - \boldsymbol{V}_\perp) = \boldsymbol{L}\boldsymbol{U}_\parallel + \boldsymbol{E}^* + \boldsymbol{S}_\parallel \tag{8.60}
$$

where

$$
\boldsymbol{L} = \begin{bmatrix} l_{11}^{ee} & l_{11}^{ei} & l_{11}^{eI} & -l_{12}^{ee} & -l_{12}^{ei} & -l_{12}^{eI} \\ l_{11}^{ie} & l_{11}^{ii} & l_{11}^{iI} & -l_{12}^{ie} & -l_{12}^{ii} & -l_{12}^{iI} \\ l_{11}^{Ie} & l_{11}^{Ii} & l_{11}^{II} & -l_{12}^{Ie} & -l_{12}^{Ii} & -l_{12}^{II} \\ -l_{21}^{ee} & -l_{21}^{ei} & -l_{21}^{eI} & l_{22}^{ee} & l_{22}^{ei} & l_{22}^{eI} \\ -l_{21}^{ie} & -l_{21}^{ii} & -l_{21}^{iI} & l_{22}^{ie} & l_{22}^{ii} & l_{22}^{iI} \\ -l_{21}^{Ie} & -l_{21}^{Ii} & -l_{21}^{II} & l_{22}^{Ie} & l_{22}^{Ii} & l_{22}^{II} \end{bmatrix}, \quad \boldsymbol{M} = \begin{bmatrix} \mu_{e1} & 0 & 0 & \mu_{e2} & 0 & 0 \\ 0 & \mu_{i1} & 0 & 0 & \mu_{i2} & 0 \\ 0 & 0 & \mu_{I1} & 0 & 0 & \mu_{I2} \\ \mu_{e2} & 0 & 0 & \mu_{e3} & 0 & 0 \\ 0 & \mu_{i2} & 0 & 0 & \mu_{i3} & 0 \\ 0 & 0 & \mu_{I2} & 0 & 0 & \mu_{I3} \end{bmatrix}
$$

$$
\boldsymbol{U}_{\parallel} = \begin{bmatrix} \langle Bu_{\parallel e}\rangle \\ \langle Bu_{\parallel i}\rangle \\ \langle Bu_{\parallel I}\rangle \\ 2\langle Bq_{\parallel e}\rangle/5P_e \\ 2\langle Bq_{\parallel i}\rangle/5P_i \\ 2\langle Bq_{\parallel I}\rangle/5P_I \end{bmatrix}, \quad \boldsymbol{V}_{\perp} = \begin{bmatrix} BV_{1e} \\ BV_{1i} \\ BV_{1I} \\ BV_{2e} \\ BV_{2i} \\ BV_{2I} \end{bmatrix}
$$

$$
\boldsymbol{E}^* = \langle BE_{\parallel}\rangle \begin{bmatrix} -en_e \\ eZ_in_i \\ eZ_In_I \\ 0 \\ 0 \\ 0 \end{bmatrix}, \quad \boldsymbol{S}_{\parallel} = \begin{bmatrix} \langle BM_e\rangle \\ \langle BM_i\rangle \\ \langle BM_I\rangle \\ \langle BQ_e\rangle \\ \langle BQ_i\rangle \\ \langle BQ_I\rangle \end{bmatrix}.
$$

Solving Equation 8.60 for $\boldsymbol{U}_{\parallel}$ gives

$$
\boldsymbol{U}_{\parallel} = (\boldsymbol{M}-\boldsymbol{L})^{-1}\boldsymbol{M}\boldsymbol{V}_{\perp} + (\boldsymbol{M}-\boldsymbol{L})^{-1}\boldsymbol{E}^* + (\boldsymbol{M}-\boldsymbol{L})^{-1}\boldsymbol{S}_{\parallel}, \tag{8.61}
$$

$$
U_{\parallel a} = \sum_b \left(\alpha_{ab}V_{\perp b} + c_{ab}(E_b^* + S_{\parallel b})\right) \tag{8.61$'$}
$$

where matrix $\boldsymbol{\alpha}$ and $\boldsymbol{c}$ are defined as $\boldsymbol{\alpha} = (\boldsymbol{M}-\boldsymbol{L})^{-1}\boldsymbol{M}$ and $\boldsymbol{c} = (\boldsymbol{M}-\boldsymbol{L})^{-1}$, respectively. Plasma current is the sum of the currents of each species as follows,

$$
\begin{aligned}
\langle \boldsymbol{B}\cdot\boldsymbol{J}\rangle &= \sum_{a=e,i,I} e_an_a\langle \boldsymbol{B}\cdot\boldsymbol{u}_a\rangle \\
&= \sum_{a=e,i,I} e_an_a\Bigg\{\sum_{b=1}^{6}\left[(\boldsymbol{M}-\boldsymbol{L})^{-1}\boldsymbol{M}\right]_{ab}V_{\perp b} \\
&\quad + \sum_{b=1}^{3}\left[(\boldsymbol{M}-\boldsymbol{L})^{-1}\right]_{ab}e_bn_b\langle BE_{\parallel}\rangle + \sum_{b=1}^{6}\left[(\boldsymbol{M}-\boldsymbol{L})^{-1}\right]_{ab}S_{\parallel b}\Bigg\} \\
&= \langle \boldsymbol{B}\cdot\boldsymbol{J}\rangle_{bs} + \sigma_{\parallel}^{\mathrm{NC}}\langle \boldsymbol{B}\cdot\boldsymbol{E}\rangle + \langle \boldsymbol{B}\cdot\boldsymbol{J}\rangle_{\mathrm{bd}} + \langle \boldsymbol{B}\cdot\boldsymbol{J}\rangle_{\mathrm{RFCD}}\,.
\end{aligned} \tag{8.62}
$$

Here, the first term of the right-hand side is the bootstrap current, the second is the conduction current, third and fourth are the beam-driven current and RF-driven current. The electrostatic potential Φ term included in $V_{\perp}$ is cancelled out due to charge neutrality and does not contribute to the bootstrap current.

8.5 Trapped Particle Effect: Electrical Conductivity

Before giving a detailed treatment of electrical conductivity in high temperature plasma, we need to discuss the basic electrical properties of fully ionized plasma. Suppose an electric field E exists in uniform plasma without a magnetic field ($B = 0$). Since ions are much heavier than electrons, electrons are much more mobile than ions (here we assume the ions are hydrogen for simplicity). The electron equation of motion is given by $en_eE = P_{ei}$, where P_{ei} is the momentum gain of the electron fluid through the collision with background ions and is given by $P_{ei} = m_en_e(u_i - u_e)\nu_{ei}$, where ν_{ei} is electron–ion collision time. Therefore, the current induced in the plasma $J = en_e(u_i - u_e)$ is given by $J = (e^2n_e/m_e\nu_{ei})E$. Thus the electrical conductivity σ is given by $\sigma = e^2n_e/m_e\nu_{ei}$. This is the simplest formula for electrical conductivity σ in fully ionized plasma. L. Spitzer Jr. (Figure 8.3) and his coworkers [5] calculated this electrical conductivity, including electron–electron collisions, and found that electrical conductivity is almost doubled. For electron–hydrogen plasma, electrical conductivity is given as follows,

$$\sigma = \frac{e^2n_e}{0.51m_e\nu_{ei}} = \frac{T_e^{3/2}(\text{keV})}{1.65\times10^{-9}\ln\Lambda}(\Omega\,\text{m})^{-1}\,. \qquad (8.63)$$

An important observation from Equation 8.63 is that σ is independent of density. This rather surprising result can be explained by the fact that the positive effect of increasing the number of current carriers ($\sim n_e$) is cancelled by the increase of the ion drag force ($\nu_{ei} \sim n_i$). Another important observation is $\sigma \sim T_e^{3/2}$. This is because the Coulomb cross section is proportional to $1/E^2$ and collision becomes rare at higher temperatures (average kinetic energy). For reactor relevant high temperature plasma,

$$\begin{array}{lll} \text{Plasma } (T_e = 10\,\text{keV}): & \sigma = 10^9 & (\Omega\,\text{m})^{-1} \\ \text{Copper } (27\,^\circ\text{C}): & \sigma = 5\times10^7 & (\Omega\,\text{m})^{-1} \\ \text{Copper } (-195\,^\circ\text{C}): & \sigma = 5\times10^8 & (\Omega\,\text{m})^{-1}\,. \end{array}$$

Figure 8.3 Lyman Spitzer Jr. (1914–1997) who gave an analytic expression of electrical conductivity in fully ionized plasma and invented the Stellarator concept (Photo by Denise Applewhite. Courtesy of Princeton Plasma Physics Laboratory)

Here, the electrical conductivity of copper at room temperature and −195 °C are shown for comparison. It is interesting to note that current flows more easily at higher temperatures in plasma, while it flows more easily at lower temperatures in metal. Spitzer and coworkers [5] evaluated the effect of impurities on σ as follows,

$$\sigma_{\parallel}^{\text{Spitzer}} = \frac{n_e e^2 \tau_{ee}}{m_e} \frac{3.4(1.13 + Z_{\text{eff}})}{Z_{\text{eff}}(2.67 + Z_{\text{eff}})} , \tag{8.64}$$

$$Z_{\text{eff}} = \frac{1}{n_e} \sum_{b=i,I} n_b Z_b^2$$

$$\tau_{ee} = \frac{6\sqrt{2}\pi^{3/2}\varepsilon_0^2 m_e^{1/2} T_e^{3/2}}{n_e e^4 \ln \Lambda} = 2.74 \times 10^{-4} \frac{T_e[\text{keV}]^{3/2}}{n_e[m^{-3}] \ln \Lambda}[\text{s}] .$$

Here, Z_{eff} is the "effective charge." From this introduction, we now proceed to the general form of electrical conductivity in collisionless plasma. The expression of electrical conductivity σ is obtained from Equation 8.62 as follows,

$$\sigma_{\parallel}^{\text{NC}} = \sum_{a=e,i,I} \sum_{b=e,i,I} e_a n_a e_b n_b \left[(\boldsymbol{M} - \boldsymbol{L})^{-1}\right]_{ab} . \tag{8.65}$$

Here, $\boldsymbol{L}$ represents the collisional friction forces among various species and $\boldsymbol{M}$ represents the effect of trapped particles. If there are no trapped particles $f_t = 0$, $\mu_{aj} = 0$, and the viscous matrix $\boldsymbol{M} = 0$. The conductivity σ is given in this case as

$$\sigma_{\parallel}^{\text{Spitzer}} = - \sum_{a=e,i,I} \sum_{b=e,i,I} e_a n_a e_b n_b \boldsymbol{L}_{ab}^{-1} . \tag{8.66}$$

This is the most exact formula of electrical conductivity in uniform plasma for multi-species plasma suitable for numerical calculation. The analytical fit of Spitzer (Equation 8.64) agrees well with that from Equation 8.66. Then, what is the effect of $\boldsymbol{M}$ on electrical conductivity in Equation 8.65? The trapped particle is trapped in a banana orbit as shown in Figure 8.1 and does not contribute to the current. It creates a frictional force through its relative velocity to the untrapped electron, which drifts through the electric field. This conductivity is typically reduced to about half of that without a trapped particle. Hirshman also gave an approximate analytic expression for σ as follows,

$$\sigma_{\parallel}^{\text{NC}} = \sigma_{\parallel}^{\text{Spitzer}} \left[1 - \frac{f_t}{1 + \xi \nu_e^*}\right] \left[1 - \frac{C_R f_t}{1 + \xi \nu_e^*}\right], \tag{8.67}$$

$$C_R(Z_{\text{eff}}) = \frac{0.56}{Z_{\text{eff}}} \frac{3 - Z_{\text{eff}}}{3 + Z_{\text{eff}}}, \xi(Z_{\text{eff}}) = 0.58 + 0.2 Z_{\text{eff}} , \tag{8.68}$$

$$f_t = 1 - \frac{(1 - \varepsilon)^2}{(1 + 1.46\varepsilon^{1/2})\sqrt{1 - \varepsilon^2}} . \tag{8.69}$$

8.6 Thermodynamic Force: Bootstrap Current

The generalized version of Ohm's law in Equation 8.62 includes current driven by the thermodynamic forces V_{1a} and V_{2a} as follows,

$$\langle \boldsymbol{B} \cdot \boldsymbol{J} \rangle_{bs} = \sum_{a=e,i,I} e_a n_a \sum_{b=1}^{6} \alpha_{ab} V_{\perp b} \,. \tag{8.70}$$

Substituting Equations 8.30 and 8.31 for thermodynamic forces into Equation 8.70, we obtain following form for the bootstrap current:

$$\langle \boldsymbol{B} \cdot \boldsymbol{J} \rangle_{bs} = -F(\psi) n_e(\psi) \sum_{a=e,i,I} \frac{1}{|Z_a|} \left[L_{31}^a \frac{1}{n_a(\psi)} \frac{\mathrm{d}P_a(\psi)}{\mathrm{d}\psi} + L_{32}^a \frac{\mathrm{d}T_a(\psi)}{\mathrm{d}\psi} \right], \tag{8.71}$$

$$L_{31}^a = \sum_{a=e,i,I} \frac{|Z_a|}{Z_a} \frac{Z_b}{n_e} \alpha_{ab} \,, \quad L_{32}^a = \sum_{a=e,i,I} \frac{|Z_a|}{Z_a} \frac{Z_b n_b}{n_e} \alpha_{a,b+3} \,. \tag{8.72}$$

Although V_{1a} includes an electrostatic potential term, the bootstrap current due to this term ($J \sim -F(\sum\sum Z_b n_b \alpha_{ba})\, \mathrm{d}\Phi/\mathrm{d}\psi$) vanishes for axisymmetric plasma due to charge neutrality.

As explained in Section 8.1, distortion of the velocity distribution function occurs in collisionless plasma. The electron distribution function drifts in the direction of $v_\parallel < 0$, while the ion velocity distribution function drifts in the direction of $v_\parallel > 0$ as seen in Figure 8.4). This produces a current in the plasma named the "bootstrap current."

We investigate major parametric dependences of the bootstrap current following Galeev [6]. Consider the case of electrons given in Section 8.1. Let the ba-

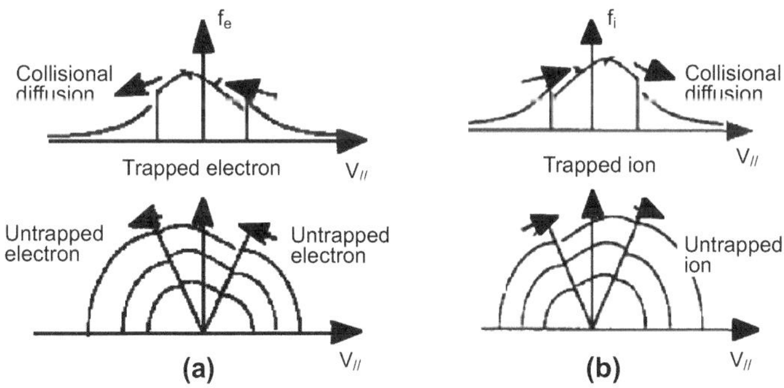

Figure 8.4 Sketch of the mechanism forming the bootstrap current in collisionless plasma for (a) electrons and (b) ions. Collisional pitch angle scattering at the trapped–untrapped boundary produces unidirectional parallel flow/momentum input and is balanced by the collisional friction force between electrons and ions. The source of the momentum input for the parallel flow drive is the radially outward motion/loss of trapped particles

nana width of the trapped electron be $\Delta_b(\sim \varepsilon^{1/2}\rho_p = \varepsilon^{1/2}m_e v_{Te}/eB_p)$. If there is a density gradient, the net particle number passing through a point is given by $n_t(v_\parallel < 0) - n_t(v_\parallel > 0) = (-\mathrm{d}n_t/\mathrm{d}r)\,\Delta_b$. For the trapped electron, $v_\parallel \sim \varepsilon^{1/2}v_{Te}$ and $n_t \sim \varepsilon^{1/2}n_e$ so the trapped electron current is given by $J_{\mathrm{bt}} \sim ev_\parallel(-\mathrm{d}n_t/\mathrm{d}r)\,\Delta_b \sim -\varepsilon^{3/2}T_e\mathrm{d}n_e/\mathrm{d}r$. Momentum supply across the trapped–untrapped boundary will balance with the momentum loss of the untrapped electron–ion collision as follows,

$$\nu_{ei}m_e u_e n_e = \frac{\nu_{ee}}{\varepsilon}\frac{J_{\mathrm{bt}}}{-e}\,. \tag{8.73}$$

Here, ν_{ee}/ε is the effective collision frequency between trapped and untrapped electrons. This equation gives the drift velocity of untrapped electrons and, hence, the current carried by the untrapped electrons is as follows,

$$J_{\mathrm{bt}} \sim -\varepsilon^{1/2}\frac{T_e}{B_p}\frac{\mathrm{d}n_e}{\mathrm{d}r} \sim -\varepsilon^{1/2}\frac{1}{B_p}\frac{\mathrm{d}P_e}{\mathrm{d}r}\,. \tag{8.74}$$

Here, T_e is included in the gradient since a similar mechanism works for $\mathrm{d}T_e/\mathrm{d}r$. The characteristic here is that both supply and loss of momentum are proportional to collision frequency, and so the resulting bootstrap current does not depend on collision frequency. Also, the electron mass is not included in the bootstrap current expression since the small banana width effect is cancelled by the large parallel velocity. A similar discussion for the ions leads to the ion bootstrap current proportional to $\varepsilon^{1/2}\mathrm{d}P_i/\mathrm{d}r/B_p$. It should be noted that the source of the power of the bootstrap current is the trapped particle, but the current is carried by untrapped particles. Now, having reviewed the principles, let us go back to the discussion on the expression of the bootstrap current from the previous section.

Let us look at the role of the viscosity coefficient in the generalized Ohm's law (the electrical conductivity and the bootstrap current) in a simple model. Ignoring the ion flow ($\mathrm{u}_i = 0$) and assuming no impurities, the electron momentum balance equation 8.53 becomes

$$\mu_{e1}\left(u_{\parallel e} - \frac{R}{en_e}\frac{\mathrm{d}P_e}{\mathrm{d}\psi}\right) = l_{11}^{ee}u_{\parallel e} - en_e E_\parallel\,. \tag{8.75}$$

So, the electron current is given by,

$$J_{\parallel e} = -en_e u_{\parallel e} = \frac{en_e}{\mu_{e1} - l_{11}^{ee}}E_\parallel - \frac{R\mu_{e1}}{\mu_{e1} - l_{11}^{ee}}\frac{\mathrm{d}P_e}{\mathrm{d}\psi}\,. \tag{8.76}$$

Using Equation 8.54, Equation 8.76 is rewritten in the following form by using the normalized viscosity coefficient $\hat{\mu}_{e1} = \mu_{e1}/m_e n_e \nu_{ei}$.

$$J_{\parallel e} = \frac{e^2 n_e}{m_e\nu_{ei}}\frac{1}{\hat{\mu}_{e1}+1}E_\parallel - \frac{\hat{\mu}_{e1}}{\hat{\mu}_{e1}+1}\frac{1}{B_p}\frac{\mathrm{d}P_e}{\mathrm{d}r}\,. \tag{8.77}$$

Here, the first term of the right-hand side of Equation 8.77 is the Ohmic current and the second term is the bootstrap current. The normalized viscosity coefficient μ_{e1} appears in both electrical conductivity and the bootstrap coefficient. This means

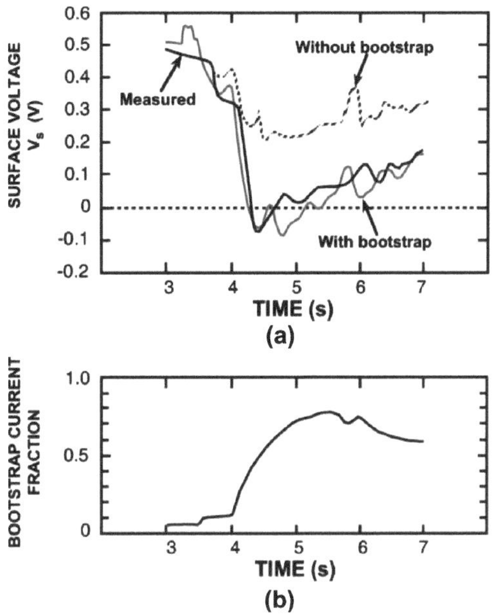

Figure 8.5 (a) Comparison of time variation of the measured surface voltage and simulation using values from JT-60 discharge [4]. (b) Up to 80% of the plasma current is carried by the bootstrap current from the simulation. Without including the bootstrap current, the measured surface voltage cannot be reproduced

that effect of trapped particle should be checked in experiments for both electrical conductivity and the bootstrap current.

Although Hinton–Hazeltine [7] gave an analytic expression of bootstrap coefficients L_{31}^a and L_{32}^a, there is a large difference between numerical coefficients and analytical expressions, except for very small values of ε without impurities [8]. Practically, it is necessary to use the Hirshman–Sigmar formulation for numerical evaluation. Figure 8.5 shows a comparison of plasma surface voltages between measurement and a 1.5-dimensional transport simulation; the existence of the bootstrap current is confirmed [4].

8.7 Momentum Input: Beam-driven Current

When a fast neutral beam is injected at a tangent to the torus, the circulating fast ion produces a fast ion current by multiple circulations around the torus. Collision with bulk electrons produces a shielding current by induced drift in the same direction as the fast ion. This shielding is not perfect due to the existence of trapped electrons and impurities. The sum of fast ion and shielding currents is the beam-driven current $\boldsymbol{J}_{\mathrm{bd}}$, which is a response to the external momentum source S_b given in Equation 8.62, but momentum balance of fast ions has to be included as follows [4],

$$\langle \boldsymbol{B} \cdot \boldsymbol{J} \rangle_{\mathrm{bd}} = \sum_{a=e,i,I,f} e_a n_a \left[(\boldsymbol{M} - \boldsymbol{L})^{-1} \right]_{af} S_{\parallel f} \,. \tag{8.78}$$

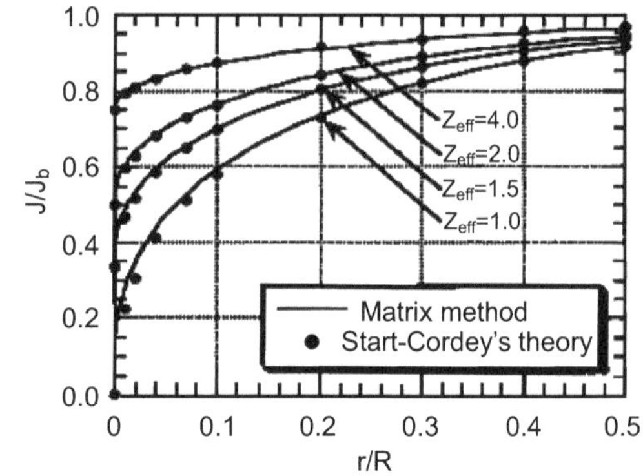

Figure 8.6 Stacking factor F as a function of r/R and Z_{eff} [4] in comparison with [9]

Here, e_a, n_a, $\boldsymbol{M}$, and $\boldsymbol{L}$ are electrical charge, density, viscous, and friction matrixes, included in the fast ion momentum balance equation, respectively, and S_f is the momentum source of the fast ion. So, if we divide the beam-driven current into fast ion current and shielding current $\langle \boldsymbol{B}\cdot\boldsymbol{J}\rangle_{\text{bd}} = \langle \boldsymbol{B}\cdot\boldsymbol{J}\rangle_{\text{fast}} + \langle \boldsymbol{B}\cdot\boldsymbol{J}\rangle_{\text{shield}}$, we obtain

$$\langle \boldsymbol{B}\cdot\boldsymbol{J}\rangle_{\text{fast}} = e_f n_f \left[(\boldsymbol{M}-\boldsymbol{L})^{-1}\right]_{ff} S_{\parallel f}\,, \tag{8.79}$$

$$\langle \boldsymbol{B}\cdot\boldsymbol{J}\rangle_{\text{shield}} = \sum_{a=e,i,I} e_a n_a \left[(\boldsymbol{M}-\boldsymbol{L})^{-1}\right]_{af} S_{\parallel f}\,, \tag{8.80}$$

$$F = \frac{\langle \boldsymbol{B}\cdot\boldsymbol{J}\rangle_{\text{bd}}}{\langle \boldsymbol{B}\cdot\boldsymbol{J}\rangle_{\text{fast}}} = \frac{\sum_{a=e,i,I,f} e_a n_a \left[(\boldsymbol{M}-\boldsymbol{L})^{-1}\right]_{af}}{e_f n_f \left[(\boldsymbol{M}-\boldsymbol{L})^{-1}\right]_{ff}} \tag{8.81}$$

where, e_b, n_b, and S_f are electrical charge, density, and momentum source density of the fast ion, respectively. F is the "stacking factor." Parametric dependences of F on Z_{eff} and ε in arbitrary aspect ratio ($0 \leq \varepsilon \leq 1$) are calculated by Equation 8.81 and shown in Figure 8.6.

The velocity distribution function of fast ions f_b is determined as a solution of the Fokker–Planck equation valid for $v_{Ti} \ll v_b \ll v_{Te}$. The fast ion current is carried by the passing ions and trapped ions do not contribute to the fast ion current since return motion cancels the current. The fast ion linear Fokker–Planck equation in a non-uniform magnetic field is give by Connor [10] as follows,

$$\tau_{se}\frac{\partial f_b}{\partial t} = v^{-2}\frac{\partial}{\partial v}\left[(v_c^3+v^3)f_b\right] + \frac{\beta v_c^3}{v^3\eta\langle v/v_\parallel\rangle}\frac{\partial}{\partial \eta}\left[\frac{(1-\eta^2)}{\eta}\left\langle\frac{v_\parallel}{v}\right\rangle\frac{\partial f_b}{\partial \eta}\right], \\ + \tau_{se}S(v,\eta) \tag{8.82}$$

$$\tau_{se} = \frac{3(2\pi)^{3/2}\varepsilon_0^2 M_b T_e^{3/2}}{e^4 Z_b^2 n_e m_e^{1/2}\ln\Lambda}\,, \quad E_c = \left(\frac{9\pi m_p}{16 m_e}\right)^{1/3}\left[\sum_j \frac{n_j Z_j^2}{n_e A_j}\right]^{2/3} A_b T_e\,.$$

where τ_{se} is the beam-electron slowing down time, v_c is critical velocity defined by $v_c = (2E_c/m_b)^{1/2}$ where E_c is the "critical energy," v_b is the beam velocity $(= (2E_b/m_b)^{1/2})$, β is defined by $\beta = Z_{\text{eff}} n_e \ln \Lambda_e / 2m_b \sum m_j^{-1} Z_j^2 \ln \Lambda_i$, $S(v, \eta)$ is the bounce averaged fast ion source rate per unit volume, Z_{eff} is the effective charge defined by $Z_{\text{eff}} = \sum_j n_j Z_j^2 / n_e$, $\langle v/v_\parallel \rangle$ is given by $\langle v/v_\parallel \rangle = (2/\pi) K[(\eta_t/\eta)^2]$ and $\langle v_\parallel / v \rangle$ is given by $\langle v_\parallel / v \rangle = (2/\pi) E[(\eta_t/\eta)^2]$ for passing ions, where K and E are complete elliptic integrals of the first and second kind, respectively. The solution of Equation 8.82 is given in Cordey [11] as follows,

$$f_b(v, \eta) = S_0 \tau_{se} \sum_n a_n(v) c_n(\eta) \tag{8.83}$$

where, $a_n(v)$ is the analytical solution for a uniform magnetic field from Gaffey [12]. $S(v, \eta) = S_0 \delta(v - v_b) k(\eta) = S_0 \delta(v - v_b) \sum k_n c_n(\eta)$ and the fast ion distribution function above beam energy $(v > v_b)$ comes from energy diffusion in velocity space. The following equation determines the eigenvalue λ_n and eigen-function $c_n(\eta)$:

$$\frac{1}{\eta \langle v/v_\parallel \rangle} \frac{\mathrm{d}}{\mathrm{d}\eta} \left[(1 - \eta^2) \left\langle \frac{v_\parallel}{v} \right\rangle \frac{\mathrm{d}c_n}{\mathrm{d}\eta} \right] + \lambda_n c_n = 0\,. \tag{8.84}$$

This equation is solved numerically by the Rayleigh–Ritz method using ACCOME code [13]. Using the above formulas, the flux surface averaged parallel fast ion current multiplied by B, $\langle \boldsymbol{B} \cdot \boldsymbol{J} \rangle_{\text{fast}}$ can be calculated as

$$\langle \boldsymbol{B} \cdot \boldsymbol{J} \rangle_{\text{fast}} = e Z_f \int_0^\infty v^3 \mathrm{d}v \int_{-1}^{1} \mathrm{d}\eta f_b(v, \eta, \rho) H(v, \eta, \rho), \tag{8.85}$$

$$H(v, \eta, \rho) = \int \frac{B}{v} \mathrm{d}s \int \frac{\mathrm{d}s}{v_\parallel}\,.$$

Note: Modification of L and M due to fast ion

With inclusion of fast ion, L and M becomes 7×7 matrices including fast ion momentum balance equation, while heat flow balance is not included. Friction coefficients are given as follows,

$$l_{11}^{fe} = l_{11}^{ef} = (n_e m_e / \tau_{ee}) \left(Z_f^2 n_f / n_e \right)$$

$$l_{11}^{fk} = l_{11}^{kf} = (3\sqrt{\pi} n_k m_k / 4\tau_{kk}) \left(Z_f^2 n_f / Z_k^2 n_k \right) (1 + m_k/m_f)$$
$$(\tau_s v_{Tk}^3 / 3\tau_{th} v_c^3) \ (k = i, I)$$

$$l_{11}^{fe} = l_{11}^{ef} = (n_e m_e/\tau_{ee}) \left(Z_f^2 n_f / n_e \right)$$
$$l_{11}^{ff} = -\left(l_{11}^{fe} + l_{11}^{fi} + l_{11}^{fI} \right),$$
$$l_{21}^{ef} = -1.5 l_{11}^{ef}, l_{21}^{if} = l_{21}^{If} = l_{12}^{fe} = l_{12}^{fi} = l_{12}^{fI} = 0$$
$$l_{11}^{kk} = l_{11}^{kk}(th) - l_{11}^{kf} \quad (k = e, i, I), \quad i_{21}^{ee} = l_{21}^{ee}(th) - l_{21}^{ef},$$

$\tau_{se} = 3(2\pi)^{3/2}\varepsilon_0^2 M_b T_e^{3/2}/e^4 Z_b^2 n_e m_e^{1/2} \ln \Lambda$: slowing down time

$\tau_{th} = (\tau_{se}/3)\ln((E_{b0}/E_c)^{1.5} + 1)$: thermalization time

$\mu_{f1} = \dfrac{f_t}{f_c}\dfrac{n_e m_e}{\tau_{ee}}\dfrac{Z_f^2 n_f}{n_e}\dfrac{\hat{Z} v_c^3}{\hat{v}_c^3}$: fast ion viscosity coefficient

$p_f = \int \dfrac{m_f v^2}{3} f_f(\boldsymbol{v})\, \mathrm{d}\boldsymbol{v}$: fast ion pressure

$$\hat{Z} = \frac{\sum Z_i^2 n_i}{\sum Z_i^2 n_i m_f/m_i}, \quad \hat{v}_c^3 = \frac{\tau_{se}}{3\tau_{th}}\frac{n_f m_f}{2p_f}\int_0^{v_b} \frac{v\, \mathrm{d}v}{v^3 + v_c^3}.$$

8.8 Rotation: Toroidal Rotation and Flow Relations among Ions, Impurities, and Electrons

One important property of the axisymmetric system is the conservation of total toroidal momentum. The net force balance equation for the plasma is obtained by summing up Equation 8.1 over all species as follows,

$$\sum_a m_a n_a \frac{\mathrm{d}\boldsymbol{u}_a}{\mathrm{d}t} = \boldsymbol{J} \times \boldsymbol{B} - \nabla P - \sum_a \nabla \cdot \boldsymbol{\Pi}_a + \sum_a \boldsymbol{M}_a \tag{8.86}$$

where P is total plasma pressure and the following momentum conservation in Coulomb collision and charge neutrality are taken into account:

$$\sum_a \boldsymbol{F}_{a1} = 0\,, \tag{8.87}$$

$$\sum_a e_a n_a = 0\,. \tag{8.88}$$

For timescales longer than the Alfven transit time $\tau_A = L/v_A$, static pressure balance valid for $O(1)$ and is given by, $\boldsymbol{J} \times \boldsymbol{B} = \nabla P$. Taking the inner product $R^2\nabla\zeta\cdot$ Equation 8.86 and the flux surface average, we obtain the following equation for slowly time-varying phenomena in response to momentum source:

$$\sum_a m_a \left\langle n_a R \frac{\mathrm{d}u_{a\zeta}}{\mathrm{d}t} \right\rangle = \left\langle \sum_a R^2\nabla\zeta \cdot \nabla \cdot \boldsymbol{\Pi}_a \right\rangle + \left\langle \sum_a R^2\nabla\zeta \cdot \boldsymbol{M}_a \right\rangle . \quad (8.89)$$

The first term of the right-hand side is the drag force due to the magnetic field variation. Since viscous tensor $\boldsymbol{\Pi}_a$ is a symmetric tensor and $\nabla(R^2\nabla\zeta)$ is anti-symmetric tensor, we obtain $\nabla(R^2\nabla\zeta)$: $\boldsymbol{\Pi}_a = 0$ and $R^2\nabla\zeta \cdot \nabla \cdot \boldsymbol{\Pi}_a = \nabla \cdot (R^2\nabla\zeta \cdot \boldsymbol{\Pi}_a) - \nabla(R^2\nabla\zeta) : \boldsymbol{\Pi}_a = \nabla \cdot (R^2\nabla\zeta \cdot \boldsymbol{\Pi}_a)$ (see Tensor Algebra in Appendix). Thus flux surface average of toroidal viscous force is given as follows,

$$\left\langle R^2\nabla\zeta \cdot \nabla \cdot \boldsymbol{\Pi}_a \right\rangle = \left\langle \nabla \cdot \left[R^2\nabla\zeta \cdot \boldsymbol{\Pi}_a \right] \right\rangle . \quad (8.90)$$

Since $\langle \nabla \cdot \boldsymbol{A} \rangle = (\mathrm{d}/\mathrm{d}V)\langle \boldsymbol{A} \cdot \nabla V \rangle$ (V: volume inside the flux surface $V(\psi)$) for any vector $\boldsymbol{A}$, the flux surface average of Equation 8.90 gives,

$$\left\langle R^2\nabla\zeta \cdot \nabla \cdot \boldsymbol{\Pi}_a \right\rangle = 0 . \quad (8.91)$$

This means that the toroidal drag force due to the magnetic field variation is zero for an axisymmetric system, while it becomes important if there is some asymmetry [14].

The general flow balance equation 8.61 allows us to derive an expression for the toroidal flows of electrons, ions, and impurities. From Equation 8.41, the local toroidal flow for species a is the summation of flux surface averaged parallel flow and Pfirsch–Schlüter flow as follows,

$$u_{a\zeta}^{(1)} = \frac{B_\zeta}{\langle B^2 \rangle} \langle \boldsymbol{B} \cdot \boldsymbol{u}_a \rangle + \left[1 - \frac{B_\zeta^2}{\langle B^2 \rangle} \right] R \left(\frac{\mathrm{d}\Phi}{\mathrm{d}\psi} + \frac{1}{e_a n_a} \frac{\mathrm{d}P_a}{\mathrm{d}\psi} \right) . \quad (8.92)$$

The flux surface averaged parallel flow for species a (electron, ion, or impurity) are given by

$$\langle \boldsymbol{B} \cdot \boldsymbol{u}_a \rangle = -F(\psi) \sum_{b=e,i,I} \left[\alpha_{ab} \left(\frac{\mathrm{d}\Phi}{\mathrm{d}\psi} + \frac{1}{e_b n_b} \frac{\mathrm{d}P_b}{\mathrm{d}\psi} \right) + \alpha_{a,b+3} \frac{1}{e_b} \frac{\mathrm{d}T_b}{\mathrm{d}\psi} \right] \quad (8.93)$$
$$+ \sum_{b=e,i,I} e_b n_b c_{ab} \langle \boldsymbol{B} \cdot \boldsymbol{E} \rangle .$$

The impurity (say, carbon) toroidal rotation can be measured by charge exchange recombination spectroscopy (CXRS), which can be used to determine the radial electric field using Equations 8.92 and 8.93 if density and temperature profiles are known. If we evaluate the toroidal rotation profiles for ions, electrons, and impurities, we observe a difference between electrons and ions which is the plasma current consisting of electric current and bootstrap current as shown in Sections 8.5 and 8.6. Interestingly, there is a difference between ion and impurity toroidal rotations, which is usually proportional to the ion temperature gradient. Since electron inertia is small, the electron contribution to ion and impurity momentum balance can be neglected and the flow relations of ions and impurities are given as follows,

$$\hat{\mu}^I \cdot \boldsymbol{u}_\theta^I = \hat{\boldsymbol{L}}_{II} \cdot \boldsymbol{u}_\parallel^I + \hat{\boldsymbol{L}}_{Ii} \cdot \boldsymbol{u}_\parallel^i \qquad \text{(impurity)}\ , \tag{8.94}$$

$$\hat{\mu}^i \cdot \boldsymbol{u}_\theta^i = \hat{\boldsymbol{L}}_{iI} \cdot \boldsymbol{u}_\parallel^I + \hat{\boldsymbol{L}}_{ii} \cdot \boldsymbol{u}_\parallel^i \qquad \text{(ion)} \tag{8.95}$$

where normalized viscosity and friction coefficient matrices, and poloidal and parallel flow vectors are defined as follows,

$$\hat{\mu}^a = \begin{bmatrix} \hat{\mu}_{a1} & \hat{\mu}_{a2} \\ \hat{\mu}_{a2} & \hat{\mu}_{a3} \end{bmatrix}, \quad \hat{\boldsymbol{L}}_{ab} = \begin{bmatrix} \hat{l}_{11}^{ab} & -\hat{l}_{12}^{ab} \\ -\hat{l}_{21}^{ab} & \hat{l}_{22}^{ab} \end{bmatrix}, \tag{8.96}$$

$$\boldsymbol{u}_\theta^a = \begin{bmatrix} \langle B^2 \rangle u_{a\theta}^*(\psi) \\ 2\langle B^2 \rangle q_{a\theta}^*(\psi)/5P_a \end{bmatrix}, \quad \boldsymbol{u}_\parallel^a = \begin{bmatrix} \langle B u_{\parallel b} \rangle \\ \langle 2 B q_{\parallel b}/5P_b \rangle \end{bmatrix}, \tag{8.97}$$

$$\hat{l}_{ij}^{ab} = \frac{\tau_{aa}}{m_a n_a} l_{ij}^{ab}\ , \quad \hat{\mu}_{ai} = \frac{\tau_{aa}}{m_a n_a} \mu_{ai}\ . \tag{8.98}$$

From Equations 8.39 and 8.40, the poloidal flow $\boldsymbol{u}_\theta^a$ is expressed as $\boldsymbol{u}_\theta^a = \boldsymbol{u}_\parallel^a - \boldsymbol{V}^a$ and $\boldsymbol{V}^a = (BV_{1a},\ BV_{2a})^t$. Since impurity collisionality is given as $\nu_I^* = (n_I Z_I^4/n_i Z_i^4)\nu_i^* \gg \nu_i^*$ (see Equation 8.51), impurities may form a Pfirsch–Schlüter regime (negligible impurity viscous force) while bulk ions form a deeply collisionless regime. Therefore, impurity parallel flow can be given as $\boldsymbol{u}_\parallel^I = -\hat{\boldsymbol{L}}_{II}^{-1}\hat{\boldsymbol{L}}_{Ii} \cdot \boldsymbol{u}_\parallel^i$. Substituting this into Equation 8.95 gives the following ion momentum balance equation:

$$\hat{\mu}^i \cdot \boldsymbol{u}_\theta^i = \left(\hat{\boldsymbol{L}}_{ii} - \hat{\boldsymbol{L}}_{iI}\hat{\boldsymbol{L}}_{II}^{-1}\hat{\boldsymbol{L}}_{Ii}\right) \cdot \boldsymbol{u}_\parallel^i\ . \tag{8.99}$$

Matrix elements of $L_{ii} - L_{iI}L_{II}^{-1}L_{Ii}$ are easily calculated using Equation 8.54 and simpler approximate expression than those given by Kim [15] is obtained for $m_i \ll m_I$ as follows,

$$\hat{\boldsymbol{L}}_{ii} - \hat{\boldsymbol{L}}_{iI}\hat{\boldsymbol{L}}_{II}^{-1}\hat{\boldsymbol{L}}_{Ii} = -\begin{bmatrix} 0 & 0 \\ 0 & \gamma \end{bmatrix} \tag{8.100}$$

where $\gamma = 2^{0.5} + \alpha$, $\alpha = n_I Z_I^2 l n_i Z_i^2$. Using $\boldsymbol{u}_\theta^a = \boldsymbol{u}_\parallel^a - \boldsymbol{V}^a$, we obtain following equation for $\boldsymbol{u}_\parallel^i$, $\boldsymbol{u}_\parallel^I$, $\boldsymbol{u}_\theta^i$,

$$\boldsymbol{u}_\parallel^i = \begin{bmatrix} 1 & K_1 \\ 0 & K_2 \end{bmatrix} \boldsymbol{V}^i\ , \quad \boldsymbol{u}_\parallel^I = \begin{bmatrix} 1 & K_1 + 1.5K_2 \\ 0 & 0 \end{bmatrix} \boldsymbol{V}^i\ , \quad \boldsymbol{u}_\theta^i = \begin{bmatrix} 0 & K_1 \\ 0 & -(\hat{\mu}_{i1}/\hat{\mu}_{i2})K_1 \end{bmatrix} \boldsymbol{V}^i \tag{8.101}$$

$$K_1 = \frac{\gamma \hat{\mu}_{i2}}{D}\ , \quad K_2 = \frac{\hat{\mu}_{i1}\hat{\mu}_{i3} - \hat{\mu}_{i2}^2}{D}\ , \quad D = \hat{\mu}_{i1}(\hat{\mu}_{i3} + \gamma) - \hat{\mu}_{i2}^2\ . \tag{8.102}$$

In this case, we obtain the following form of $u_{\theta i}$, $u_{\theta I}$, $u_{\parallel i}$, and $u_{\parallel I}$ by neglecting poloidal variation and approximating $B \sim B_\zeta$,

$$u_{\parallel i} \approx -\frac{1}{B_\theta}\left[\frac{\mathrm{d}\Phi}{\mathrm{d}r} + \frac{1}{eZ_i n_i}\frac{\mathrm{d}P_i}{\mathrm{d}r} + \frac{K_1}{eZ_i}\frac{\mathrm{d}T_i}{\mathrm{d}r}\right], \tag{8.103}$$

$$u_{\parallel I} \approx -\frac{1}{B_\theta}\left[\frac{\mathrm{d}\Phi}{\mathrm{d}r} + \frac{1}{eZ_i n_i}\frac{\mathrm{d}P_i}{\mathrm{d}r} + \frac{K_1 + 1.5K_2}{eZ_i}\frac{\mathrm{d}T_i}{\mathrm{d}r}\right], \tag{8.104}$$

$$u_{i\theta} \approx -\frac{B_\zeta}{e\langle B^2\rangle}\frac{K_1}{Z_i}\frac{\mathrm{d}T_i}{\mathrm{d}r}, \tag{8.105}$$

$$u_{I\theta} \approx -\frac{B_\zeta}{e\langle B^2\rangle}\left(\frac{1}{Z_I n_I}\frac{\mathrm{d}P_I}{\mathrm{d}r} - \frac{1}{Z_i n_i}\frac{\mathrm{d}P_i}{\mathrm{d}r} + \frac{K_1 + 1.5K_2}{Z_i}\frac{\mathrm{d}T_i}{\mathrm{d}r}\right). \tag{8.106}$$

From (8.103) and (8.104), we obtain the expression of a difference between impurity and ion toroidal rotation as $\Delta u_{//} = u_{//I} - u_{//i} = -1.5K_2(\mathrm{d}T_i/\mathrm{d}r)/(eZ_i B_\theta)$, which is proportional to ion temperature gradient and inversely proportional to poloidal magnetic field.

8.9 Neoclassical Transport: Transport Across the Magnetic Field

To determine collisional transport across the magnetic field, higher order terms must be retained when the perpendicular flux is obtained from Equation 8.1 and 8.2. Neglecting source terms, we obtain the following perpendicular fluxes from $\boldsymbol{B}\times$ Equations 8.1 and 8.2.

$$\boldsymbol{u}_{\perp a} = \frac{\boldsymbol{E}\times\boldsymbol{B}}{B^2} + \frac{\boldsymbol{b}\times\nabla P_a}{m_a n_a \Omega_a} + \frac{\boldsymbol{b}\times(\nabla\cdot\boldsymbol{\Pi}_a - \boldsymbol{F}_{a1})}{m_a n_a \Omega_a}, \tag{8.107}$$

$$\boldsymbol{q}_{\perp a} = \frac{5}{2}P_a\frac{\boldsymbol{b}\times\nabla T_a}{m_a\Omega_a} + \frac{T_a}{m_a\Omega_a}\boldsymbol{b}\times(\nabla\cdot\boldsymbol{\Theta}_a - \boldsymbol{F}_{a2}). \tag{8.108}$$

The lowest order $O(\delta)$ of Equation 8.107 and 8.108 is the first order flow on the flux surface given in Section 8.2. Collisional transport across the flux surface is second order in δ. Taking the flux surface average of the inner product with Equations 8.107 and 8.108 gives,

$$\begin{bmatrix}\langle n_a \boldsymbol{u}_{a\perp}\cdot\nabla\psi\rangle \\ \left\langle \frac{\boldsymbol{q}_{a\perp}}{T_a}\cdot\nabla\psi\right\rangle\end{bmatrix} = \begin{bmatrix}\Gamma_a^{cl} \\ \frac{q_a^{cl}}{T_a}\end{bmatrix} + \begin{bmatrix}\Gamma_a^{\mathrm{NC}} \\ \frac{q_a^{\mathrm{NC}}}{T_a}\end{bmatrix}. \tag{8.109}$$

Here,

$$\begin{bmatrix}\Gamma_a^{cl} \\ \frac{q_a^{cl}}{T_a}\end{bmatrix} = \left\langle\frac{\boldsymbol{B}\times\nabla\psi}{e_a B^2}\cdot\begin{bmatrix}\boldsymbol{F}_{a1} \\ \boldsymbol{F}_{a2}\end{bmatrix}\right\rangle, \tag{8.110}$$

$$\begin{bmatrix}\Gamma_a^{\mathrm{NC}} \\ \frac{q_a^{\mathrm{NC}}}{T_a}\end{bmatrix} = \left\langle\frac{\boldsymbol{B}\times\nabla\psi}{e_a B^2}\cdot\begin{bmatrix}\nabla P_a + \nabla\cdot\boldsymbol{\Pi}_a - e_a n_a\boldsymbol{E} \\ \frac{5}{2}n_a\nabla T_a + \nabla\cdot\boldsymbol{\Theta}_a\end{bmatrix}\right\rangle. \tag{8.111}$$

The fluxes given by Equations 8.110 and 8.111 are the classical transport flux, and neoclassical transport flux, respectively. If the mean free path $\lambda = \tau_{aa}v_{Ta}$ is much shorter than the connection length Rq ($\lambda < Rq$), inhomogeneity of the friction force along the magnetic field on the magnetic surface produces Pfirsch–Schlüter diffusion. Also, if $\lambda \gg Rq$ (collisionless regime), the anisotropic velocity distribution function formed by the non-uniform magnetic field (see Section 8.1) produces viscous forces ($\nabla \cdot \boldsymbol{\Pi}_a$ and $\nabla \cdot \boldsymbol{\Theta}_a$) and hence banana-plateau diffusion. Neoclassical transport fluxes can be rewritten using the axisymmetric equilibrium relation $\boldsymbol{b} \times \nabla\psi = F(\psi)\boldsymbol{b} - R^2 B \nabla\zeta$ derived from $\nabla\psi = -R^2\nabla\zeta \times \boldsymbol{B}$ as follows,

$$\Gamma_a^{\mathrm{NC}} = -\left\langle \frac{F(\psi)}{e_a B^2} \boldsymbol{B} \cdot (\nabla P_a + \nabla \cdot \boldsymbol{\Pi}_a) \right\rangle + \left\langle \frac{n_a}{B^2} \boldsymbol{E} \cdot (\nabla\psi \times \mathbf{B}) \right\rangle$$
$$= -\frac{F(\psi)}{e_a} \left\langle \left(\frac{1}{B^2} - \frac{1}{\langle B^2 \rangle} \right) \boldsymbol{B} \cdot \mathbf{F}_{a1} \right\rangle - \frac{F(\psi)}{e_a \langle B^2 \rangle} \langle \boldsymbol{B} \cdot \nabla \cdot \boldsymbol{\Pi}_a \rangle + \Gamma_a^E \,, \tag{8.112}$$

$$\frac{q_a^{\mathrm{NC}}}{T_a} = -\frac{F(\psi)}{e_a} \left\langle \left(\frac{1}{B^2} - \frac{1}{\langle B^2 \rangle} \right) \mathbf{B} \cdot \boldsymbol{F}_{a2} \right\rangle - \frac{F(\psi)}{e_a \langle B^2 \rangle} \langle \boldsymbol{B} \cdot \nabla \cdot \boldsymbol{\Theta}_a \rangle \,. \tag{8.113}$$

Here,

$$\Gamma_a^E = \frac{F(\psi) \langle n_a \boldsymbol{B} \cdot \mathbf{E} \rangle}{\langle B^2 \rangle} - \left\langle R^2 \nabla\zeta \cdot n_a \boldsymbol{E}_A \right\rangle , \quad \boldsymbol{E}_A = -\frac{\partial \boldsymbol{A}}{\partial t} \,. \tag{8.114}$$

The flux Γ_a^E includes the movement of the toroidal flux, $E_\zeta \times B_\theta$ pinch and flux induced by the poloidal variation of the electrostatic potential Φ [1]. The first term of the right-hand side of Equations 8.112 and 8.113 is the Pfirsch–Schlüter transport flux and the second term is the banana-plateau transport flux. Substituting Equation 8.43 into the classical transport flux expression Equation 8.110, we obtain

$$\begin{bmatrix} \Gamma_a^{cl} \\ \frac{q_a^{cl}}{T_a} \end{bmatrix} = \left\langle \frac{|\nabla\psi|^2}{B^2} \right\rangle \sum_b \frac{1}{e_a e_b} \begin{bmatrix} l_{11}^{ab} & -l_{12}^{ab} \\ -l_{21}^{ab} & l_{22}^{ab} \end{bmatrix} \begin{bmatrix} P_b'(\psi)/n_b \\ T_b'(\psi) \end{bmatrix} . \tag{8.115}$$

Pfirsch–Schlüter transport flux can be expressed as follows,

$$\begin{bmatrix} \Gamma_a^{ps} \\ \frac{q_a^{ps}}{T_a} \end{bmatrix} = -\frac{F(\psi)}{e_a} \left\langle \left(\frac{1}{B^2} - \frac{1}{\langle B^2 \rangle} \right) \begin{bmatrix} \boldsymbol{B} \cdot \boldsymbol{F}_{a1} \\ \boldsymbol{B} \cdot \boldsymbol{F}_{a2} \end{bmatrix} \right\rangle . \tag{8.116}$$

Using Equations 8.39 and 8.40, the inner product of $\boldsymbol{B}$ and the friction forces $\boldsymbol{F}_{ai}$ above can be expressed as follows,

$$\begin{bmatrix} \boldsymbol{B} \cdot \boldsymbol{F}_{a1} \\ \boldsymbol{B} \cdot \boldsymbol{F}_{a2} \end{bmatrix} = \sum_b \begin{pmatrix} l_{11}^{ab} & -l_{12}^{ab} \\ -l_{21}^{ab} & l_{22}^{ab} \end{pmatrix} \begin{bmatrix} B^2 u_{a\theta}^*(\psi) + B V_{1a} \\ B^2 q_{a\theta}^*(\psi) + \frac{5}{2} P_a B V_{2a} \end{bmatrix} . \tag{8.117}$$

So, the Pfirsch–Schlüter transport flux can be expressed as

$$\begin{bmatrix} \Gamma_a^{ps} \\ \frac{q_a^{ps}}{T_a} \end{bmatrix} = \frac{F(\psi)^2}{e_a} \left(\left\langle \frac{1}{B^2} \right\rangle - \frac{1}{\langle B^2 \rangle} \right) \sum_b \frac{1}{e_b} \begin{bmatrix} l_{11}^{ab} & -l_{12}^{ab} \\ -l_{21}^{ab} & l_{22}^{ab} \end{bmatrix} \begin{bmatrix} P_b'(\psi)/n_b \\ T_b'(\psi) \end{bmatrix} . \tag{8.118}$$

The banana plateau transport fluxes of Equations 8.112 and 8.113 can be written in the following matrix form,

$$\begin{bmatrix} \Gamma_a^{\mathrm{bp}} \\ \frac{q_a^{\mathrm{bp}}}{T_a} \end{bmatrix} = -\frac{F(\psi)}{e_a \langle B^2 \rangle} \begin{pmatrix} \langle \boldsymbol{B} \cdot \nabla \cdot \boldsymbol{\Pi}_a \rangle \\ \langle \boldsymbol{B} \cdot \nabla \cdot \boldsymbol{\Theta}_a \rangle \end{pmatrix} . \tag{8.119}$$

Substituting Equation 8.44 into this equation and utilizing Equations 8.39 and 8.40 gives following expression for banana-plateau transport flux,

$$\begin{bmatrix} \Gamma_a^{\mathrm{bp}} \\ \frac{q_a^{\mathrm{bp}}}{T_a} \end{bmatrix} = \frac{F(\psi)}{e_a \langle B^2 \rangle} \begin{bmatrix} \mu_{a1} & \mu_{a2} \\ \mu_{a2} & \mu_{a3} \end{bmatrix} \begin{bmatrix} \langle B u_{\|a} \rangle + F(\psi) \left(\Phi'(\psi) + P_a'(\psi)/e_a n_a \right) \\ 2 \langle B q_{\|a} \rangle / 5 P_a + F(\psi) T_a'(\psi)/e_a \end{bmatrix} . \tag{8.120}$$

Substituting Equation 8.61 into $\langle B u_{\|a} \rangle$ and $2\langle B q_{\|a} \rangle / 5 P_a$, we can readily obtain the following form of banana-plateau flux, where the expression for K and g are left for the reader to derive,

$$\begin{bmatrix} \Gamma_a^{\mathrm{bp}} \\ \frac{q_a^{\mathrm{bp}}}{T_a} \end{bmatrix} = -\sum_{b=e,i,I} \begin{bmatrix} K_{11}^{ab} & K_{12}^{ab} \\ K_{21}^{ab} & K_{22}^{ab} \end{bmatrix} \begin{bmatrix} P_a'(\psi)/n_a \\ T_a'(\psi) \end{bmatrix} + \begin{bmatrix} g_{1a} \\ g_{2a} \end{bmatrix} \langle B E_{\|} \rangle . \tag{8.121}$$

8.10 Neoclassical Ion Thermal Diffusivity: Ion Thermal Diffusivity by Coulomb Collisions

As an application of previous discussion, we derive an expression for neoclassical ion thermal diffusivity. At first, the label on the flux surface is changed from ψ to the equivalent radius defined by $\rho = (\phi/\phi_a)^{1/2} a$, and the dimensionless friction coefficient is introduced, as follows,

$$\hat{l}_{ij}^{ab} = \frac{\tau_{aa}}{m_a n_a} l_{ij}^{ab} . \tag{8.122}$$

Then the classical and Pfirsch–Schlüter transport fluxes is obtained as follows,

$$\begin{bmatrix} \Gamma_a^{cl} \\ \frac{q_a^{cl}}{T_a} \end{bmatrix} = \left\langle \frac{R^2 B_p^2}{B^2} \right\rangle \frac{m_a n_a}{e^2 \psi'(\rho) \tau_{aa}} \sum_b \frac{1}{Z_a Z_b} \begin{bmatrix} \hat{l}_{11}^{ab} & -\hat{l}_{12}^{ab} \\ -\hat{l}_{21}^{ab} & \hat{l}_{22}^{ab} \end{bmatrix} \begin{bmatrix} P_b'(\rho)/n_b \\ T_b'(\rho) \end{bmatrix} , \tag{8.123}$$

$$\begin{bmatrix} \Gamma_a^{ps} \\ \frac{q_a^{ps}}{T_a} \end{bmatrix} = F(\psi)^2 \left(\left\langle \frac{1}{B^2} \right\rangle - \frac{1}{\langle B^2 \rangle} \right) \frac{m_a n_a}{e^2 \psi'(\rho) \tau_{aa}} \times \sum_b \frac{1}{Z_a Z_b} \begin{bmatrix} \hat{l}_{11}^{ab} & -\hat{l}_{12}^{ab} \\ -\hat{l}_{21}^{ab} & \hat{l}_{22}^{ab} \end{bmatrix} \begin{bmatrix} (\mathrm{d}P_b/\mathrm{d}\rho)/n_b \\ \mathrm{d}T_b/\mathrm{d}\rho \end{bmatrix} . \tag{8.124}$$

Here, $|\nabla\psi| = RB_p$ and $\langle 1/B^2 - 1/\langle B^2\rangle\rangle \sim 2\varepsilon^2/B_0^2$ in cylindrical approximation and $\Gamma_{ps}/\Gamma_c \sim 2\varepsilon^2(B_0/B_p)^2 \sim 2q^2$ is obtained. The combined flux of classical and Pfirsch–Schlüter transport is given by following form,

$$\begin{bmatrix} \Gamma_a^{c+ps} \\ \frac{q_a^{c+ps}}{T_a} \end{bmatrix} = \left(\langle R^2\rangle - \frac{F(\psi)^2}{\langle B^2\rangle}\right) \frac{m_a n_a}{e^2\psi'(\rho)\tau_{aa}} \times \sum_b \frac{1}{Z_a Z_b} \begin{bmatrix} \hat{l}_{11}^{ab} & -\hat{l}_{12}^{ab} \\ -\hat{l}_{21}^{ab} & \hat{l}_{22}^{ab} \end{bmatrix} \begin{bmatrix} (\mathrm{d}P_b/\mathrm{d}\rho)/n_b \\ \mathrm{d}T_b/\mathrm{d}\rho \end{bmatrix} . \tag{8.125}$$

Collecting the q_a terms in proportion to $\mathrm{d}T_b/\mathrm{d}r$ from Equation 8.125, the classical and Pfirsch–Schlüter thermal transport coefficient χ_a^{cps} is given as follows,

$$\chi_a^{\mathrm{cps}} = -\frac{\langle \boldsymbol{q}_{a\perp}\cdot\nabla\rho\rangle}{\left\langle|\nabla\rho|^2\right\rangle n_a \mathrm{d}T_a/\mathrm{d}\rho} = \sqrt{\varepsilon}\rho_{pa}^2 \nu_{aa}\hat{K}_{2a}^{\mathrm{cps}} . \tag{8.126}$$

Here, $\nu_{aa} = 1/\tau_{aa}$ and

$$\rho_{pa} = \frac{m_a v_{Ta}}{e_a B_{p1}} , \quad B_{p1}^2 = \frac{B_0^2}{F(\psi)^2}|\nabla\psi|^2 ,$$

$$\hat{K}_{2a}^{\mathrm{cps}} = \frac{B_0^2\left(\hat{l}_{21}^{aa} - \hat{l}_{22}^{aa}\right)}{2\sqrt{\varepsilon}\langle B^2\rangle}\left(\frac{\langle R^2\rangle\langle B^2\rangle}{F(\psi)^2} - 1\right) . \tag{8.127}$$

Assuming that the temperature equipartition time between the main ions and impurity ions is short and the contribution of the electron to ion heat flux (Equation 8.125) is small ($\hat{l}_{21}^{ae} \sim \hat{l}_{22}^{ae} \sim 0$; $a \neq e$), we can derive heat flux, taking into account the contribution of the heat flux produced by different ion temperature gradients,

$$\hat{K}_{2a}^{\mathrm{cps}} = \frac{B_0^2}{2\sqrt{\varepsilon}\langle B^2\rangle}\left(\frac{\langle R^2\rangle\langle B^2\rangle}{F(\psi)^2} - 1\right)\sum_{b=i,I}\frac{Z_a}{Z_b}\left(\hat{l}_{21}^{ab} - \hat{l}_{22}^{ab}\right) . \tag{8.128}$$

Including the heat flux through the impurity ion channel, ion thermal diffusivity $\chi_{i(\mathrm{tot})}^{\mathrm{cps}}$ is given by

$$\chi_{i(\mathrm{tot})}^{\mathrm{cps}} = \frac{\langle(\boldsymbol{q}_{i\perp} + \boldsymbol{q}_{I\perp})\cdot\nabla\rho\rangle}{\left\langle|\nabla\rho|^2\right\rangle(n_i + n_I)\mathrm{d}T_a/\mathrm{d}\rho} = \sqrt{\varepsilon}\rho_{pi}^2\nu_{ii}\left[f_i\hat{K}_{2i}^{\mathrm{cps}} + (1 - f_i)\alpha\hat{K}_{2I}^{\mathrm{cps}}\right] . \tag{8.129}$$

Here, $f_i = n_i/(n_i + n_I)$ and $\alpha = Z_I^2 n_I/Z_i^2 n_i$. Collecting the q_a terms in proportion to $\mathrm{d}T_b/\mathrm{d}\rho$ from Equation 8.121, the banana-plateau thermal transport coefficient χ_a^{bp} is given by

$$\chi_a^{\mathrm{bp}} = -\frac{\left\langle \boldsymbol{q}_{a\perp}^{\mathrm{bp}}\cdot\nabla\rho\right\rangle}{\left\langle|\nabla\rho|^2\right\rangle n_a\mathrm{d}T_a/\mathrm{d}\rho} = \sqrt{\varepsilon}\rho_{pa}^2\nu_{aa}\hat{K}_{2a}^{\mathrm{bp}} \tag{8.130}$$

where

$$\hat{K}_{2a}^{\text{bp}} = \frac{B_0^2 \left(\hat{\mu}_{3a}(1 - \alpha_{a+3,a} + \alpha_{a+3,a+3}) + \hat{\mu}_{2a}(1 + \alpha_{a,a} - \alpha_{a,a+3})\right)}{2\sqrt{\varepsilon}\langle B^2 \rangle}, \tag{8.131}$$

$$\hat{\mu}_{3a} = \frac{\tau_{aa}}{m_a n_a}\mu_{3a}, \hat{\mu}_{2a} = \frac{\tau_{aa}}{m_a n_a}\mu_{2a}, \alpha_{ij} = \left[(\boldsymbol{M} - \boldsymbol{L})^{-1}\boldsymbol{M}\right]_{ij} . \tag{8.132}$$

From Equations 8.129 and 8.130, ion thermal diffusivity χ_a^{tot} is given by

$$\chi_a^{\text{NC}} = \sqrt{\varepsilon}\rho_{pa}^2 \nu_{aa} \hat{K}_{2a}^{\text{NC}}, \quad \hat{K}_{2a}^{\text{NC}} = \hat{K}_{2a}^{\text{cps}} + \hat{K}_{2a}^{\text{bp}} . \tag{8.133}$$

The effective ion thermal diffusivity including the impurity ion channels is

$$\chi_{i\,(\text{tot})}^{\text{NC}} = \frac{\langle (\boldsymbol{q}_{i\perp} + \boldsymbol{q}_{I\perp}) \cdot \nabla\rho \rangle}{\left\langle |\nabla\rho|^2 \right\rangle (n_i + n_I)\, \mathrm{d}T_i/\mathrm{d}\rho} = \sqrt{\varepsilon}\rho_{pi}^2 \nu_{ii} \left[f_i \hat{K}_{2i}^{\text{NC}} + (1 - f_i)\alpha \hat{K}_{2I}^{\text{NC}} \right] . \tag{8.134}$$

Application of this form to pure electron–ion plasma without impurities confirms that χ_i^{NC} given by Chang–Hinton is correct [16], while application of this formula to electron–ion–impurity plasma [17] gives χ_i^{NC} lower than that given by Chang–Hinton with impurities [18].

References

1. Hirshman SP and Sigmar DJ (1981) Nucl. Fusion, 21, 1079–1201.
2. Chew GF, Goldberger ML, Low FE (1956) Proc. Roy. Soc., A236, 112.
3. Hirshman SP (1977) Phys. Fluids, 20, 589.
4. Kikuchi M, Azumi M (1995) Plasma Phys. Contr. Fusion, 37, 1215–1238.
5. Spitzer L Jr. (1962) Physics of Fully Ionized Gases. 2nd edn. Interscience Publishers.
6. Galeev AA and Sagdeev RZ (1972) Nuclear Fusion Supplement, p. 45.
7. Hinton FL, Hazeltine RD (1976) Rev. Mod. Phys., 48, 239–308.
8. Kikuchi M, Azumi M, et al. (1990) Nucl. Fusion, 30, 343–355.
9. Start DFH, Cordey J (1980) Phys. Fluids, 23, 1477.
10. Connor JW, Cordey JG (1974) Nucl. Fusion, 14. 185.
11. Cordey J (1976) Nucl. Fusion, 16, 499.
12. Gaffey JD Jr. (1976) J. Plasma Phys., 16, 149
13. Tani K, Azumi M, Devoto RS (1992) J. Comp. Phys., 98, 332.
14. Zhu W, Sabbagh SA, et al. (2006) Phys. Rev. Lett., 96, 225002.
15. Kim YB, Diamond P, Groebner RJ (1991) Phys. Fluids, B3, 2050–2060.
16. Chang CS, Hinton FL (1982) Phys. Fluids, 25, 1494.
17. Kikuchi M, Shirai H, Itakura H, Azumi M (1998) JAERI-Research, 98-039, 43–46.
18. Chang CS, Hinton FL (1986) Phys Fluids, 29, 3314.

Chapter 9
Turbulence in the Plasma: Self-organized Criticality and Its Local Breakdown

Turbulent transport is at the frontier of plasma confinement research and is still under development. Here, the basic concepts of the dynamical system are introduced as a useful fundamental discipline. Thermal diffusion of the low confinement state can be explained by the self-organized criticality examined by complexity science and the transport barrier is formed by the local breakdown of self-organized criticality when a turbulent cell is torn by the flow shear.

9.1 Concepts of Nonlinear Dynamics: Dynamical System and Attractor

The concept of "attractor" in the dynamical system is useful in understanding the turbulence generated in high temperature plasma. A system evolving according to the deterministic law is called a "dynamical system." The motion of the dynamical system is expressed in the phase space, which consists of the variables of the equation of motion. Even if the system is not related to physical dynamics, it is still called a dynamical system if the state follows the deterministic law. A dynamical system with some conserved quantity is called a "conservative system," while a system with some quantity being dissipated is called a "dissipative system." The motion of a dissipative system settles down to a specific trajectory or point in time. The stable state after the transition is called the attractor. There are four types of attractors (Figure 9.1):

1. Equilibrium point: the motion converges to a point in phase space.
2. Limit cycle: repeat the periodic motion in phase space.
3. Torus: Motion in phase space around the torus.
4. Chaotic attractor: phase space orbit never closes.

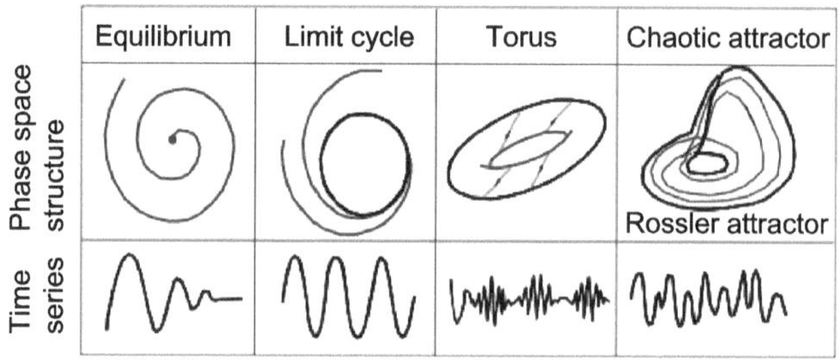

Figure 9.1 Phase space structures and time waveforms of attractors in the dynamical system

Even in the following simple linear oscillation system, the motion around the equilibrium point shows interesting variations in the parameter space (b, c), [1] (Figure 9.2).

$$\frac{\mathrm{d}^2 X}{\mathrm{d}t^2} + b\frac{\mathrm{d}X}{\mathrm{d}t} + cX = 0\,. \tag{9.1}$$

The phase diagram $(X, \dot{X})$ of this system has a saddle point at $(0, 0)$ (unstable equilibrium point) in the case where $c < 0$ and the damping term $b = 0$. Adding the nonlinear term $eX^3 (e > 0)$ to above equation, two equilibrium points (point attractors) appear at $X = \pm(-c/e)^{1/2}$ and motion around each one and encircling the two develops. If the damping term is added, the system converges asymptotically toward one of equilibrium points. Addition of the nonlinear term $e(\mathrm{d}X/\mathrm{d}t)^3$ also produces the limit cycle in the region $c > 0, b < 0$. Negative linear damping ($b < 0$) produces an outward spiral orbit from $(0, 0)$ and reaches a stable limit cycle orbit dominated by the nonlinear term. In the planar phase space, the equilibrium and the limit cycle are the only two final states of the dynamical system. This comes from the simple fact that the plane is divided into an inner region and an outer region by a closed curve.

An important 2-dimensional phase space structure that differs from the planar phase space is the torus. A torus naturally occurs due to the forced vibration when the external force acts on the system. There are two quasi-periodic motions caused by the two low frequencies. Such a system is structurally unstable for small perturbations by the frequency locking when the ratio of two frequencies becomes an integer.

The stability of the equilibrium point can be described by "Lyapunov stability," mentioned in Section 6.1. According to Lyapunov, a system is stable if the neighborhood of the equilibrium remains as the neighborhood forever. The system is called "asymptotically stable" if the solution converges to an equilibrium point as $t \to \infty$. The system is said to be "structurally stable" when the topology of the flow does not change when a small perturbation is applied.

When a system parameter μ changes, the eigenvalue of the linearized evolution equation $\dot{\boldsymbol{x}} = F(\boldsymbol{x}, \mu)$ draw a trajectory in the complex plane. If the eigenvalue stays in the left half of the plane for $\mu = \mu_0$, "bifurcation" occurs when the eigenvalue crosses the imaginary axis.

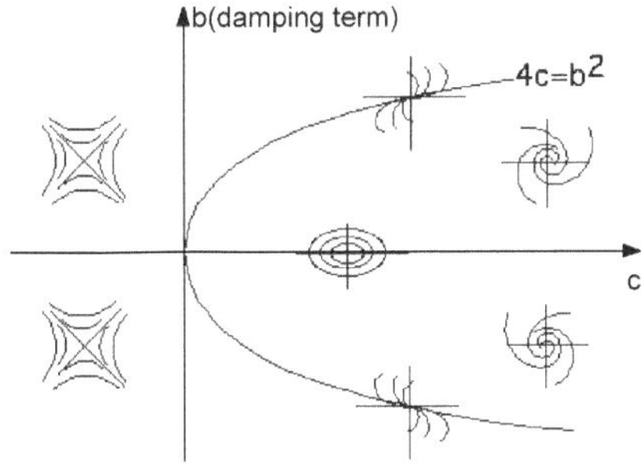

Figure 9.2 Phase-space structure on the parameter space (c, b) for the linear dynamic system Equation 9.1

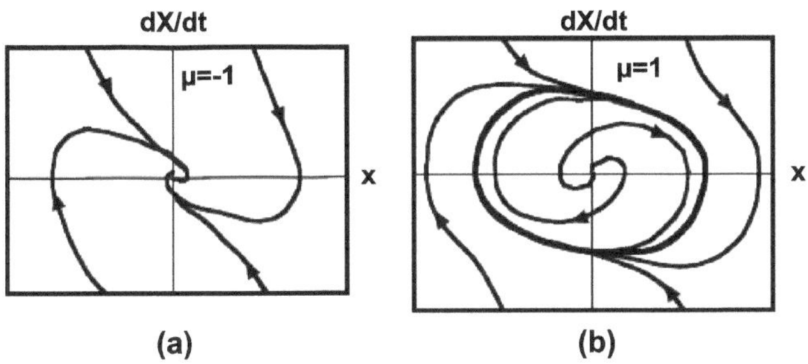

Figure 9.3 Phase diagrams of Equation 9.2 (a) before ($\mu = -1$) and (b) after ($\mu = 1$) Hopf bifurcation. The case of $\mu = 1$ is the limit cycle

If the system has only one eigenvalue, it moves along the real axis. Bifurcation occurs when one crosses the imaginary axis and is called "fold bifurcation." Fold bifurcation is a structurally stable bifurcation. If the system has two eigenvalues, they are complex conjugate and the bifurcation occurs when they cross the imaginary axis; this bifurcation is called "Hopf bifurcation." As an example of Hopf bifurcation, consider the following nonlinear damped oscillation system:

$$\frac{\mathrm{d}^2 X}{\mathrm{d}t^2} - \mu \frac{\mathrm{d}X}{\mathrm{d}t} + X + \left(\frac{\mathrm{d}X}{\mathrm{d}t}\right)^3 = 0 \,. \tag{9.2}$$

The origin is asymptotically stable when $\mu < -2$ and is a stable spiral when $-2 < \mu < 0$. Furthermore, the origin is an unstable spiral when $0 < \mu$ and the flow is attracted by the newly formed limit cycle (see Figure 9.3).

9.2 Self-organized Criticality: Turbulent Transport and Critical Temperature Gradient

The study of heat transport across magnetic surfaces has been ongoing since the start of the fusion research. But, mechanisms governing the process have not been

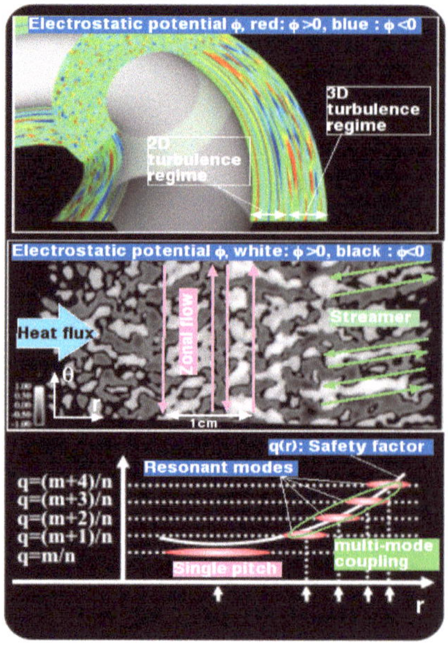

Figure 9.4 Variation of the spatial structure of turbulence with magnetic shear in the electron channel [2]. In the low shear regime, zonal flow is formed by the inverse cascade of 2-dimensional turbulence. In the finite shear regime, streamer is formed due to radial mode coupling

fully clarified since turbulent transport in plasma is a complex and nonlinear process. In recent years, fundamental processes have been elucidated since zonal flow and streamers produced by the wave–wave interaction is clarified through the gyrokinetic and gyrofluid simulations. The existence of the critical temperature gradient (see Section 7.5) has also been shown by experiments using localized electron cyclotron heating.

In plasma confined in a closed magnetic configuration, a large spatial temperature gradient occurs since the plasma core is heated to a few 100 million degrees while the temperature at the boundary of the closed magnetic field is low enough (but still a few million degrees). In high-temperature plasma, the ion temperature gradient drift wave mentioned in Section 7.5 (ITG-drift wave), electron temperature gradient drift wave (ETG-drift wave), and trapped electron mode (TEM) can be unstable. These modes become unstable when the temperature gradient exceeds a critical temperature gradient $(\mathrm{d}T/\mathrm{d}r)_\mathrm{c}$ similar to the case of thermal convection (see Salon 1). This is called the "critical temperature gradient."

An interesting observation in the gyrokinetic simulation of ETG turbulence [2] is that the structure of the turbulence is strongly dependent on magnetic shear as shown in Figure 9.4). In the zero magnetic shear regime (regime around q_{min} in the negative magnetic shear configuration), plasma turbulence has a 2-dimensional structure as described by the Hasegawa–Mima equations in Section 7.3 and excites zonal flow by the inverse cascade of the turbulent spectrum [3]. Meanwhile, a radially stretched "streamer" is created through radial mode coupling and intermittent heat transport is produced in the positive shear region. This situation is exactly the same as avalanches of sand piles.

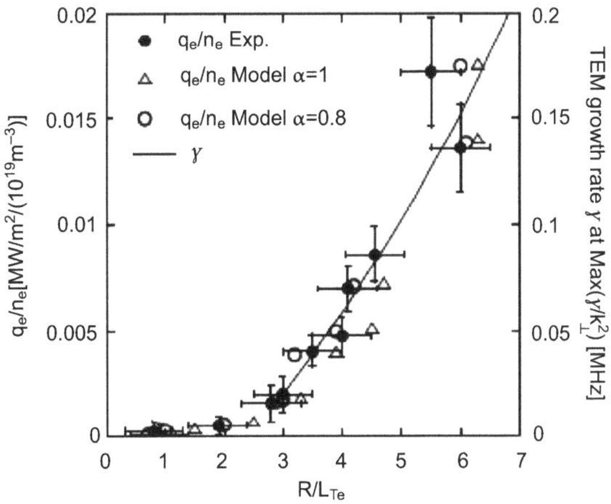

Figure 9.5 Experimental verification of critical temperature gradient transport in a tokamak in $(R/L_{Te}, q_e/n_e)$ plane [4]. At low R/L_{Te}, heat flux is quite small indicating that ETG turbulence is suppressed, while ETG turbulence grows at high R/L_{Te} [4]

In the electron channel, the ETG-drift wave becomes unstable if the electron temperature is close to or lower than the ion temperature, while TEM becomes unstable if the electron temperature is higher than the ion temperature. Figure 9.5 shows experimental observations of electron heat transport [4]. There is a critical temperature gradient $R/L_{Te} \sim 2.5$ $(L_{Te} = -T_e/\mathrm{d}T_e/\mathrm{d}r)$ and transport is very small in $R/L_{Te} < 2.5$, while the transport rate grows rapidly in $R/L_{Te} > 2.5$.

Note: Dissipative Structure and Bernard Convective Cell

A closed magnetic configuration for controlled fusion is a "closed system" in terms of equilibrium but is an "open system" in terms of heat. In the closed system, the equilibrium structure is determined in line with thermal equilibrium. On the other hand, the open system can maintain an off-equilibrium state as long as a driving force exists. The various shapes and motions arising from the driving force of open systems are named "dissipative structures" by I. Prigogine [5]. As a typical example, he uses "Bernard cells." In thermal convection under gravity, the system becomes linearly unstable when a non-dimensional "Rayleigh number" R related to the temperature difference ΔT exceeds a critical value R_c and a convective cell structure appears [6]:

$$\frac{\partial u}{\partial t} + \frac{1}{P_r}(u \cdot \nabla u) = -\frac{1}{P_r}\nabla p + \nabla^2 u + RkT \, , \tag{9.3}$$

$$P_r \frac{\partial T}{\partial t} + (u \cdot \nabla)T = \nabla^2 T \nabla \cdot u = 0 \, . \tag{9.4}$$

Here, $R = (g\alpha d^4/\kappa\nu)|\Delta T/\Delta z|$ is the Rayleigh number, $P_r = \kappa/\nu$ is the Prandtl number, g gravitational acceleration, α thermal expansion coefficient, κ thermal conductivity, ν dynamic viscosity, d the vertical thickness, and $\Delta T/\Delta z = (T_1 - T_2)/d$.

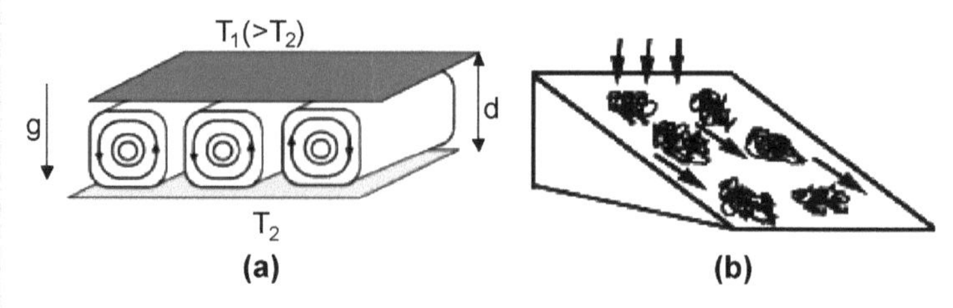

Figure 9.6 (a) Bernard convective cell as dissipative structure and (b) sand pile as self-organized criticality

Critical Phenomena and Self-organized Criticality

Critical phenomena are observed in the phase transition of thermal equilibrium, for example, the paramagnetic-ferromagnetic transition in magnetism. Magnetization defined in the crystal lattice u_j is given by $u_j = 0$ at $T > T_c$ and $u_j = a(T - T_c)$ at $T < T_c$. The spatial correlation length of the magnetization is $\sim 1/\kappa$, while $\kappa = \kappa_0((T - T_c)/T_c)^{1/2}$ and is divergent if $T \to T_c$. Namely, the effect of fluctuation in the critical state extends over a long distance. The statistical physics tells us that fluctuation in thermal equilibrium follows a Gaussian distribution. If you check the process that leads to the Gaussian from a random walk, it can be seen that Gaussian distribution $P(x) = \exp(-bx^2)$ is a special case.

The power law $P(x) = x^{-n}$ frequently appears in non-equilibrium open systems. Sand pile avalanches occur when the sand pile reaches a certain critical slope and the relation between the magnitude and frequency is given by the power law x^{-2}. Per Bak (1948–2002) [7] called this state as "self-organized criticality." The frequency and scale of earthquakes (relation between magnitude M and frequency (occurrence/year) of earthquakes above M) follows the famous Gutenberg–Richter law f^{β} ($\beta \sim 0.94$).

9.3 Chaotic Attractor: Three-wave Interaction in Drift Wave Turbulence

As mentioned in Section 7.3, drift wave turbulence can be regarded as the result of many three-wave interactions. Plasma turbulence is a natural (or settled) state with oscillations due to nonlinear processes rather than growing instability. It is important to examine the phase space structure to understand its behavior. The saturated state of plasma turbulence dominated by the three-wave interaction has a "chaotic attractor" structure. Fourier expansion of the electrostatic potential is given by

$$\tilde{\Phi}(\boldsymbol{x}, t) = \sum_{\boldsymbol{k}} \tilde{\Phi}_{\boldsymbol{k}}(t) \exp(\mathrm{i}\boldsymbol{k} \cdot \boldsymbol{x}) \,. \tag{9.5}$$

Then, the nonlinearity is given by the interaction of the wave with wave number $\boldsymbol{k}_1$ and $\boldsymbol{k}_2$ as follows,

$$\boldsymbol{k} = \boldsymbol{k}_1 + \boldsymbol{k}_2 \,. \tag{9.6}$$

Although $k_{//} \neq 0$ is assumed, the Hasegawa–Mima equation 7.80 is essentially 2-dimensional turbulence and can be rewritten by transforming $\boldsymbol{k}_\perp \to \nabla_\perp$ as follows,

$$(1 - \rho_s^2\nabla^2)\frac{\partial\tilde{\Phi}}{\partial t} + v_{de}\frac{\partial\tilde{\Phi}}{\partial y} - \left[\tilde{\Phi}, \rho_s^2\nabla^2\tilde{\Phi}\right] = 0 \tag{9.7}$$

$$\rho_s^2 = \frac{m_i T_e}{e^2 B^2}\,, \quad \left[\tilde{\Phi}, \tilde{\Psi}\right] = \boldsymbol{e}_z \cdot \nabla\tilde{\Phi} \times \nabla\tilde{\Psi} \,. \tag{9.8}$$

The adiabatic response (Boltzmann distribution) for electrons (Equation 7.68) is assumed in the Hasegawa–Mima equation and is an essentially conservative system. The mass $\mathcal{M}$, energy $\mathcal{W}$, and potential enstrophy $\mathcal{U}$ of the turbulence are conserved [8].

$$\mathcal{M} = \int [\tilde{\Phi} - \rho_s^2\nabla_\perp^2\tilde{\Phi}]\mathrm{d}x\mathrm{d}y \,, \tag{9.9}$$

$$W = \frac{1}{2}\int \left[\tilde{\Phi}^2 + \rho_s^2\left(\nabla_\perp\tilde{\Phi}\right)^2\right]\mathrm{d}x\mathrm{d}y \,, \tag{9.10}$$

$$U = \frac{1}{2}\int \left[\left(\nabla_\perp\tilde{\Phi}\right)^2 - \rho_s^2\left(\nabla_\perp^2\tilde{\Phi}\right)^2\right]\mathrm{d}x\mathrm{d}y \,. \tag{9.11}$$

The 2-dimensional turbulence equation with dissipation is given by Hasegawa–Wakatani [9]. Here, we show this effect according to Horton–Ichikawa [10]. Two-dimensional drift wave turbulence including dissipation is expressed by adding a dissipation term to the Hasegawa–Mima equation (9.7).

$$(1 + L)\frac{\partial\tilde{\Phi}}{\partial t} + v_{de}\frac{\partial\tilde{\Phi}}{\partial y} + \hat{\gamma}_i\tilde{\Phi} + \left[\tilde{\Phi}, L\tilde{\Phi}\right] = 0 \tag{9.12}$$

$$L = L_h + L_{ah} = -\rho_s^2\nabla^2 + \delta_0(c_0 + \rho_s^2\nabla^2)\frac{\partial}{\partial y} \,.$$

Here, $\hat{\gamma}_i$ is the ion Landau damping and the wave decay rate by the coupling to the decay wave. $L_h = -\rho_s^2\nabla^2$ is the Hermitian operator, and the effect of dissipation is given by the non-Hermitian operator $L_{ah} = \delta_0(c_0 + \rho_s^2\nabla^2)\partial/\partial y$.

We introduce the electrostatic potentials of interacting three-drift waves $\varphi_j(t) = \varphi_{\boldsymbol{k}_j}(t)$ $(j = 1, 2, 3)$ by using amplitude a_j and phase ζ_j as follows,

$$(1 + k_j^2)^{1/2}\varphi_{\boldsymbol{k}_j}(t) = a_j(t)\exp[i\zeta_j(t)] \,. \tag{9.13}$$

Then, the following equations for three-wave amplitudes a_1, a_2, a_3 and total phase $\zeta = \zeta_1 + \zeta_2 + \zeta_3$ can be obtained for (j, k, l) 3 wave interaction.

$$\frac{\mathrm{d}a_j(t)}{\mathrm{d}t} = \gamma_j a_j - A a_k a_l (F_j \cos\zeta + G_j \sin\zeta) \,, \tag{9.14}$$

$$\frac{d\zeta}{dt} = -\Delta\omega + A \sum_{jkl} \frac{a_k a_l}{a_j} (F_j \sin\zeta - G_j \cos\zeta) . \tag{9.15}$$

Here, $\Delta\omega = \omega_1 + \omega_2 + \omega_3$. If there is no dissipation, Equations 9.14 and 9.15 are integrable (conserved quantity exists). If there is dissipation, they are not integrable and become chaotic. If there is a fixed point at which the system settles, it is determined by $\dot{a}_j = 0$ and $\dot{\zeta}_j = 0$.

The above discussion is the case for 2-dimensional turbulence. As is well know, there is an essential difference between 2-dimensional and 3-dimensional turbulence. As introduced in Section 9.2, a 2-dimensional turbulence structure is observed at zero magnetic shear for the electron. With magnetic shear, turbulence becomes 3-dimensional due to poloidal coupling of the modes and creates the unique structure of a streamer.

Note: 2- and 3-dimensional isotropic turbulence

Drift wave turbulence in the plasma does not have a single wave number k in the wave number space but has a spectrum in k space. If there is a driving force of the turbulence at some k region, energy flow occurs in the direction of a larger k in k space. The plasma has the characteristics of a fluid and the nonlinear term $\boldsymbol{u} \cdot \nabla \boldsymbol{u}$ in the fluid equation produces a wave with a wave number twice that of initial wave and produces energy flow toward the high k region.

The turbulence energy entering the small wave number region passes through the inertia region where dissipation is negligible and reaches the high wave number region where wave energy is dissipated as thermal energy, the dissipation region. Homogeneous isotropic turbulence is well known with a known spectrum which does not vary with translation, rotation and reflection. The spectrum of the inertia region of homogeneous isotropic turbulence is given by Kolmogorov [11] and is called the "Kolmogorov spectrum":

$$F(k) = Ck^{-5/3} . \tag{9.16}$$

In confined plasma, turbulence may develop perpendicular to the magnetic field because of the magnetic field. In this case, turbulence may become 2-dimensional and show quite a different character compared with 3-dimensional homogeneous isotropic turbulence:

1. Energy flow is from high k to low k, the opposite of 3-dimensional turbulence;
2. k spectrum of homogeneous isotropic 2-dimensional turbulence is $\sim k^{-3}$.

Mode Selection Rule in Three-wave Coupling [12]

If we consider the coherent three-wave interaction, it should satisfy following selection rules:

$$\omega_{k_1} + \omega_{k_2} = \omega_k \,, \tag{9.17}$$

$$k_1 + k_2 = k \,. \tag{9.18}$$

This coherent three-wave interaction becomes possible only when the structure of the dispersion curve $((k, \omega)$ is convex upward since $|\boldsymbol{k}_1 + \boldsymbol{k}_2| = |\boldsymbol{k}| \le |\boldsymbol{k}_1| + |\boldsymbol{k}_2|$ as shown in Figure 9.7.

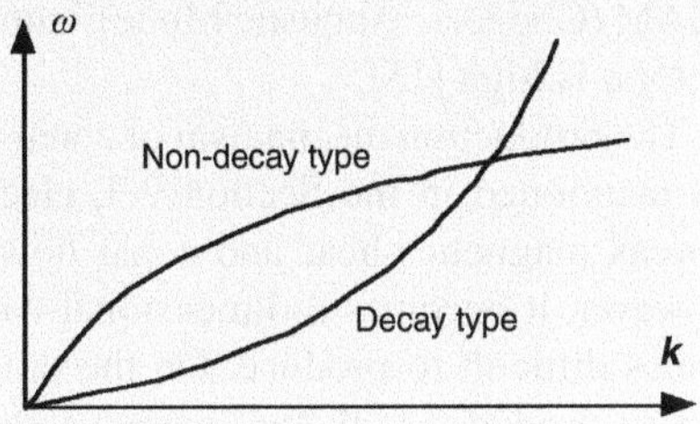

Figure 9.7 Dispersion curve of decay-type and non-decay type waves

9.4 Structure Formation: Turbulence Suppression by Shear Flow and Zonal Flow

Convective cells produced by drift wave turbulence induce heat and particle transport with a step length of the cell size. This convection cell experiences $v_E = E_r/B$ drift due to radial electric field E_r. If the magnitude of v_E changes radially, the displacement of the two edges of the cell is different and the cell is distorted and breaks down by the sheared flow as shown in Figure 9.8. The shearing rate is proportional to dv_E/dr. Poloidal rotation in a tokamak is kept small due to neoclassical viscosity, but v_E itself is not subject to this constraint and the turbulence can be suppressed by the radial electric field shear [13].

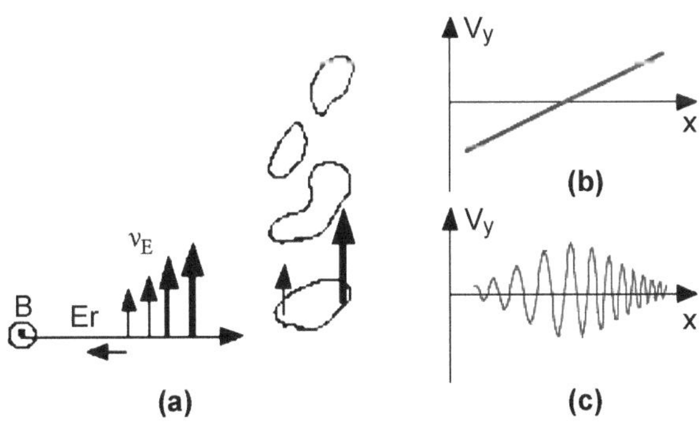

Figure 9.8 (a) Schematics of break up of the cell by the radial electric shear. (b) Schematics of the neoclassical Er shear flow balanced to pressure gradient, and (c) the zonal flow produced by turbulence [16]

While drift wave turbulence is formed by the interaction of a broad wave spectrum, a mode with $m = n = 0$, $k_{//} = 0$ can be created in ITG turbulence or ETG/TEM turbulence in the case of zero magnetic shear through the interaction of similar wave numbers. This is called "zonal flow." According to Rosenbluth–Hinton [14], the zonal flow can remain in a collisionless plasma and suppress the level of turbulence. When potential fluctuation in an $m = n = 0$ mode coupled with $m = 1, n = 0$ mode of the toroidal effect, a low frequency oscillation called "GAM (Geodesic Acoustic Mode)" appears. This mode is observed when the safety factor q is high [15].

The production mechanism of zonal flow is different between ions and electrons. As mentioned in the Section 9.3, electron turbulence becomes 2-dimensional in a weak magnetic shear and zonal flow is created if a density gradient exists [3]. However, it becomes 3-dimensional with a magnetic shear and the zonal flow becomes difficult to produce. On the other hand, in ITG turbulence, the zonal flow is also produced with the magnetic shear. In a tokamak, the drift wave also has the structure of the ballooning eigen function from the constraints of the double-periodic boundary conditions. Uniform amplitude approximation (see Section 6.6) holds for the pump wave of drift wave turbulence as follows,

$$\tilde{\Phi}_0(\boldsymbol{x},t) = \exp(-in\zeta - \mathrm{i}\omega_0 t)\sum_m \tilde{\varphi}_0(m-nq)\exp(im\theta) + \text{c.c.} \tag{9.19}$$

Then, zonal flow ($\tilde{\Phi}_{\text{ZF}}$) and side bands ($\tilde{\Phi}_+$ and $\tilde{\Phi}_-$) are created by "modulational instability" [16]. Let q_r be the radial wave number,

$$\tilde{\Phi}_{\text{ZF}}(\boldsymbol{x},t) = \exp(\mathrm{i}q_r r - \mathrm{i}\Omega t)\tilde{\varphi}_{\text{ZF}} + \text{c.c.}\,, \tag{9.20}$$

$$\tilde{\Phi}_+(\boldsymbol{x},t) = \exp(-in\zeta - i\omega_0 t + \mathrm{i}q_r r - \mathrm{i}\Omega t)\sum_m \tilde{\varphi}_+(m-nq)\exp(\mathrm{i}m\theta) + \text{c.c.} \tag{9.21}$$

$$\tilde{\Phi}_-(\boldsymbol{x},t) = \exp(-in\zeta + \mathrm{i}\omega_0 t + \mathrm{i}q_r r - \mathrm{i}\Omega t)\sum_m \tilde{\varphi}_-(m-nq)\exp(\mathrm{i}m\theta) + \text{c.c.} \tag{9.22}$$

Growth of zonal flow is governed by the following equation:

$$\frac{\partial V_{\theta,\text{ZF}}}{\partial t} = \frac{\partial}{\partial r}\langle \tilde{v}_\theta \tilde{v}_r\rangle - \gamma_{\text{damp}} V_{\theta,\text{ZF}}\,. \tag{9.23}$$

Here, $V_{q\text{ZF}}$ is the zonal flow velocity, $\tilde{v}_\theta$ and $\tilde{v}_r$ are the poloidal and radial velocity fluctuations, γ_{damp} is the damping rate of the zonal flow. The first term of right-hand side of Equation 9.23 is called "Reynolds stress." Zonal flow and GAM are observed experimentally in toroidal plasma considered to be the main mechanism for improved confinement [17]. Since the first observation of the internal transport barrier (ITB) in JT-60 [18], formation of various ITB are observed as shown in

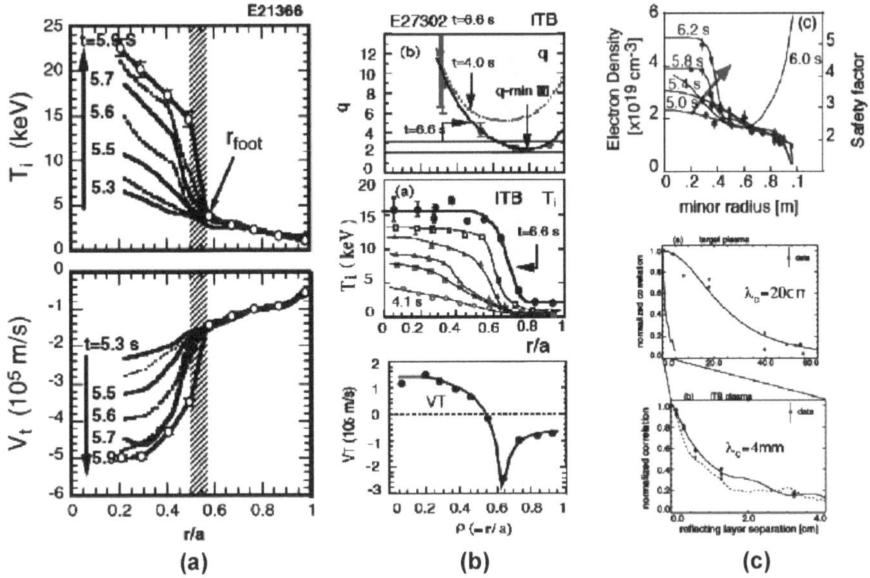

Figure 9.9 (a) Formation of ITB in the positive magnetic shear in JT-60 [19]. (b) Formation of ITB in negative magnetic shear in JT-60 [20]. (c) Measured radial correlation length before and after ITB formation [21]

Figure 9.9 [19, 20]. The long correlation length of L-mode is observed as a common characteristic of self-organized criticality [21]. It is also observed that correlation length becomes quite short due to the ITB formation. Thus, self-organized criticality is locally relaxed due to the formation of E_r shear and "structure formation" ITB occurs in the plasma.

References

1. Thompson JM and Stewart HB (1986) Nonlinear Dynamics and Chaos: Geometrical Methods for Engineers and Scientists. John Wiley & Sons Ltd.
2. Idomura Y, Tokuda S, Kishimoto Y (2005) Nucl. Fusion, 45, 1571–1581; also Idomura Y, Wakatani M and Tokuda S (2000) Phys. Plasmas, 7, 3551–3556.
3. Idomura Y (2006) Phys. Plasmas, 13, 080701.
4. Ryter F, Angioni C, et al. (2005) Phys. Rev. Lett. 95, 085001; also Ryter F, Tardini G et al. (2005) Nucl. Fusion, 43, 1396–1404.
5. Nicolis G and Prigogine I (1977) John Wiley and Sons.
6. Chandrasekhar S (1961) Hydrodynamics and Hydromagnetic Stability. Dover Books.
7. Bak P et al. (1988) Phys. Rev., A38, 364–374; also Bak P et al. (1987) Phys. Rev. Lett., 59, 381–384.
8. Hasegawa A, Mima T (1978) Phys. Fluids, 21, 87–92.
9. Hasegawa A, Wakatani M (1987) Phys. Rev. Lett., 59, 1581–1584.

10. Horton W, Ichikawa Y-H (1996) Chaos and Structures in Nonlinear Plasmas. World Scientific.
11. Kolmogorov AN (1941) Doklady Akad. Nauk SSSR, 30, 9.
12. Hasegawa A (1975) Plasma Instabilities and Nonlinear Effects. Springer-Verlag Berlin, Section 4.3 b.
13. Biglari H, Diamond PH, Terry PW (1990) Phys. Fluids, B2, 1.
14. Rosenbluth MN, Hinton F (1998) Phys. Rev. Lett., 80, 724.
15. Miyato N, Kishimoto Y, Li YJ (2007) Nucl. Fusion, 47, 929.
16. Diamond PH, Itoh S-I, Itoh K, Hahm TS (2005) PPCF, 47, R35.
17. Fujisawa A (2009) Nucl. Fusion, 49, 1.
18. Koide Y, Kikuchi M et al. (1994) Phys. Rev. Lett., 72, 3662.
19. Koide Y, Ishida S, M. Kikuchi M et al. (1994) Proc. 15th FEC, Seville, 1994, Vol. 1, 199–210.
20. Fujita T et al. (1998) Nucl. Fusion, 38, 207.
21. Nazikian R, Shinohara K, Kramer GJ et al. (2005) Phys. Rev. Lett., 94, 135002.

Chapter 10
Towards the Realization of Fusion Energy

In this chapter, energy research and development in terms of fusion is overviewed, including energy and environment issues, progress of plasma confinement in the three major areas of fusion research, namely tokamak and helical by magnetic confinement, and laser fusion by inertial confinement. The experimental reactor ITER is aiming at DT fusion energy production and its control via the tokamak concept is described. A the broader approach is supplementing ITER towards DEMO. The possible role of fusion is also considered in terms of transforming the energy supply and demand structure with the goal of reducing carbon dioxide emissions during this century.

10.1 Energy, Environmental Problems, and Fusion Energy

Today, fusion energy is under the spotlight again as a carbon-free energy source. Planet Earth is the only planet on which life exists, as far as we know. Earth was formed 4.6 billion years ago, the geomagnetic field and the ozone layer were created shielding most of the harmful radiation from the Sun and realizing an environment where only the energy of light reaching the ground was useful for life. On the other hand, 4.6 billion years ago, Earth was at a high temperature due to the greenhouse effect of carbon dioxide at a few tens of atmospheres, which was reduced to 0.0003 atmospheres and produces the temperature environment of living plants and animals (present atmospheric carbon is only 750 billion tons).

Fossil fuel resources are from carbon fixed into the earth over billions of years. In the history of human energy use, the twentieth century was a critical turning point, as shown in Figure 10.1. Mechanized civilization initiated by the Industrial Revolution (1760–1800) resulted in large-scale energy consumption and caused the expansion of energy usage from energy such as firewood and charcoal to coal, oil, and natural gas. As the use of fossil fuel resources has expanded significantly in developed countries, large amounts of carbon dioxide (6.32 billion tons in 1999) are emitted, as shown in Figure 10.2. This is just the beginning. An era of huge energy consumption is beginning in the twenty-first century because the developing

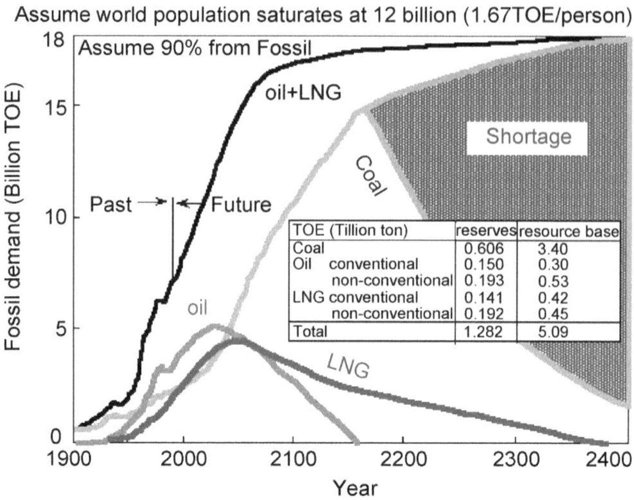

Figure 10.1 Energy supply scenario by fossil resources [1]. Table shows remaining resources (reserves and resource base)

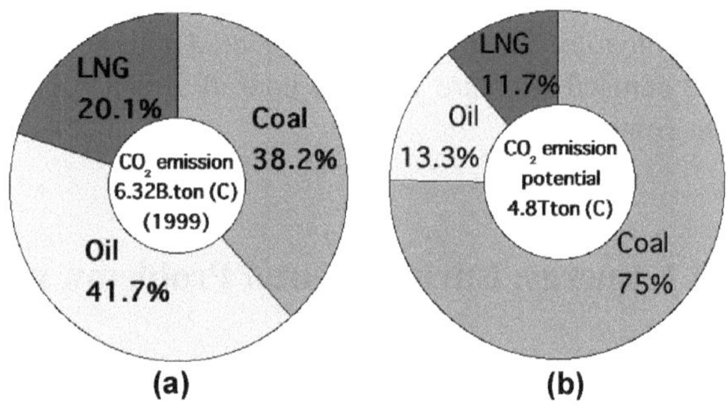

Figure 10.2 (a) Carbon dioxide emissions in 1999 and (b) potential carbon dioxide emissions of remaining resources [2]

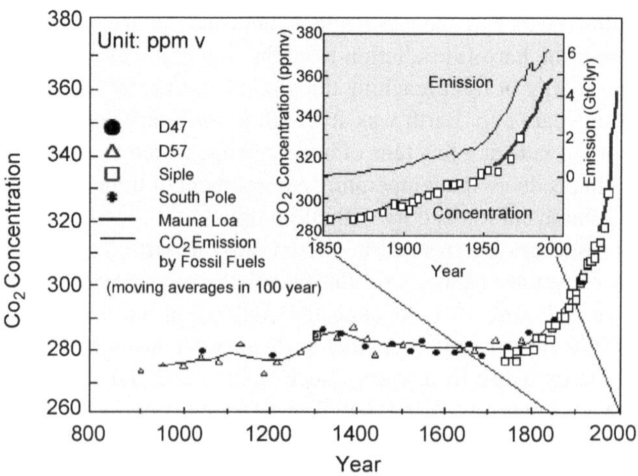

Figure 10.3 Changes of carbon dioxide concentration in the atmosphere for past 1,000 years

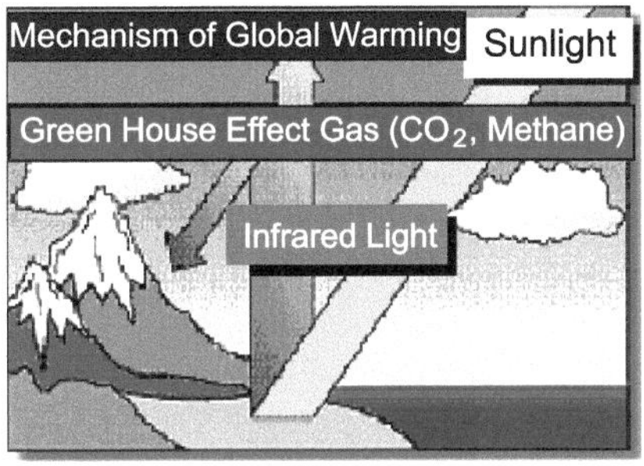

Figure 10.4 Schematics of the mechanism of global warming by greenhouse effect

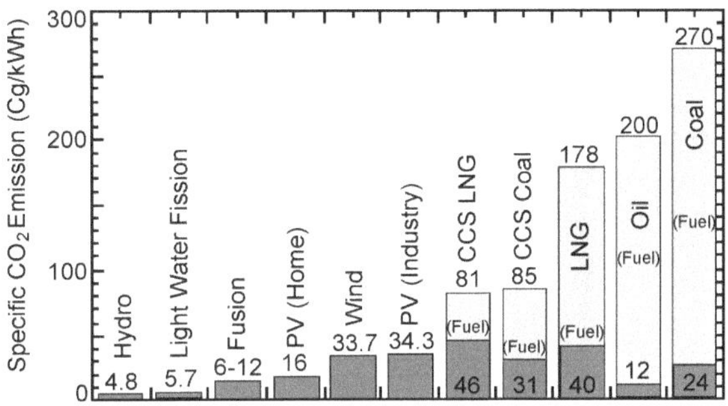

Figure 10.5 Specific carbon dioxide emissions of fired (coal, oil, natural gas), renewables (solar, wind), fission and fusion plants (*PV*: photovoltaics; *CCS*: carbon capture and storage; *LNG*: liquefied natural gas)

countries that are home to three fourths of the world's population will increase their energy use pursuing the rich living of developed countries. Fossil energies, including conventional and unconventional sources, measure about five trillion tons, and will produce 6.4 times atmospheric carbon dioxide if they are burned.

Emission of carbon dioxide into the atmosphere due to huge energy consumption is influencing the environment of Earth. About half of the carbon dioxide emissions to the atmosphere remain in the atmosphere. Charles David Keeling started taking measurements of atmospheric carbon dioxide concentrations at the Mauna Loa Observatory at the summit of Mt. Mauna Loa, Hawaii, in 1957, as shown in Figure 10.3. These showed that carbon dioxide concentration was rising steadily. He was a researcher at the Scripps Institute of Oceanography, La Jolla, near San Diego, California, and started these observations to get basic data to investigate the absorption of carbon dioxide by the ocean. This led to this important discovery.

In 1988, the Intergovernmental Panel on Climate Change (IPCC) was established under the United Nations (UN) as an intergovernmental organization monitoring global warming. It published four reports in 1990, 1995, 2001, and 2007. Figure 10.3 shows the variation of carbon dioxide concentration in the air over the past 1,000 years from the second IPCC report. Furthermore, it was agreed in the fourth report in 2007 that the increase in greenhouse gas emissions associated with the expansion of human economic activity is a major cause of climate change The effect of greenhouse gases on global warming is shown schematically in Figure 10.4.

To prevent global warming, a low carbon society free from fossil energy must be achieved during the twenty-first century. Although attention was drawn to electricity

Figure 10.6 Comparison of fired, fission and fusion electricity plants in terms of radiation hazard potential and specific carbon dioxide emission

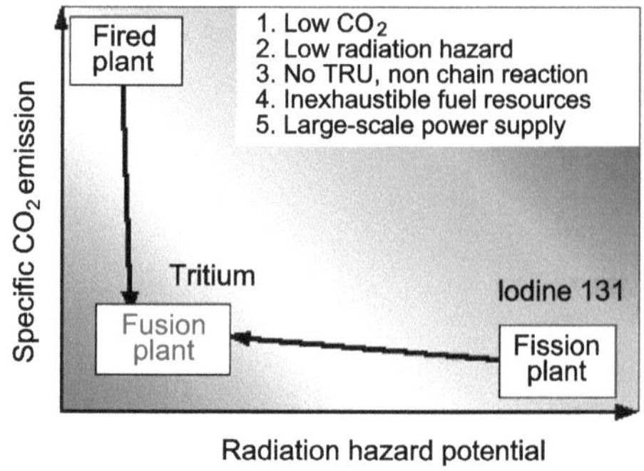

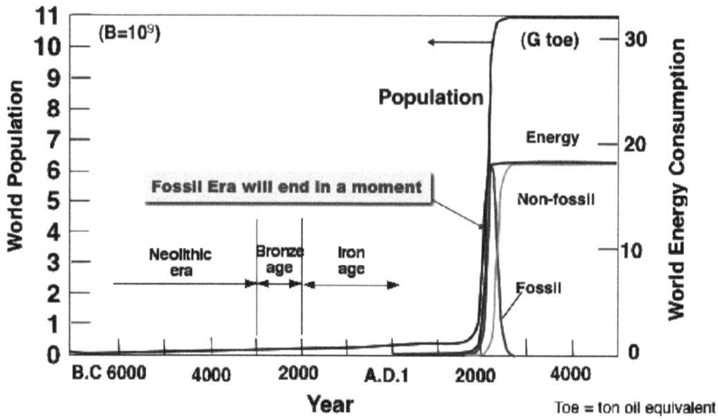

Figure 10.7 Energy consumption and population trends since the Neolithic era [5]

and hydrogen power as clean energy, low carbon dioxide emission energy sources must be used to create the electricity and hydrogen. Carbon dioxide emissions per electric power kilowatt · hour is an important consideration and is called "specific carbon dioxide emission." As shown in Figure 10.5, fusion is one of the energy sources that has a low specific carbon dioxide emission next to hydroelectricity and light water reactors.

In light water reactors and fast reactors, iodine 131 (^{131}I) is produced in the reaction. This tends to accumulate in certain organs of the human body, and the concentration limit in the air (tolerance) is set to a very low level. The hazard potential of ^{131}I in 1 GW fission power station is about 1,000 times larger than that of tritium (T) in a 1 GW fusion power station (Figure 10.6). Therefore, fusion energy has favorable characteristics in terms of reducing the radiation hazard.

In the long history of humankind, the current era of mass energy consumption is only a moment (Figure 10.7). The transition from the fossil energy era to a non-fossil energy era must be made now while fossil energy still exists. To rely on a single energy source is not desirable for energy security as experienced in the case of oil shock. The source of energy supply must be diversified to ensure energy security. In this regard, it is desirable to achieve a long-term energy supply for humankind by promoting the use of renewable energy, fission energy, and fusion energy.

10.2 Status of Major Three Fusion Concepts and the Progress of Plasma Confinement

Plasma temperature must be increased to a few hundred million degrees for the DT fusion reaction mentioned in Chapter 2. Not only temperature but also the energy gain as a ratio of the fusion reaction energy to input energy from outside to maintain a high temperature (energy multiplication factor) must be large enough for fusion plasma conditions. In this regard, the time constant of the thermal insulation of the plasma thermal energy (τ_E: the energy confinement time) multiplied by the fuel density (ion density) ($n_i \tau_E$) affecting reaction rate becomes an important parameter. Since reactions tend to occur in the plasma core, it can be diagrammatically represented with the central ion temperature as the horizontal axis and product of confinement time and central ion density as the vertical axis to show the fusion plasma condition. This figure is called a "Lawson diagram" as introduced by British physicist J. D. Lawson. The condition to obtain $Q = 1$ or reactor relevant Q is different between magnetic confinement fusion (MCF) and inertial confinement fusion (ICF) since power production for the latter is essentially pulsed in nature while magnetic fusion is diffusively quasi-steady. So, the required $n_i \tau_E$ is somewhat higher in ICF as shown in Figure 10.8. Also, the energy gain necessary for the power reactor is higher in ICF ($Q \sim 100$), while $Q \sim 30$ is required for MCF. Right now, both helical and laser fusion still require breakthroughs to achieve high $n_i \tau_E$ at reactor relevant high temperatures. Tokamak has already achieved the plasma condition re-

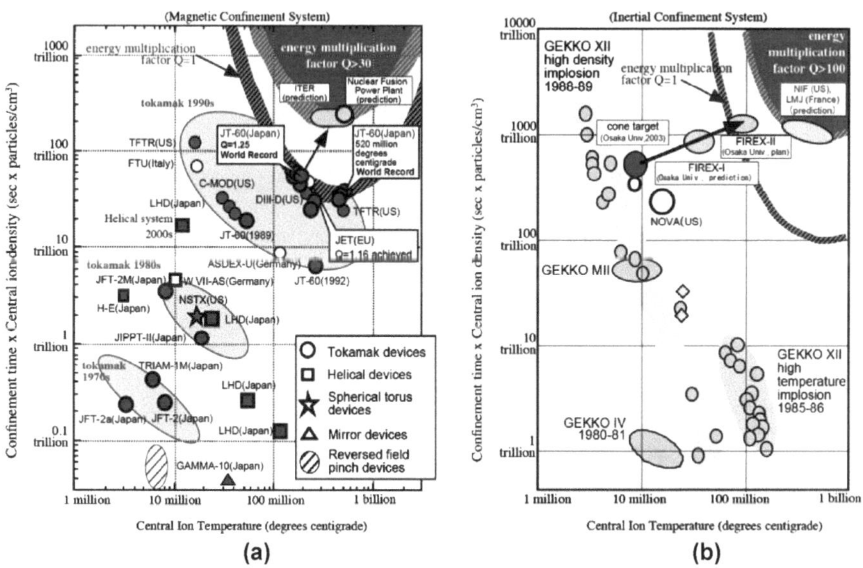

Figure 10.8 Major three schemes of fusion (tokamak, helical, laser) and progress in Lawson diagram for (a) MCF (tokamak, helical, *etc.*) and (b) ICF [6]

(a) (b) (c)

Figure 10.9 Large tokamaks, (a) JET (EU) (Courtesy of EFDA-JET), (b) JT-60 (Japan), and (c) TFTR (US) (Courtesy of Dr. R. Hawryluk, Princeton Plasma Physics Laboratory)

(a) (b)

Figure 10.10 Inside views of (a) DIII-D (Courtesy of Dr. T. Taylor, General Atomics) and (b) ASDEX-U (Courtesy of Prof. H. Zohm, Max Planck Institute for Plasma Physics) tokamak vacuum vessel

Figure 10.11 EAST tokamak (Courtesy of Institute of Plasma Physics, Chinese Academy of Sciences, Hefei, China) [7]

Figure 10.12 KSTAR tokamak (Courtesy of National Fusion Research Institute, Daejon, Korea) [8]

quired for $Q \sim 1$ and has a good prospect for achieving reactor relevant high Q in ITER.

As seen from Figure 10.8, the performance of tokamak plasma confinement improved rapidly with increasing device size during the 1970s, 1980s, and 1990s reaching the equivalent energy multiplication factor $Q^{eq} = 1$ (equivalent value because there is no actual fuel). The Japanese large tokamak device JT-60 remains at the forefront in terms of both Q^{eq} and Ti(0). There is the JET (Joint European Torus) in the UK and TFTR (Tokamak Fusion Test Reactor) in Princeton Plasma Physics Laboratory in the United States, which have a similar performance to JT-60. Together these are referred to as the three large tokamaks and are shown in Figure 10.9.

Tokamak devices following the large tokamaks are the DIII-D device of General Atomics (constructed under US–Japan cooperation and modified in the United States) and ASDEX-U of the Max Planck Institute für Plasma Physics (Garching), shown in Figure 10.10, and the Alcator C-MOD at MIT in the United States. The plasma performance of DIII-D is actually quite high and high quality R&D is being done for ITER and advanced tokamak.

Based on these results, the ITER project, in which 500 MW DT fusion heat will be produced, has begun. In recent years, China, Korea, and India subsequently constructed superconducting tokamaks for long pulse operation (China: EAST, Figure 10.11; Korea: KSTAR, Figure 10.12; India: SST-1) to contribute to the ITER project.

The helical system has shown good confinement, second to tokamak. Based on sequential works on heliotron lines, the world largest helical device LHD was build at National Institute for Fusion Science of NINS and has achieved world best performance among helical devices (Figure 10.13 (a)).

In Germany, advanced helical device Wendelstein VII-X using optimization methodology different from LHD (Large Helical Device) is being constructed at the Max Planck Institute für Plasma Physics, Greifswald, and is expected to start

(a) (b)

Figure 10.13 (a) Large Helical Device (LHD) (National Institute for Fusion Science, Toki, Japan) [9] and (b) Wendelstein VII-X (Courtesy of Max Planck Institute für Plasma Physics Laboratory, Germany) [10]

(a)

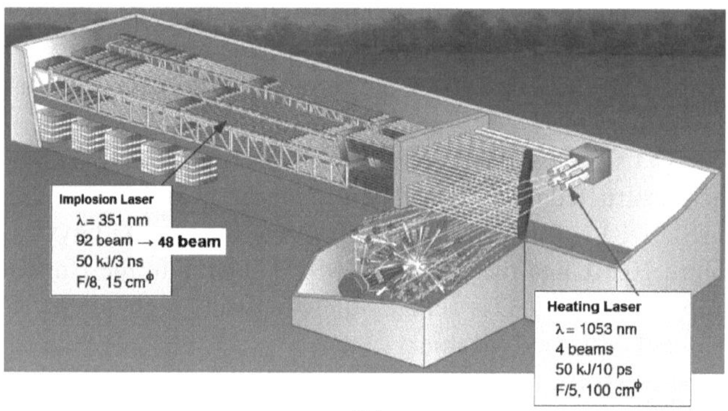

(b)

Figure 10.14 (a) Gekko XII and (b) the next generation ICF program FIREX-II [11]

operating in 2014 (Figure 10.13 (b)). Advanced helical devices such as Heliotron-J at Kyoto University and HSX at Wisconsin University are being used to research the optimization of the helical magnetic configuration.

Laser fusion is major line in inertial confinement. The ILE at Osaka University achieved an implosion of 600 times solid density using GEKKO XII in 1980 (Figure 10.14 (a)). Now they are conducting fast ignition experiments to heat the center

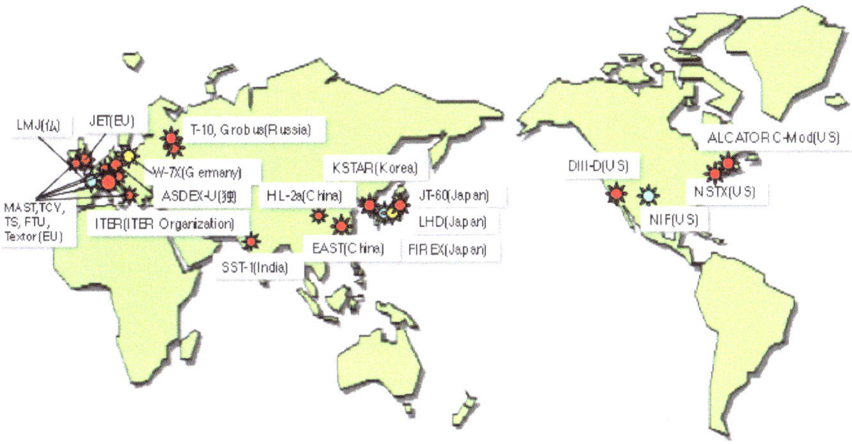

Figure 10.15 Locations of the world's leading fusion device aimed at the Sun on Earth (*red* is tokamak device, *yellow* is helical device, and *blue* is laser facility)

using an ultra-short laser pulse to achieve higher plasma temperatures. If FIREX-I achieves its programmatic goal, FIREX-II will be promoted (Figure 10.14 (b)). A similar program (HiPER program) is also proposed in Britain. Construction of large laser fusion facility NIF (USA) and LMJ (France) using the traditional implosion scheme (central ignition) is close to completion and it is thought that they will reach ignition (start to burn).

In Figure 10.15, the locations of world's major fusion facilities including ITER are shown.

Many tokamaks are actively studying tokamak physics in support of ITER, not only the above mentioned tokamaks but also Tore Supra in France, TCV in Switzerland, TEXTOR in Germany, FTU in Italy, T-10 in the Russian Federation, and HL-2a in China. Spherical torus (ST) is one of the most important research lines and has made a significant contribution to tokamak physics with its unique low aspect ratio and commonality as an axisymmetric torus. Major ST devices are NSTX in the United States and MAST in the UK. There are also smaller STs in Japan (such as QUEST) and in the Russian Federation (such as Globus).

10.3 ITER Project and the Broader Approach

The ITER project is an international project among Europe, Japan, the United States, Russia, China, Korea, and India, whose total population exceeds half of the world's population. Construction of ITER is jointly performed by the ITER Organization as an international organization and the domestic agencies of participating parties (countries). The first director general of the ITER Organization is Mr. Kaname Ikeda from Japan. Construction of ITER has started at a construction site adjacent to the

Figure 10.16 Birds eye view of ITER site [12]

1. Programmatic objective
   - Demonstration of scientific and technological feasibility of fusion power for peaceful purposes.
2. Detailed technical objectives
   - Extended burn with energy multiplication factor $Q \geq 10$ for 300-500s.
   - Aim at achieve $Q \geq 5$ in non-inductive current drive operation.
   - Not preclude controlled ignition
   - Average neutron flux $\geq 0.5 MW/m^2$ and average fluence $\geq 0.3 MWa/m^2$.
   - Test tritium breeding module.
3. Main parameters

Item	Value
Plasma current	15 MA
Toroidal magnetic field	5.3 Tesla
Plasma major radius	6.2 m
Plasma minor radius	2.0 m
Nominal fusion power	0.5GW

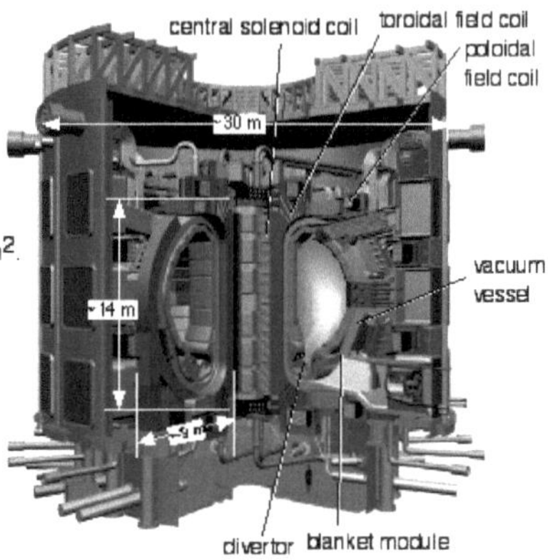

Figure 10.17 Program objectives, technical objectives, major parameters of ITER and view of the ITER main body [6]

Cadarache Research Center, near Aix-en-Provence in southern France. Figure 10.16 is a composite aerial photograph of the proposed building. The development in the foreground is the fusion facilities of the Cadarache Center and beyond the road is the ITER building.

Aix-en-Provence in southern France is near the port of Marseilles and is a central city in Provence, which has rich natural surroundings, such as olive groves and fragrant lavender. It is known for its culinary culture producing a variety of foods and wines, as introduced by Peter Mayle's best-selling book *A Year in Provence*.

ITER put cites its program objective as to "demonstrate the scientific and technical feasibility of fusion energy for peaceful purposes," and states its technical objectives as to "achieve energy multiplication factor Q of more than 10 by inductive operation, and aim at Q of more than 5 by non-inductive operation, realize average neutron rate $0.5\ MW/m^2$ and average fluence $0.3\ MWa/m^2$ and test tritium breeding modules" (see Figure 10.17).

ITER itself is giant high-technology device as shown in Figure 10.18 and its superconducting coil is a large device for generating magnetic fields to confine hot

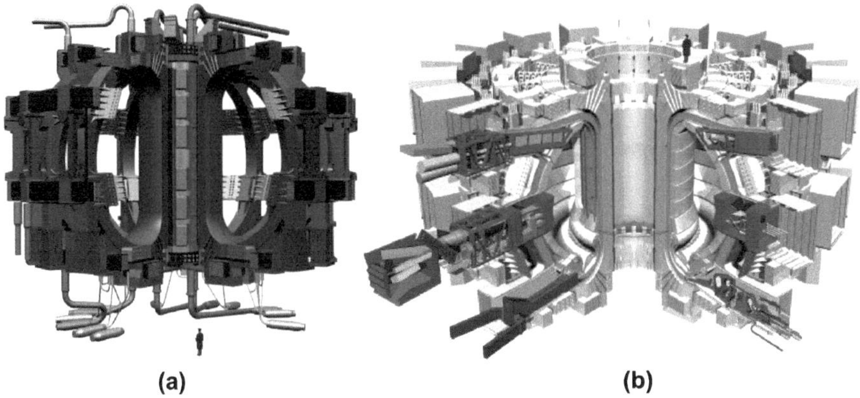

Figure 10.18 Views of (a) ITER superconducting coils and (b) ITER main body [11]

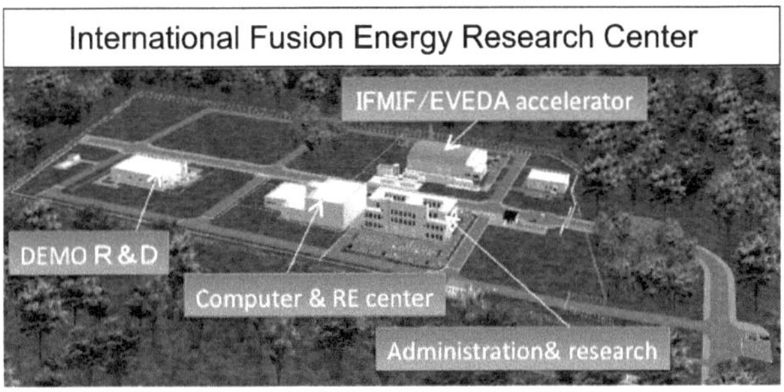

Figure 10.19 International Fusion Energy Research Center under construction at Rokkasho village, Aomori prefecture, Japan [13]

plasma. With the vacuum vessel to keep the hot plasma (a strong vacuum is necessary to prevent impurity contamination), the radiation shield, and plasma measurement system, it is a complex structure. The ITER project is a long-term project taking 35 years for construction, operation, and decontamination. Site preparation and procurement of components are ongoing and it is hoped that it will be completed in $\sim$9 years.

To allow a smooth transition to the stage of DEMO reactor when ITER has achieved its technical missions, a broader approach (BA) program has been launched by Japan and Europe with the idea of supporting and supplementing the program as necessary in parallel with ITER program for the early realization of fusion energy [13]. In the BA program, engineering demonstration of the engineering design activities (EVEDA) for the International Fusion Materials Irradiation Facility (IFMIF), accelerator facilities for testing materials for DEMO, and International Fusion Energy Research Center (IFERC) program consisting of DEMO design and an R&D center, ITER Remote Experimental Center, and a Fusion Computational Simulation Center will be developed at the Rokkasho site. Site preparation is rapidly progressing (Figure 10.19).

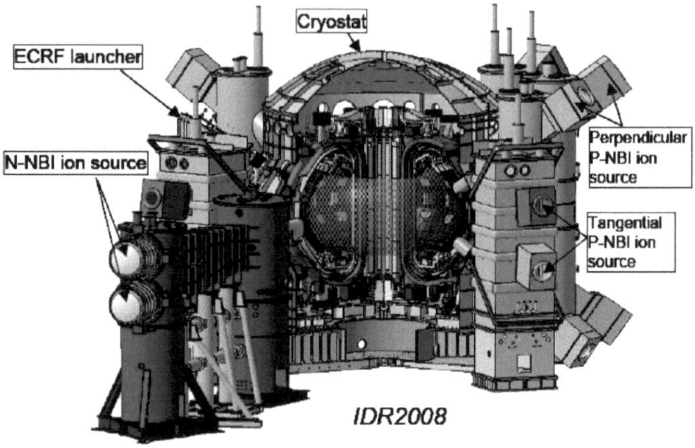

Figure 10.20 View of the large superconducting tokamak JT-60SA device, a modification of the JT-60 at Naka city in Ibaraki Prefecture, Japan [14]

Another BA program is modification of the JT-60 to a superconducting tokamak JT-60SA (see Figure 10.20). The plan is to improve the performance of the tokamak system as a magnetic bottle, to strengthen the physical basis for continuous operation and for high power density, to bridge to the high-performance DEMO reactor, and to perform supporting research flexibly according to ITER's needs.

10.4 Fusion Energy as an Energy Option for a Low Carbon Society

As mentioned in Section 10.1, realization of a low carbon society by suppressing the use of fossil fuels is necessary to prevent global warming and emissions of greenhouse gases such as carbon dioxide and methane. The 2100 Atomic Vision proposed by the Office of Strategy Research, Japan Atomic Energy Agency [15], aims to utilize previous R&D results and atomic energy technologies under development to reduce national carbon dioxide emissions to 10% of the present value, while maintaining a stable energy supply by 2100 (Figure 10.21 (a)). In that scenario, use of fossil energy currently providing 85% of primary energy is reduced to 30% by the end of this century and other 70% will be provided by non-fossil energy with a dominant contribution from atomic energy (Figure 10.21 (b)).

In this scenario, the use of electricity and hydrogen is promoted as an energy source. A significant reduction in energy consumption is realized by improved energy efficiency in the transport sector through the increased use of electric and fuel cell vehicles by way of hybrid vehicles. Coal and oil use in the industrial sector is abolished by substituting coke as a reducing agent in the steel industry and naphtha in the chemical industry with hydrogen. Energy use in civilian areas becomes electricity-based, except for solar heat.

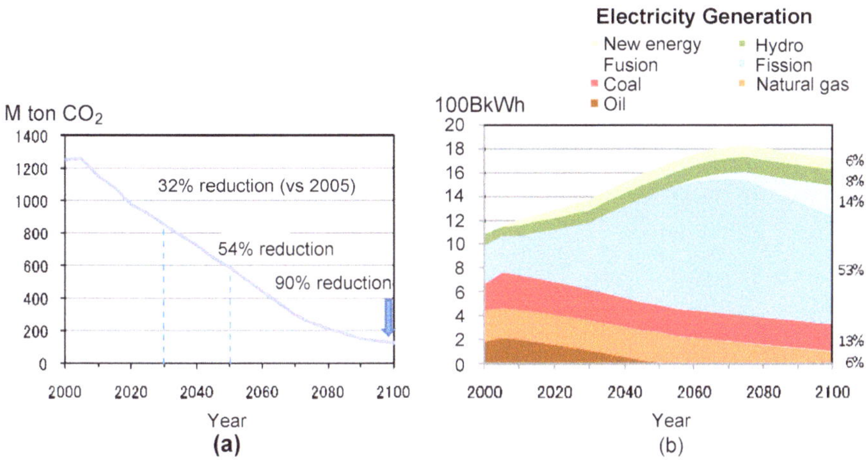

Figure 10.21 (a) carbon dioxide reduction scenario and (b) electricity supply scenario from various sources in 2100 Atomic Vision [15]

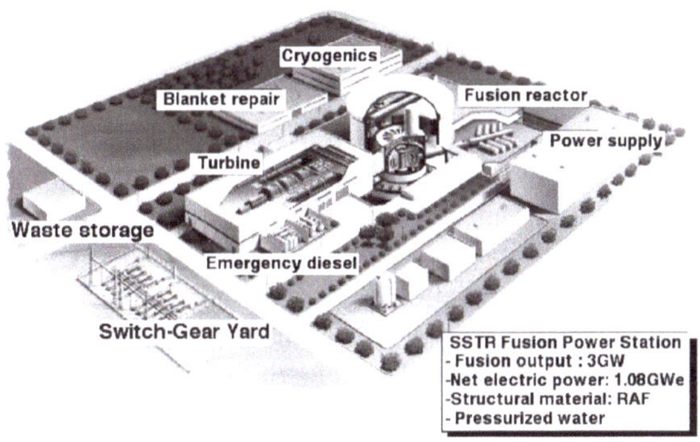

Figure 10.22 View of SSTR (Steady State Tokamak Reactor) plant layout [16]

In this way, 60% of the final energy demand in the year 2100 will be electricity. It is difficult to respond to such a huge power demand with only renewable energy and a large-scale stable supply from atomic energy is the most promising. In the vision, it is expected that ~ 30 fusion plants will be operational in Japan at the end of this century, assuming the scientific and technical feasibility of fusion energy is demonstrated in ITER and the construction and operation of DEMO is advanced to start construction of commercial fusion plants in late 2050s. It should be remembered that it is possible fusion energy may not be in the energy market by the end of the twenty-first century if fusion R&D does not go well or the economic efficiency and reliability of operation is not good enough.

The tokamak system, adopted in ITER, shows the best performance in high temperature plasma confinement. The operation of the tokamak stops when an inductive electric field cannot be supplied, since the confinement field is created by the inductive plasma current. The power is, therefore, generated in pulses.

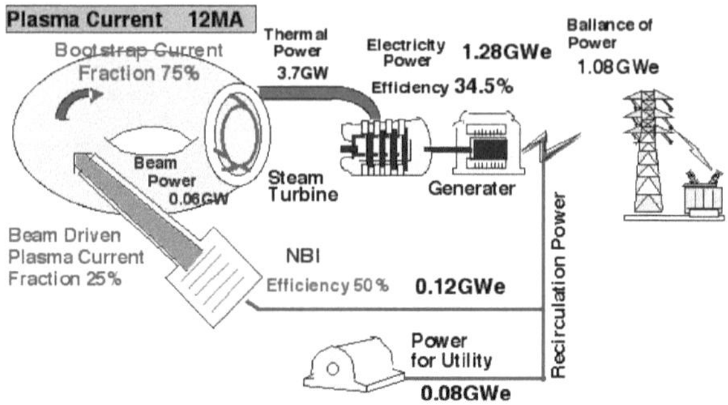

Figure 10.23 Principle to achieve high net generated power by achieving efficient continuous operation using bootstrap current [16]

As a method to overcome this drawback and to produce energy continuously, the use of the bootstrap current mentioned in Section 8.5 is being considered. Since 80% of the plasma current is shed by the bootstrap current in the JT-60 (Figure 8.5), a steady-state tokamak reactor (SSTR) is being developed in cooperation with industries [16]. The proposed site plan is shown in Figure 10.22.

To achieve continuous operation using the bootstrap current, the majority of plasma current has to be driven by the bootstrap current and the rest by beam for example, as shown in Figure 10.23. The heat including that generated inside the blanket is removed and used to produce electricity in a steam turbine and this fraction is used as recirculating power for beam generation and other plants. The achievement of continuous operation with $Q > 5$ given in ITER technical objectives is important. In this case, about half of the plasma current should be driven by the bootstrap current while the rest should be driven by the active current drive methods.

References

1. CRIEPI (1998) Trilemma as Crisis for Human. In Japanese.
2. Produced from Key World Energy Statistics (2001), 18th World Energy Congress, International Energy Agency (IEA).
3. Climate Change (1995) The Science of Climate Change, Contribution of Working Group I to the Second Assessment Report of IPCC. Cambridge University Press.
4. The Subcommittee of the Fusion Council for Fusion Development Strategy (2000) Report on technical feasibility of fusion energy and extension of the fusion program and basic supporting researches. May 17, 2000.
5. CRIEPI, IMIDAS (2000) Next Generation Energy Plan.
6. ITER Council, ITER Final Report, http://www-naweb.iaea.org/napc/physics/PS/publications/pub_iter.htm.
7. Wan Y Li J Weng P, GA, PPPL Team (2006)Proc. 21st IAEA FEC OV/1-1.
8. Bak JS et al. (2008) Proc. 22nd IAEA FEC FT/1-1.
9. Courtesy of Professor H. Yamada, NIFS, Japan.

10. Klinger T et al. (2008) Proc. 22nd IAEA FEC FT/1-4.
11. Courtesy of Professor Azechi, Osaka University, Japan.
12. From homepage of ITER Organization (http://www.ITER.org).
13. Matsuda S (2007) Fusion Eng. & Design, 82, 435–442.
14. Ishida S, Barabasci P, Kamada Y et al. (2010) Proc. 23nd IAEA Fusion Energy Conference, IAEA-CN-180/OV/P-4
15. Office of Strategy Research-JAEA (2008) 2100 Atomic Vision: A Proposal for Low Carbon Society, http://www.jaea.go.jp/.
16. Kikuchi M (1990) Nucl. Fusion, 30, 265; also Seki Y, Kikuchi M et al. (1990) Proc. 13th IAEA Fusion Energy Conference, IAEA-CN-53/G-1-2.

1.2 Sun: Gravitationally Confined Fusion Reactor

1. Einstein's equation

$$E = mc^2$$

2. Fusion reactions in the Sun

$$\begin{aligned} 4\mathrm{p} + 2\mathrm{e}^- &\longrightarrow {}^4_2\mathrm{He} + 26.72\,\mathrm{MeV}\,, \\ \mathrm{p} + \mathrm{p} &\longrightarrow \mathrm{D} + \mathrm{e}^+ + \nu_e + 0.42\,\mathrm{MeV}\,, \\ \mathrm{p} + \mathrm{D} &\longrightarrow {}^3_2\mathrm{He} + \gamma + 5.49\,\mathrm{MeV}\,, \\ {}^3_2\mathrm{He} + {}^3_2\mathrm{He} &\longrightarrow {}^4_2\mathrm{He} + \mathrm{p} + 12.86\,\mathrm{MeV} \end{aligned}$$

3. Action integral for special relativity

$$S = -m_{a0}c \int_1^2 \mathrm{d}s = -m_{a0}c^2 \int_{t_1}^{t_2} \mathrm{d}\tau$$

4. World interval

$$\mathrm{d}s \equiv (c^2 \mathrm{d}t^2 - \mathrm{d}\boldsymbol{x} \cdot \mathrm{d}\boldsymbol{x})^{1/2}$$

5. Proper time

$$\tau = t[1 - (\mathrm{d}\boldsymbol{x}/\mathrm{d}t \cdot \mathrm{d}\boldsymbol{x}/\mathrm{d}t)/c^2]^{1/2}$$

6. Action integral for special relativity

$$S = \int_{t_1}^{t_2} L\,\mathrm{d}t$$

$$L = -m_{a0}c^2 \sqrt{1 - (\mathrm{d}\boldsymbol{x}/c\,\mathrm{d}t)^2}$$

7. Relativistic momentum

$$\boldsymbol{p} = m_{a0}\boldsymbol{v}/\sqrt{1 - v^2/c^2}$$

8. Relativistic energy

$$E = m_{a0}c^2/\sqrt{1 - v^2/c^2}$$

9. Metric in weak gravitational field

$$g_{00} = 1 + 2\phi/c^2$$

1.3 Fusion: Challenge of the Sun on Earth

1. DT fusion reaction

$$D + T \rightarrow {}^4\mathrm{He}(3.52\,\mathrm{MeV}) + \mathrm{n}(14.06\,\mathrm{MeV})$$

1.4 Plasma: Fourth State of Matter

1. Saha's ionization rate

$$\alpha = \frac{\rho}{\rho + 1}, \quad \rho = \sqrt{\frac{3 \times 10^{27}}{n_0(m^{-3})} T(\mathrm{eV})^{3/4} \exp\left(-\frac{V_i}{2kT}\right)}$$

Chapter 2
Hydrogen Fusion: Light Nuclei and Theory of Fusion Reactions

2.1 Fusion: Fusion of Little Nuts

1. DT fusion reaction

$$D + T \rightarrow {}^5_2\mathrm{H}^*_e \rightarrow {}^4_2\mathrm{He} + \mathrm{n}$$

2. Planck relation

$$E = \hbar\omega$$

3. de Broglie relation

$$\boldsymbol{p} = \hbar\boldsymbol{k}$$

4. Equivalence relation

$$\mathrm{i}\boldsymbol{k} = \partial/\partial\boldsymbol{x}$$
$$-\mathrm{i}\omega = \partial/\partial t$$

5. Energy conservation relation

$$\hbar^2\boldsymbol{k}^2/2m + V(\boldsymbol{x}) = \hbar\omega$$

6. Time-dependent Schrödinger equation

$$\left[-(\hbar^2/2m)\partial^2/\partial\boldsymbol{x}^2 + V(\boldsymbol{x})\right]\psi = \mathrm{i}\hbar(\partial/\partial t)\psi$$

7. Time-independent Schrödinger equation

$$\left[-(\hbar^2/2m)\partial^2/\partial \boldsymbol{x}^2 + V(\boldsymbol{x})\right]u = Eu$$

2.2 Deuterium: Nucleus Loosely Bound by a Neutron and Proton

1. Yukawa's nuclear meson equation

$$\left[\hbar^2 \frac{\partial^2}{\partial \boldsymbol{x}^2} - m^2c^2 - \hbar^2 \frac{1}{c^2}\frac{\partial^2}{\partial t^2}\right]U = 0$$

2.3 Tritium: Nucleus Emitting an Electron and Neutrino

1. Neutron decay into proton

$$\mathrm{n} \to \mathrm{p} + \mathrm{e} + \nu$$

2. Li reactions

$$^6\mathrm{Li} + \mathrm{n} \longrightarrow {}^3\mathrm{T} + {}^4\mathrm{He} + 4.8\,\mathrm{MeV}$$
$$^7\mathrm{Li} + \mathrm{n} \longrightarrow {}^3\mathrm{T} + {}^4\mathrm{He} + \mathrm{n}' - 2.5\,\mathrm{MeV}$$

2.6 Fusion Cross-Section: Tunneling Effect and Resonance

1. Wave front equation in Coulomb potential

$$z + b_0 \ln k(r - z) = \text{const.}$$

2. Coulomb barrier penetration probability

$$P(E/E_\mathrm{c}) = \frac{\sqrt{E_\mathrm{c}/E}}{\exp\sqrt{E_\mathrm{c}E} - 1}$$
$$E_\mathrm{c} = \frac{m_r e^4}{8\varepsilon_0^2\hbar^2} = 0.98A_r\,(\mathrm{MeV})$$

3. Analytical form of fusion cross-section

$$\sigma_r = \pi\lambdabar^2 P(E/E_\mathrm{c})\frac{\Gamma_i\Gamma_f}{(E - E_r)^2 + \Gamma^2/4}$$

4. Empirical form of fusion cross-section

$$\sigma_r = \sigma_0 \frac{E_{cl}}{E_l[\exp\sqrt{E_{cl}/E_l} - 1]}\left[\frac{1}{1 + 4(E_l - E_{rl})^2/\Gamma_l^2} + \alpha\right]$$

Chapter 3
Confinement Bottle:
Topology of Closed Magnetic Field and Force Equilibrium

3.1 Field: Magnetic Field and Closed Magnetic Configuration

1. Variational principle of field line

$$\delta \int \boldsymbol{A} \cdot \mathrm{d}\boldsymbol{x} = 0$$

2. Integrable system (manifold with constant H)

$$H(\boldsymbol{q}, \boldsymbol{p}) = \text{constant}$$

3.2 Topology: Closed Surface without a Fixed Point

1. Euler's index

$$K = p - q + r$$

(p, q, r: numbers of vertexes, sides, and polygons)

2. Euler index of sphere (S^2)

$$K = p - q + r = 2$$

3. Euler index of torus (T^2)

$$K = p - q + r = 0$$

3.3 Coordinates: Analytical Geometry of the Torus

1. Descartes coordinates

$$\boldsymbol{x} = x\boldsymbol{e}_x + y\boldsymbol{e}_y + z\boldsymbol{e}_z$$

2. General coordinates (u^1, u^2, u^3)

$$\boldsymbol{x}(u^1, u^2, u^3) = x(u^1, u^2, u^3)\boldsymbol{e}_x + y(u^1, u^2, u^3)\boldsymbol{e}_y + z(u^1, u^2, u^3)\boldsymbol{e}_z$$

3. Jacobian J (defined by volume element)

$$\mathrm{d}V = \frac{\partial \boldsymbol{x}}{\partial u^1} \cdot \left(\frac{\partial \boldsymbol{x}}{\partial u^2} \times \frac{\partial \boldsymbol{x}}{\partial u^3} \right) \mathrm{d}u^1 \mathrm{d}u^2 \mathrm{d}u^3$$
$$J \equiv 1/\nabla u^1 \cdot (\nabla u^2 \times \nabla u^3)$$

4. Gradient vector

$$\nabla u^i = \frac{\partial u^i}{\partial x}\boldsymbol{e}_x + \frac{\partial u^i}{\partial y}\boldsymbol{e}_y + \frac{\partial u^i}{\partial z}\boldsymbol{e}_z$$

5. Tangent vector

$$\frac{\partial \boldsymbol{x}}{\partial u^i} = \frac{\partial x}{\partial u^i}\boldsymbol{e}_x + \frac{\partial y}{\partial u^i}\boldsymbol{e}_y + \frac{\partial z}{\partial u^i}\boldsymbol{e}_z$$

6. Orthogonality relation

$$\nabla u^i \cdot \frac{\partial \boldsymbol{x}}{\partial u^j} = \delta_{ij}$$

7. Dual relations (i, j, k); right handed

$$\nabla u^i = \frac{1}{J}\left(\frac{\partial \boldsymbol{x}}{\partial u^j} \times \frac{\partial \boldsymbol{x}}{\partial u^k}\right)$$
$$\frac{\partial \boldsymbol{x}}{\partial u^i} = J\nabla u^j \times \nabla u^k$$

8. Contravariant form (expansion by tangent vector)

$$\boldsymbol{a} = \sum_i a^i \frac{\partial \boldsymbol{x}}{\partial u^i} = \sum_i \left(\boldsymbol{a}\cdot\nabla u^i\right)\frac{\partial \boldsymbol{x}}{\partial u^i}$$

9. Covariant form (expansion by gradient vector)

$$\boldsymbol{a} = \sum_i a_i \nabla u^i = \sum_i \left(\boldsymbol{a}\cdot\frac{\partial \boldsymbol{x}}{\partial u^i}\right)\nabla u^i$$

10. Metric

$$g_{ij} = \frac{\partial \boldsymbol{x}}{\partial u^i}\cdot\frac{\partial \boldsymbol{x}}{\partial u^j},$$
$$g^{ij} = \nabla u^i \cdot \nabla u^j,$$
$$[g_{ij}] = [g^{ij}]^{-1}$$

11. Differential length (definition of metric)

$$\mathrm{d}s^2 = \mathrm{d}\boldsymbol{x}\cdot\mathrm{d}\boldsymbol{x} = \sum_{i,j}\frac{\partial \boldsymbol{x}}{\partial u^i}\cdot\frac{\partial \boldsymbol{x}}{\partial u^j}\mathrm{d}u^i\mathrm{d}u^j = \sum_{i,j} g_{ij}\mathrm{d}u^i\mathrm{d}u^j$$

12. Covariant component a_i

$$a_i = \boldsymbol{a} \cdot \frac{\partial \boldsymbol{x}}{\partial u^i} = \sum_j g_{ij} a^j$$

13. Contravariant component a^i

$$a^i = \boldsymbol{a} \cdot \nabla u^i = \sum_j g^{ij} a_j$$

14. Inner product

$$\boldsymbol{a} \cdot \boldsymbol{b} = \sum_i a_i b^i = \sum_i a^i b_i$$

15. Outer product

$$\boldsymbol{a} \times \boldsymbol{b} = J \sum_i (a^j b^k - a^k b^j) \frac{\partial \boldsymbol{x}}{\partial u^i} = J^{-1} \sum_i (a_j b_k - a_k b_j) \nabla u^i$$

16. Gradient

$$\nabla f = \sum_i \frac{\partial f}{\partial u^i} \nabla u^i$$

17. Rotation (i, j, k); right handed

$$\nabla \times \boldsymbol{a} = \sum_{i=1,3} \sum_{j=1,3} \frac{\partial a_i}{\partial u^j} \nabla u^j \times \nabla u^i = J^{-1} \sum_{k=1,3} \left[\frac{\partial a_j}{\partial u^i} - \frac{\partial a_i}{\partial u^j} \right] \frac{\partial \boldsymbol{x}}{\partial u^k}$$

18. Divergence

$$\nabla \cdot \boldsymbol{a} = \nabla \cdot \sum_i a^i \frac{\partial \boldsymbol{x}}{\partial u^i} = J^{-1} \sum_i \frac{\partial J a^i}{\partial u^i}$$

3.4 Field Line Dynamics: Hamilton Dynamics of the Magnetic Field

1. General form of vector potential (G: gauge term)

$$\boldsymbol{A} = \phi \nabla \theta - \psi \nabla \zeta + \nabla G$$

2. Simplectic form of $\boldsymbol{B}$

$$\boldsymbol{B} = \nabla \phi \times \nabla \theta - \nabla \psi \times \nabla \zeta$$

3. Magnetic coordinates

$$(\phi, \theta, \zeta)$$

4. Orbit equation of field line

$$\frac{d\theta}{d\zeta} = \frac{\partial \psi}{\partial \phi} ,$$
$$\frac{d\phi}{d\zeta} = -\frac{\partial \psi}{\partial \theta}$$

5. Action integral of magnetic field line

$$S = \int \boldsymbol{A} \cdot d\boldsymbol{x}$$

6. Lagrangian (T: kinetic energy, V: potential energy)

$$L = T - V$$

7. Canonical momentum

$$p_i \equiv \frac{\partial L}{\partial \dot{q}_i}$$

8. Hamilton equation of motion

$$d\boldsymbol{q}/dt = \partial H / \partial \boldsymbol{p} ,$$
$$d\boldsymbol{p}/dt = -\partial H / \partial \boldsymbol{q}$$

9. Action integral in Hamilton form

$$S(\boldsymbol{q}, \boldsymbol{p}) = \int_{t_1}^{t_2} \left[\sum p_i \dot{q}_i - H(\boldsymbol{q}, \boldsymbol{p}, t) \right] dt$$

3.5 Magnetic Surface: Integrable Magnetic Field and Hidden Symmetry

1. Equilibrium equation

$$\boldsymbol{J} \times \boldsymbol{B} = \nabla P$$

2. Magnetic field stays on constant P surface

$$\boldsymbol{B} \cdot \nabla P = 0$$

3. Current stays on constant P surface

$$\boldsymbol{J} \cdot \nabla P = 0$$

4. Stream function for magnetic field

$$\boldsymbol{B} = \nabla\phi \times \nabla h = \nabla\phi \times \left[\frac{\partial h}{\partial \theta}\nabla\theta + \frac{\partial h}{\partial \zeta}\nabla\zeta\right]$$

5. Toroidal flux inside the flux surface

$$2\pi\phi(u) = \int \boldsymbol{B} \cdot d\boldsymbol{a}_\zeta$$

6. Poloidal flux outside of flux surface

$$2\pi\psi(u) = \int \boldsymbol{B} \cdot d\boldsymbol{a}_\theta$$

7. Clebsch form of magnetic field (ϕ: toroidal flux, α: surface function)

$$\boldsymbol{B} = \nabla\phi \times \nabla\alpha$$

8. Flux coordinates

$$(\phi, \theta_m, \zeta)$$

9. Jacobian of flux coordinates

$$J = 1/(2\pi\boldsymbol{B} \cdot \nabla\zeta)$$

10. Field line is straight line

$$\frac{\mathrm{d}\theta_m}{\mathrm{d}\zeta} = \frac{1}{q(\phi)}$$

11. Safety factor

$$q = \mathrm{d}\phi/\mathrm{d}\psi$$

12. Surface function

$$\alpha = \theta_m - \zeta/q$$

13. Action integral for field line in flux coordinates

$$S = \int \boldsymbol{A} \cdot \mathrm{d}\boldsymbol{x} = \int [\phi \mathrm{d}\theta_m - \psi \mathrm{d}\zeta]$$

14. Field line Lagrangian in flux coordinates:

$$\mathcal{L} = \phi \frac{d\theta_m}{d\zeta} - \psi(\phi)$$

3.6 Flux Coordinates: Hamada and Boozer Coordinates

1. Toroidal current inside the flux surface

$$2\pi f(u) = \int \boldsymbol{J} \cdot d\boldsymbol{a}_\zeta$$

2. Poloidal current flux

$$2\pi g(u) = \int \boldsymbol{J} \cdot d\boldsymbol{a}_\theta$$

3. Current safety factor

$$q_J = -f'(\phi)/g'(\phi)$$

4. Kruskal–Kulsrud equilibrium equation

$$g'(\phi) + f'(\phi)/q(\phi) = -V'(\phi)P'(\phi)$$

5. Hamada coordinates

$$(v, \theta_h, \zeta_h)$$

6. Jacobian of Hamada coordinates

$$J = 1$$

7. Equilibrium $\boldsymbol{B}$ in Hamada coordinates

$$\boldsymbol{B} = \nabla\zeta_h \times \nabla\psi + \nabla\phi \times \nabla\theta_h = \nabla\phi \times \nabla(\theta_h - \zeta_h/q)$$

8. Equilibrium $\boldsymbol{J}$ in Hamada coordinates

$$\boldsymbol{J} = \nabla\zeta_h \times \nabla g + \nabla f \times \nabla\theta_h = \nabla f \times \nabla(\theta_h - \zeta_h/q_J)$$

9. Equilibrium relation in Hamada coordinates

$$f'(v)\psi'(v) - \phi'(v)g'(v) = P'(v)$$

10. Boozer coordinates

$$(\phi, \theta_b, \zeta_b)$$

11. Jacobian of Boozer coordinates

$$J = (g + f/q)/2\pi B^2$$

12. Equilibrium $\boldsymbol{B}$ in Boozer coordinates

$$\boldsymbol{B} = g(\phi)\nabla\zeta_b + f(\phi)\nabla q_b + \beta_*(\phi, q_b, \zeta_b)\nabla\phi$$
$$\boldsymbol{B} = \nabla\phi \times \nabla\alpha \qquad (\alpha = \theta_b - \zeta_b/q)$$

13. Equilibrium $\boldsymbol{J}$ in Boozer coordinates

$$\mu_0 \boldsymbol{J} = \nabla\phi \times \nabla\beta \qquad (\beta = \beta_* - g'(\phi)\zeta_b - f'(\phi)\theta_b)$$

14. Boozer–Grad coordinates

$$(\phi, \alpha, \chi)$$

15. Jacobian of BG coordinates

$$J = 1/B^2$$

16. Equilibrium $\boldsymbol{B}$ in BG coordinates

$$\boldsymbol{B} = \nabla\chi + \beta\nabla\phi(\beta = \beta_* - g'(\phi)\zeta_b - f'(\phi)\theta_b)$$
$$\boldsymbol{B} = \nabla\phi \times \nabla\alpha(\alpha = \theta_b - \zeta_b/q)$$

3.7 Ergodicity: A Field Line Densely Covers the Torus

1. Poincaré mapping of magnetic field line

$$\theta_0, g\theta, g^2\theta_0, g^3\theta_0, g^4\theta_0, \ldots, g^j\theta_0$$
$$g : \Theta \longrightarrow \Theta\ , \quad g\theta_0 = \theta_0 + 2\pi/q \text{ for } \theta_0 \in \Theta$$

3.8 Apparent Symmetry: Force Equilibrium of Axisymmetric Torus

1. Cylindrical coordinates

$$(R, \zeta, Z)$$

2. Flux functions

$$P(\psi), RB_\zeta(\psi)$$

3. Magnetic field

$$\boldsymbol{B} = \nabla\psi \times \nabla\zeta + F\nabla\zeta\ ,$$
$$RB_R = -\frac{\partial\psi}{\partial Z}\ ,$$
$$RB_Z = \frac{\partial\psi}{\partial R}$$

4. Collorary 1

$$\nabla\psi = -R^2\nabla\zeta \times \boldsymbol{B}$$

5. Collorary 2

$$\boldsymbol{B} \times \nabla\psi = -R^2 B^2 \nabla\zeta + F(\psi)\boldsymbol{B}$$

6. Current density

$$\boldsymbol{J} = \mu_0^{-1}[\nabla F \times \nabla\zeta + \Delta^*\psi\nabla\zeta]$$

7. Grad–Shafranov equation

$$[R\partial/\partial R(R^{-1}\partial/\partial R) + \partial^2/\partial R^2]\psi = -\mu_0 R^2 P'(\psi) + FF'(\psi)$$

8. Action integral for equilibrium

$$S = \int L\, \mathrm{d}R\, \mathrm{d}Z = \int R\left(\frac{B_p^2}{2\mu_0} - \frac{B_\zeta^2}{2\mu_0} - P\right)\mathrm{d}R\, \mathrm{d}Z$$

9. Euler–Lagrange equation

$$\frac{\partial L}{\partial\psi} - \frac{\partial}{\partial R}\frac{\partial L}{\partial\psi_R} - \frac{\partial}{\partial Z}\frac{\partial L}{\partial\psi_Z} = 0$$

10. Flux surface average

$$\langle A\rangle = \frac{\int_\psi^{\psi+\mathrm{d}\psi} AJ\, \mathrm{d}\psi\, \mathrm{d}\theta\, \mathrm{d}\zeta}{\int_\psi^{\psi+\mathrm{d}\psi} J\, \mathrm{d}\psi\, \mathrm{d}\theta\, \mathrm{d}\zeta} = \frac{\int_0^{2\pi} \frac{A\, \mathrm{d}\theta}{\boldsymbol{B}_p\cdot\nabla\theta}}{\int_0^{2\pi} \frac{\mathrm{d}\theta}{\boldsymbol{B}_p\cdot\nabla\theta}}$$

11. Jacobian

$$J = 1/(\nabla\zeta \times \nabla\psi)\cdot\nabla\theta = 1/\boldsymbol{B}_p\cdot\nabla\theta$$

12. Differential operator along $\boldsymbol{B}$

$$\boldsymbol{B}\cdot\nabla = J^{-1}\partial/\partial\theta$$

13. Magnetic differential equation (MDE)

$$\boldsymbol{B}\cdot\nabla h = S$$

14. Solvable condition of MDE

$$\int_0^{2\pi} \frac{S}{\boldsymbol{B}_p\cdot\nabla\theta}\mathrm{d}\theta = 0$$

15. Property of flux surface average 1 ($\boldsymbol{A}$: arbitrary vector, V: volume enclosed by flux surface)

$$\langle \nabla \cdot \boldsymbol{A} \rangle = (\mathrm{d}/\mathrm{d}V)\langle \boldsymbol{A} \cdot \nabla V \rangle$$

16. Property of flux surface average 2 (F: arbitrary function)

$$\langle \boldsymbol{B} \cdot \nabla F \rangle = 0$$

17. Property of flux surface average 3 ($\boldsymbol{A}$: arbitrary vector)

$$\langle \nabla \psi \cdot \nabla \times \boldsymbol{A} \rangle = \langle \nabla \cdot (\boldsymbol{A} \times \nabla \psi) \rangle = 0$$

18. Cyclic coordinate

$$\frac{\partial L}{\partial q_s} = 0$$

19. Conservation of conjugate momentum

$$\frac{\mathrm{d}}{\mathrm{d}t}\left(\frac{\partial L}{\partial \dot{q}_s}\right) = 0$$

3.9 3-dimensional Force Equilibrium: Search for Hidden Symmetry

1. Action integral for 3-dimensional equilibrium

$$S = \int L \,\mathrm{d}V = \int \left[\frac{B^2}{2\mu_0} - P\right] \mathrm{d}V$$

2. Magnetic field

$$\boldsymbol{B} = \nabla P \times \nabla \omega$$

3. Current density

$$\boldsymbol{J} = \nabla P \times \nabla \omega_J$$

4. Euler–Lagrange equation

$$\boldsymbol{J} \cdot \nabla \omega = 1\,, \quad \boldsymbol{J} \cdot \nabla P = 0$$

5. Equilibrium relation

$$(\nabla\omega_J \times \nabla\omega)\cdot\nabla P = 1\,,$$
$$\boldsymbol{B}\cdot\nabla\omega_J = -1$$

Chapter 4
Charged Particle Motion: Lagrange–Hamilton Orbit Dynamics

4.1 Variational Principle: Hamilton's Principle

1. Lagrange action integral

$$S(\boldsymbol{x}) = \int_{t_1}^{t_2} \mathrm{d}t L(\boldsymbol{x}(t), \dot{\boldsymbol{x}}(t), t)$$

2. Lagrange equation of motion

$$\frac{\mathrm{d}}{\mathrm{d}t}\left(\frac{\partial L}{\partial \dot{q}_i}\right) - \frac{\partial L}{\partial q_i} = 0$$

3. Gauge freedom for Lagrangian

$$L(\boldsymbol{q}, \dot{\boldsymbol{q}}, t) + \frac{\mathrm{d}W(\boldsymbol{q}, t)}{\mathrm{d}t}$$

4. Hamilton action integral

$$S(\boldsymbol{x}, \boldsymbol{p}) = \int_{t_1}^{t_2} \left[\sum p_i \dot{q}_i - H(\boldsymbol{q}, \boldsymbol{p}, t)\right] \mathrm{d}t$$

5. Hamilton equation

$$\frac{\mathrm{d}\boldsymbol{q}}{\mathrm{d}t} = \frac{\partial H}{\partial \boldsymbol{p}}\,,$$
$$\frac{\mathrm{d}\boldsymbol{p}}{\mathrm{d}t} = -\frac{\partial H}{\partial \boldsymbol{q}}$$

6. Noether's theorem

$$I = \sum_{i=1}^{n} \frac{\partial L(\boldsymbol{q}, \dot{\boldsymbol{q}}, t)}{\partial \dot{q}_i} S_i - W(\boldsymbol{q}, t) = \text{constant}$$

6.1 Condition for Noether's theorem

$$\delta L = L(\boldsymbol{q}', \dot{\boldsymbol{q}}', t) - L(\boldsymbol{q}, \dot{\boldsymbol{q}}, t) = \varepsilon\, \mathrm{d}W(\boldsymbol{q}, t)/\mathrm{d}t$$

4.2 Lagrange–Hamilton Mechanics: Motion in an Electromagnetic Field

1. Charged particles Lagrangian

$$L_a(\boldsymbol{x}, \dot{\boldsymbol{x}}, t) = \frac{1}{2} m_a \dot{\boldsymbol{x}}^2 + e_a \boldsymbol{A}(\boldsymbol{x}, t) \cdot \dot{\boldsymbol{x}} - e_a \Phi(\boldsymbol{x}, t)$$

2. Canonical momentum

$$\boldsymbol{p} = \frac{\partial L}{\partial \dot{\boldsymbol{x}}} = m_a \boldsymbol{v} + e_a \boldsymbol{A}$$

3. Hamilton form of Lagrangian

$$L(\boldsymbol{p}, \boldsymbol{x}, t) = \boldsymbol{p} \cdot \dot{\boldsymbol{x}} - H(\boldsymbol{p}, \boldsymbol{x}, t)$$

4. Hamiltonian

$$H(\boldsymbol{p}, \boldsymbol{x}, t) = \frac{1}{2m_a} (\boldsymbol{p} - e_a \boldsymbol{A})^2 + e_a \Phi(\boldsymbol{x}, t)$$

5. Canonical angular momentum in tokamak

$$p_\zeta = \frac{\partial L}{\partial \zeta} = m_a R^2 \dot{\zeta} + e_a R A_\zeta = \text{constant}$$

6. Relativistic Lagrangian

$$L_a(\boldsymbol{x}, \boldsymbol{v}, t) = -m_{a0} c^2 \sqrt{1 - \left(\frac{v}{c}\right)^2} + e_a (\boldsymbol{v} \cdot \boldsymbol{A} - \Phi)$$

7. Relativistic canonical momentum

$$\boldsymbol{p} = \frac{m_{a0} \boldsymbol{v}}{\sqrt{1 - (v/c)^2}} + e_a \boldsymbol{A}$$

8. Relativistic Hamiltonian

$$H(\boldsymbol{p}, \boldsymbol{q}, t) = \sqrt{m_{a0} c^4 + c^2 (\boldsymbol{p} - e_a \boldsymbol{A})^2} + e_a \Phi(\boldsymbol{x}, t)$$

9. Action for free particle

$$S_{\text{FreeParticle}} = -m_{a0} c \int_1^2 \mathrm{d}s = -m_{a0} c^2 \int_{t_1}^{t_2} \sqrt{1 - \frac{v^2}{c^2}} \, \mathrm{d}t$$

10. Action for field particle

$$S_{\text{Field-Particle}} = -e_a \int_1^2 A_\mu \mathrm{d}x^\mu = -\int_{t_1}^{t_2} e_a \Phi \, \mathrm{d}t + \int_{t_1}^{t_2} e_a \boldsymbol{A} \cdot \mathrm{d}\boldsymbol{x}$$

11. Action for field

$$S_{\text{Field}} = \int_{t_1}^{t_2} \left[\frac{\varepsilon_0 \boldsymbol{E}^2}{2} - \frac{\boldsymbol{B}^2}{2\mu_0} \right] \mathrm{d}V \, \mathrm{d}t$$

4.3 Littlejohn's Variational Principle: Orbital Mechanics of the Guiding Center

1. Charged particle Lagrangian

$$L(\boldsymbol{x}, \boldsymbol{v}, t) = (e_a \boldsymbol{A} + m_a \boldsymbol{v}) \cdot \dot{\boldsymbol{x}} - H(\boldsymbol{x}, \boldsymbol{v}, t)$$

2. Hamiltonian

$$H(\boldsymbol{x}, \boldsymbol{v}, t) = \frac{1}{2} m_a \boldsymbol{v}^2 + e_a \Phi(\boldsymbol{x}, t)$$

3. Charged particle position

$$\boldsymbol{x}(t) = \boldsymbol{r}(t) + \rho[\boldsymbol{e}_x \cos\theta + \boldsymbol{e}_y \sin\theta] \,, \quad \rho = \frac{v_\perp}{\Omega}$$

4. Guiding center Lagrangian

$$L(\boldsymbol{r}, \dot{\boldsymbol{r}}, v_\parallel, \mu, \dot{\theta}, t) = e_a \boldsymbol{A}^*(\boldsymbol{r}, v_\parallel, t) \cdot \dot{\boldsymbol{r}} - \frac{m_a}{e_a} \mu \dot{\theta} - H(\boldsymbol{r}, v_\parallel, \mu, t)$$

5. Modified vector potential

$$\boldsymbol{A}^* = \boldsymbol{A} + (m_a/e_a) v_\parallel \boldsymbol{b}$$

6. Magnetic moment

$$\mu = \frac{m_a v_\perp^2}{2B}$$

7. Guiding center Hamiltonian

$$H(\boldsymbol{r}, v_\parallel, \mu, t) = \frac{1}{2} m_a v_\parallel^2 + \mu B(\boldsymbol{r}) + e_a \Phi(\boldsymbol{r}, t)$$

8. Guiding center equation of motion

$$\frac{\mathrm{d}v_\parallel}{\mathrm{d}t} = -\frac{1}{B_\parallel^*} \boldsymbol{B}^* \cdot (\mu \nabla B - e_a \boldsymbol{E}^*) \,,$$

$$\frac{\mathrm{d}\boldsymbol{r}}{\mathrm{d}t} = \frac{1}{B_\parallel^*} [v_\parallel \boldsymbol{B}^* + \boldsymbol{b} \times ((\mu/e_a) \nabla B - \boldsymbol{E}^*)]$$

9. Modified magnetic field

$$\boldsymbol{B}^* = \boldsymbol{B} + (m_a v_\parallel / e_a)\nabla \times \boldsymbol{b}$$

10. Modified electric field

$$\boldsymbol{E}^* = \boldsymbol{E} - (m_a v_\parallel / e_a)\frac{\partial \boldsymbol{b}}{\partial t}$$

11. Guiding center velocity for static field

$$\frac{\mathrm{d}\boldsymbol{r}}{\mathrm{d}t} = \frac{v_\parallel}{\boldsymbol{b} \cdot \boldsymbol{B}^*}\nabla \times \left(\boldsymbol{A} + \frac{m_a v_\parallel}{e_a}\boldsymbol{b}\right)$$

12. Morosov-Solovev guiding center velocity

$$\frac{\mathrm{d}\boldsymbol{r}}{\mathrm{d}t} = \frac{v_\parallel}{B}\nabla \times (\boldsymbol{A} + \rho_\parallel \boldsymbol{B})$$

13. Parallel Larmor radius

$$\rho_\parallel = \frac{m_a v_\parallel}{e_a B}$$

4.4 Orbital Dynamics: Hamiltonian Orbit Dynamics in Flux Coordinates

General formula in magnetic coordinates

1. Guiding center Lagrangian

$$L = e_a(\phi\dot{\theta} - \psi\dot{\zeta}) + \frac{m_a}{2B^2}(B_\phi\dot{\phi} + B_\theta\dot{\theta} + B_\zeta\dot{\zeta})^2 - \mu B - e_a\Phi$$

2. Canonical momentum

$$P_\theta = e_a(\phi + \rho_\parallel B_\theta)\ ,$$
$$P_\zeta = e_a(-\psi + \rho_\parallel B_\zeta)$$

3. Hamiltonian

$$H = \frac{e_a^2}{2m_a}\rho_\parallel^2 B^2 + \mu B + e_a\Phi$$

4. Parallel Larmor radius

$$\rho_\parallel = \frac{m_a v_\parallel}{e_a B} = \frac{m_a}{e_a B^2}(B_\phi\dot{\phi} + B_\theta\dot{\theta} + B_\zeta\dot{\zeta})$$

5. Magnetic field

$$\boldsymbol{B} = B_\phi \nabla\phi + B_\theta \nabla\theta + B_\zeta \nabla\zeta$$

Hamilton orbit dynamics in Boozer–Grad coordinates

1. Morosov–Solovev guiding center velocity

$$\begin{aligned}
&: \frac{\mathrm{d}\phi}{\mathrm{d}t} = \frac{\mathrm{d}\boldsymbol{r}}{\mathrm{d}t} \cdot \nabla\phi = v_\parallel B \frac{\partial\rho_\parallel}{\partial\alpha} \,, \\
&: \frac{\mathrm{d}\alpha}{\mathrm{d}t} = \frac{\mathrm{d}\boldsymbol{r}}{\mathrm{d}t} \cdot \nabla\alpha = v_\parallel B \left(\frac{\partial\rho_\parallel\beta}{\partial\chi} - \frac{\partial\rho_\parallel}{\partial\phi} \right) \,, \\
&: \frac{\mathrm{d}\chi}{\mathrm{d}t} = \frac{\mathrm{d}\boldsymbol{r}}{\mathrm{d}t} \cdot \nabla\chi = v_\parallel B \left(1 - \frac{\partial\rho_\parallel\beta}{\partial\alpha} \right) \,, \\
&: \frac{\mathrm{d}\rho_\parallel}{\mathrm{d}t} = \frac{\mathrm{d}\boldsymbol{r}}{\mathrm{d}t} \cdot \nabla\rho_\parallel = v_\parallel B \left[\frac{\partial\rho_\parallel}{\partial\chi} - \rho_\parallel \left(\frac{\partial\beta}{\partial\chi}\frac{\partial\rho_\parallel}{\partial\alpha} - \frac{\partial\beta}{\partial\alpha}\frac{\partial\rho_\parallel}{\partial\chi} \right) \right]
\end{aligned}$$

2. Guiding center Lagrangian

$$L = e_a \phi \dot{\alpha} + \frac{m_a}{2B^2}(\dot{\chi} + \beta\dot{\phi})^2 - \mu B - e_a \Phi$$

3. Canonical momentum

$$\begin{aligned}
P_\alpha &= e_a \phi \,, \\
P_\chi &= \frac{m_a}{B^2}(\dot{\chi} + \beta\dot{\phi}) = e_a \rho_\parallel
\end{aligned}$$

4. Hamiltonian

$$H = \frac{B^2}{2m_a} P_\chi^2 + \mu B + e_a \Phi$$

5. Taylor Lagrangian guiding center velocity

$$\begin{aligned}
\frac{\mathrm{d}\alpha}{\mathrm{d}t} &= \frac{\partial H}{e_a \partial\phi} = \Phi'(\phi) + \left(\frac{\mu}{e_a} + \frac{e_a \rho_\parallel^2 B}{m_a} \right) \frac{\partial B}{\partial\phi} \,, \\
\frac{\mathrm{d}\chi}{\mathrm{d}t} &= \frac{\partial H}{e_a \partial\rho_\parallel} = \frac{e_a \rho_\parallel B^2}{m_a} \,, \\
\frac{\mathrm{d}\phi}{\mathrm{d}t} &= -\frac{\partial H}{e_a \partial\alpha} = -e_a^{-1} \left(\mu + \frac{e_a^2 \rho_\parallel^2 B}{m_a} \right) \frac{\partial B}{\partial\alpha} \,, \\
\frac{\mathrm{d}\rho_\parallel}{\mathrm{d}t} &= -\frac{\partial H}{e_a \partial\chi} = -e_a^{-1} \left(\mu + \frac{e_a^2 \rho_\parallel^2 B}{m_a} \right) \frac{\partial B}{\partial\chi}
\end{aligned}$$

Hamilton orbit dynamics in Boozer coordinates

1. Morosov-Solovev guiding center velocity

$$\frac{d\phi}{dt} = \frac{d\boldsymbol{r}}{dt} \cdot \nabla\phi = \frac{v_\parallel B}{g + f/q}\left(f\frac{\partial\rho_\parallel}{\partial\zeta} - g\frac{\partial\rho_\parallel}{\partial\theta}\right),$$

$$\frac{d\theta}{dt} = \frac{d\boldsymbol{r}}{dt} \cdot \nabla\theta = \frac{v_\parallel B}{g + f/q}\left(\frac{\partial\beta_*\rho_\parallel}{\partial\zeta} - \psi'(\phi) - \frac{\partial g\rho_\parallel}{\partial\phi}\right),$$

$$\frac{d\zeta}{dt} = \frac{d\boldsymbol{r}}{dt} \cdot \nabla\zeta = \frac{v_\parallel B}{g + f/q}\left(1 + \frac{\partial f\rho_\parallel}{\partial\phi} - \frac{\partial\beta_*\rho_\parallel}{\partial\theta}\right),$$

$$\frac{d\rho_\parallel}{dt} = \frac{d\boldsymbol{r}}{dt} \cdot \nabla\rho_\parallel = \frac{v_\parallel B}{g + f/q}\left[2\frac{\partial\rho_\parallel}{\partial\phi}\left(f\frac{\partial\rho_\parallel}{\partial\zeta} - g\frac{\partial\rho_\parallel}{\partial\theta}\right) - \psi'(\phi) - \rho_\parallel\left(g'(\phi)\frac{\partial\rho_\parallel}{\partial\theta} - f'(\phi)\frac{\partial\rho_\parallel}{\partial\zeta}\right)\right]$$

2. Guiding center Lagrangian

$$L = e_a(\phi\dot{\theta} - \psi\dot{\zeta}) + \frac{m_a}{2B^2}(g\dot{\zeta} + f\dot{\theta})^2 - \mu B - e_a\Phi$$

3. Canonical momentum

$$P_\theta = e_a(\phi + f(\phi)\rho_\parallel),$$
$$P_\zeta = e_a(-\psi(\phi) + g(\phi)\rho_\parallel)$$

4. Hamiltonian

$$H = \frac{e_a^2}{2m_a}\rho_\parallel^2 B^2 + \mu B + e_a\Phi$$

5. Taylor Lagrangian guiding center velocity

$$\frac{d\phi}{dt} = \frac{\mu_1}{D}\left(f\frac{\partial B}{\partial\zeta} - g\frac{\partial B}{\partial\theta}\right),$$

$$\frac{d\theta}{dt} = \frac{g}{D}\left(\mu_1\frac{\partial B}{\partial\zeta} + e_a\frac{\partial\Phi}{\partial\phi}\right) - \frac{e_a^2 B^2}{m_a D}(\rho_\parallel g'(\phi) - 1/q(\phi))\rho_\parallel,$$

$$\frac{d\zeta}{dt} = -\frac{f}{D}\left(\mu_1\frac{\partial B}{\partial\zeta} + e_a\frac{\partial\Phi}{\partial\phi}\right) + \frac{e_a^2 B^2}{m_a D}(\rho_\parallel f'(\phi) + 1)\rho_\parallel,$$

$$\frac{d\rho_\parallel}{dt} = \frac{\mu_1}{D}\left[(\rho_\parallel g'(\phi) - 1/q(\phi))\frac{\partial B}{\partial\theta} - (\rho_\parallel f'(\phi) + 1)\frac{\partial B}{\partial\zeta}\right],$$

$$D = e_a[g + f/q + \rho_\parallel(gf'(\phi) - fg'(\phi))],$$

$$\mu_1 = \mu + e_a^2\rho_\parallel^2 B/m_a = \left(1 + 2v_\parallel^2/v_\perp^2\right)\mu$$

4.5 Periodicity and Invariants: Magnetic Moment and Longitudinal Adiabatic Invariant

1. Adiabatic invariant

$$J = \oint \boldsymbol{p} \cdot \mathrm{d}\boldsymbol{q}$$

2. Magnetic moment

$$J = \mu(2\pi m_a/e_a)$$

3. Longitudinal adiabatic invariant

$$J = m_a \oint v_\parallel \mathrm{d}l_\parallel$$

4. Longitudinal adiabatic invariant in Clebsch Coordinate

$$J = e_a \oint \rho_\parallel \mathrm{d}\chi$$

5. Bounce time of banana orbit

$$\frac{\partial J}{\partial H} = \oint \frac{\mathrm{d}l_\parallel}{v_\parallel} = \tau_b$$

6. Change in toroidal flux per bounce of banana orbit

$$\Delta\phi = \oint \frac{\partial \rho_\parallel}{\partial \alpha} \mathrm{d}\chi = e_a^{-1} \frac{\partial J}{\partial \alpha}$$

7. Change in Clebsch angle per bounce of banana orbit

$$\Delta\alpha = -\oint \frac{\partial \rho_\parallel}{\partial \phi} \mathrm{d}\chi = -e_a^{-1} \frac{\partial J}{\partial \phi}$$

8. Drift velocity of banana orbit

$$\frac{\mathrm{d}\phi}{\mathrm{d}t} = \frac{1}{e_a} \frac{\partial J/\partial \alpha}{\partial J/\partial H},$$

$$\frac{\mathrm{d}\alpha}{\mathrm{d}t} = -\frac{1}{e_a} \frac{\partial J/\partial \phi}{\partial J/\partial H}$$

4.6 Coordinate Invariance: Non-canonical Variational Principle and Lie Transformation

1. Canonical Lagrangian and its differential form

$$L = \boldsymbol{p} \cdot \dot{\boldsymbol{q}} - H(\boldsymbol{q}, \boldsymbol{p}, t)$$
$$\gamma = \boldsymbol{p} \cdot d\boldsymbol{q} - H\mathrm{d}t$$

2. Non-canonical Lagrangian and its differential form

$$L(\boldsymbol{z}, \dot{\boldsymbol{z}}, t) = \sum_{i=1}^{6} \gamma_i \dot{z}^i - h$$
$$\gamma = \gamma_\mu \mathrm{d}z^\mu = \gamma_i \mathrm{d}z^i - h\mathrm{d}t$$

3. Lagragian transformation law (($\boldsymbol{p}, \boldsymbol{q}$) to $\boldsymbol{z}$)

$$\gamma_i(\boldsymbol{z}, t) = \boldsymbol{p} \cdot \frac{\partial \boldsymbol{q}}{\partial z^i},$$
$$h(\boldsymbol{z}, t) = H(\boldsymbol{q}(\boldsymbol{z}, t), \boldsymbol{p}(\boldsymbol{z}, t), t) - \boldsymbol{p} \cdot \frac{\partial \boldsymbol{q}}{\partial t}$$

4. Equation of motion in non-canonical coordinates

$$\frac{\mathrm{d}z^i}{\mathrm{d}t} = \{z^i, h\}$$

5. Lagrange bracket

$$\omega_{ij} = [z^i, z^j] \equiv (\partial \boldsymbol{p}/\partial z^i) \cdot (\partial \boldsymbol{q}/\partial z^j) - (\partial \boldsymbol{p}/\partial z^j) \cdot (\partial \boldsymbol{q}/\partial z^i)$$

6. Poisson bracket

$$\pi_{ij} = \{z^i, z^j\} \equiv (\partial z^i/\partial \boldsymbol{q}) \cdot (\partial z^j/\partial \boldsymbol{p}) - (\partial z^i/\partial \boldsymbol{p}) \cdot (\partial z^j/\partial \boldsymbol{q})$$

7. Non-canonical coordinates

$$\boldsymbol{z} = \{z^\mu\} = \{t, z^i\}$$
$$\bar{\boldsymbol{z}} = \{\bar{z}^\mu\} = \{t, \bar{z}^i\}$$

8. Lagrangian 1 forms in $\boldsymbol{z}$ and $\bar{\boldsymbol{z}}$

$$\gamma = \gamma_\mu \mathrm{d}z^\mu, \ \Gamma = \Gamma_\nu \mathrm{d}\bar{z}^\nu$$

9. Transformation law between Γ and γ

$$\Gamma_\mu = \frac{\partial z^\nu}{\partial \bar{z}^\mu} \gamma_\nu$$

Lie transformation

1. Definition of Lie transformation

$$\partial \bar{z}^{\mu}(\boldsymbol{z}, \varepsilon)/\partial \varepsilon = g^{\mu}(\bar{\boldsymbol{z}})$$

2. Identity relation

$$z^{\mu}(\bar{z}^{\mu}(\boldsymbol{z}, \varepsilon), \varepsilon) = z^{\mu}$$

3. Lie transformation relation

$$\partial z^{\mu}(\bar{\boldsymbol{z}}, \varepsilon)/\partial \varepsilon = -g^{\nu}(\bar{\boldsymbol{z}}) \partial z^{\mu}(\bar{\boldsymbol{z}}, \varepsilon)/\partial \bar{z}^{\nu}$$

4. Lie transformation relation for scalar

$$\begin{aligned} S(\bar{\boldsymbol{z}}, \varepsilon) &= s(\boldsymbol{z}) \ , \\ \partial S(\bar{\boldsymbol{z}}, \varepsilon)/\partial \varepsilon &= -g^{\mu}(\bar{\boldsymbol{z}}) \partial S(\bar{\boldsymbol{z}}, \varepsilon)/\partial \bar{z}^{\mu} \ , \\ S(\boldsymbol{y}, \varepsilon) &= \exp(-\varepsilon L) s(\boldsymbol{y}) \ , \\ S(\boldsymbol{y}, \varepsilon) &= \exp(-\varepsilon L) s(\boldsymbol{y}, \varepsilon) \end{aligned}$$

5. Lie transformation relation for differential form

$$\begin{aligned} \partial \Gamma_{\mu}(\boldsymbol{y}, \varepsilon)/\partial \varepsilon &= -g^{\lambda}(\boldsymbol{y}) \left[\partial \Gamma_{\mu}(\boldsymbol{y}, \varepsilon)/\partial y^{\lambda} - \partial \Gamma_{\lambda}(\boldsymbol{y}, \varepsilon)/\partial y^{\mu} \right] \ , \\ \Gamma_{\mu}(\bar{\boldsymbol{z}}, \varepsilon) &= (\partial z^{\nu}(\bar{\boldsymbol{z}}, \varepsilon)/\partial \bar{z}^{\mu}) \gamma_{\nu}(\boldsymbol{z}(\bar{\boldsymbol{z}}, \varepsilon)) \ , \\ \partial \Gamma_{\mu}(\bar{\boldsymbol{z}}, \varepsilon)/\partial \varepsilon &= -L \Gamma_{\mu} \end{aligned}$$

6. Operator L for differential form

$$(L\omega)_{\mu} = g_{n}^{\lambda} \left(\partial \omega_{\mu}/\partial y^{\lambda} - \partial \omega_{\lambda}/\partial y^{\mu} \right)$$

7. Lie transformation relation

$$\Gamma = T\gamma + \mathrm{d}S$$

4.7 "Lie perturbation theory": Gyro Center Orbit Dynamics

1. Lagrangian differential 1 form in $\boldsymbol{z} = \{z^{\mu}\} = \{t, z^{i}\}$

$$L\,\mathrm{d}t = \gamma = \gamma_{i}(\boldsymbol{z}, \varepsilon)\,\mathrm{d}z^{i}(\bar{\boldsymbol{z}}, \varepsilon) - h(\boldsymbol{z}, \varepsilon)\,\mathrm{d}t$$

2. Lagrangian differential 1 form in $\bar{\boldsymbol{z}} = \{\bar{z}^{\mu}\} = \{t, \bar{z}^{i}\}$

$$L\,\mathrm{d}t = \Gamma = \Gamma_{i}(\bar{\boldsymbol{z}})\,\mathrm{d}\bar{z}^{i} - H(\bar{\boldsymbol{z}})\,\mathrm{d}t + \mathrm{d}S(\bar{\boldsymbol{z}})$$

3. Taylor expansion of T

$$T = \ldots \exp(-\varepsilon^2 L_2)\exp(-\varepsilon L_1) = 1 - \varepsilon L_1 + \varepsilon^2((1/2)L_1^2 - L_2) + \ldots$$

4. Lie perturbation relations

$$\Gamma_0 = \mathrm{d}S_0 + \gamma_0\,, \qquad (\varepsilon^0)$$
$$\Gamma_1 = \mathrm{d}S_1 - L_1\gamma_0 + \gamma_1\,, \qquad (\varepsilon^1)$$
$$\Gamma_2 = \mathrm{d}S_2 - L_2\gamma_0 + \gamma_2 - L_1\gamma_1 + (1/2)L_1^2\gamma_0 \qquad (\varepsilon^2)$$

5. Lagrangian 1 forms of a charged particle

$$\gamma(t, \boldsymbol{x}, \boldsymbol{v}) = (e_a \boldsymbol{A}(\boldsymbol{x}, t) + m\boldsymbol{v}) \cdot \mathrm{d}\boldsymbol{x} - \left[m_a v^2/2 + e_a \varphi(\boldsymbol{x}, t)\right] \mathrm{d}t$$

6. Zero, first, and second order Lagrangian 1 forms of a charged particle

$$\gamma_0(t, \boldsymbol{r}, v_\parallel, \mu, \theta) = (e_a \boldsymbol{A} + m_a v_\parallel \boldsymbol{b}) \cdot \mathrm{d}\boldsymbol{r} - (m_a/e_a)\mu \mathrm{d}\theta - \left[m_a v_\parallel^2/2 + \mu B(\boldsymbol{r})\right] \mathrm{d}t\,,$$
$$\gamma_1 = -e_a \varphi(\boldsymbol{r} + \boldsymbol{\rho}, t)\mathrm{d}t\,,$$
$$\gamma_2 = 0$$

7. First order generating function of Lie transform

$$g_1^i = \{S_1, z^i\}$$

8. First order Perturbed Hamiltonian in $\bar{z}$

$$H_1 = h_1 - \frac{\mathrm{d}S_1}{\mathrm{d}t} = e_a\varphi - \frac{\partial S_1}{\partial t} - \{S_1, h_0\}$$

9. Gyro phase averaged Hamiltonian

$$\langle H_1 \rangle = H_1 = \langle e_a \varphi \rangle$$

10. First order gauge

$$S_1 = -e_a \int \varphi \mathrm{d}t \approx -\frac{e_a}{\Omega_a} \int \varphi \mathrm{d}\theta$$

11. Second order generating function of Lie transform

$$g_2^i = \{S_2, z^i\}$$

12. Second order perturbed Hamiltonian in $\bar{z}$

$$H_2 = \langle H_2 \rangle = -\frac{1}{2} \langle \{S_1, h_1\} \rangle$$

13. Coordinate transformation relation

$$\bar{z}^\mu = z^\mu + \varepsilon \{S_1, z^\mu\} + O(\varepsilon^2)$$

Chapter 5
Plasma Kinetic Theory: Collective Equation in Phase Space

5.1 Phase Space: Liouville Theorem and Poincare Recurrence Theorem

1. Hamilton equation

$$\frac{\mathrm{d}q_j}{\mathrm{d}t} = \frac{\partial H}{\partial p_j}, \quad \frac{\mathrm{d}p_j}{\mathrm{d}t} = -\frac{\partial H}{\partial q_j}$$

2. Liouville theorem

$$\frac{\mathrm{d}D}{\mathrm{d}t} = \frac{\partial D}{\partial t} + \{D, H\} = 0$$

3. Poisson bracket

$$\{D, H\} = \sum_{j=1}^{3N} \left[\frac{\partial D}{\partial q_j}\frac{\partial H}{\partial p_j} - \frac{\partial D}{\partial p_j}\frac{\partial H}{\partial q_j} \right]$$

5.2 Dynamics and Kinetics: Individual Reversible and Collective non-reversible Equations

1. Klimontovich equation

$$\frac{\mathrm{d}F}{\mathrm{d}t} = \frac{\partial F}{\partial t} + \boldsymbol{v} \cdot \frac{\partial F}{\partial \boldsymbol{x}} + \boldsymbol{a} \cdot \frac{\partial F}{\partial \boldsymbol{v}} = 0$$

2. N-body distribution function

$$F(\boldsymbol{x}, \boldsymbol{v}, t) = \sum_{i=1}^{N} \delta(\boldsymbol{x} - \boldsymbol{x}_i(t))\delta(\boldsymbol{v} - \boldsymbol{v}_i(t))$$

3. Acceleration by electromagnetic force

$$\boldsymbol{a} = \frac{e_a}{m_a}(\boldsymbol{E} + \boldsymbol{v} \times \boldsymbol{B})$$

4. Ensemble averaged velocity distribution function

$$f = \langle F(\boldsymbol{x}, \boldsymbol{v}, t)\rangle_{\mathrm{ensemble}}$$

5. Fluctuating velocity distribution function

$$F = f + \tilde{F}$$

6. Ensemble averaged EM acceleration

$$\bar{\boldsymbol{a}} = \frac{e_a}{m_a}(\bar{\boldsymbol{E}} + \boldsymbol{v} \times \bar{\boldsymbol{B}})$$

7. Fluctuating EM acceleration

$$\tilde{\boldsymbol{a}} = \boldsymbol{a} - \bar{\boldsymbol{a}}$$

8. Boltzmann equation

$$\frac{\mathrm{d}f}{\mathrm{d}t} = \frac{\partial f}{\partial t} + \boldsymbol{v} \cdot \frac{\partial f}{\partial \boldsymbol{x}} + \bar{\boldsymbol{a}} \cdot \frac{\partial f}{\partial \boldsymbol{v}} = C(f)$$

9. General collision term

$$C(f) = -\left\langle \tilde{\boldsymbol{a}} \cdot \frac{\partial \tilde{F}}{\partial \boldsymbol{v}} \right\rangle_{\text{ensemble}}$$

10. Boltzmann collision integral

$$C(f) = \int \left[f(\boldsymbol{v}') f(\boldsymbol{v}_1') - f(\boldsymbol{v}) f(\boldsymbol{v}_1) \right] |\boldsymbol{v}_1 - \boldsymbol{v}| \, \sigma \, \mathrm{d}\Omega \, \mathrm{d}\boldsymbol{v}_1$$

11. Boltzmann's H theorem

$$\frac{\mathrm{d}H}{\mathrm{d}t} = \frac{\mathrm{d}}{\mathrm{d}t} \int f(\boldsymbol{v}) \ln f(\boldsymbol{v}) \mathrm{d}\boldsymbol{v} \leq 0$$

5.3 Vlasov Equation: Invariants, Time Reversibility, and Continuous Spectrum

1. Vlasov equation ($\bar{\boldsymbol{a}} = (e_s/m_s)(\boldsymbol{E} + \boldsymbol{v} \times \boldsymbol{B})$)

$$\frac{\mathrm{d}f}{\mathrm{d}t} = \frac{\partial f}{\partial t} + \boldsymbol{v} \cdot \frac{\partial f}{\partial \boldsymbol{x}} + \bar{\boldsymbol{a}} \cdot \frac{\partial f}{\partial \boldsymbol{v}} = 0$$

2. Generalized entropy conservation law (G: arbitrary function)

$$\frac{\mathrm{d}H}{\mathrm{d}t} = \frac{\mathrm{d}}{\mathrm{d}t} \int G(f_s) \mathrm{d}\boldsymbol{x} \mathrm{d}\boldsymbol{v} = 0$$

3. Linearized Vlasov–Poisson equation for Langmuir wave

$$\frac{\partial f_{a1}}{\partial t} + \boldsymbol{v} \cdot \frac{\partial f_{a1}}{\partial \boldsymbol{x}} = \frac{e_a}{m_a} \nabla \varphi \cdot \frac{\partial f_{a0}}{\partial \boldsymbol{v}} ,$$

$$\varepsilon_0 \nabla^2 \varphi = -e_a \int_{-\infty}^{\infty} f_{a1} \mathrm{d}\boldsymbol{v}$$

7. Fourier transformed Vlasov equation

$$(\omega - \boldsymbol{k} \cdot \boldsymbol{v}) f_{a1\boldsymbol{k}\omega} = -\frac{e_a}{m_a} \varphi_{\boldsymbol{k}\omega} \boldsymbol{k} \cdot \frac{\partial f_{a0}}{\partial \boldsymbol{v}}$$

8. Solution of Fourier transformed Vlasov equation

$$f_{a1\boldsymbol{k}\omega} = \left[-\frac{e_a}{m_a} \boldsymbol{k} \cdot \frac{\partial f_{a0}}{\partial \boldsymbol{v}} P \frac{1}{w - \boldsymbol{k} \cdot \boldsymbol{v}} + \lambda \delta (w - \boldsymbol{k} \cdot \boldsymbol{v}) \right] \varphi_{\boldsymbol{k}\omega}$$

9. Dispersion relation for Vlasov-Poisson equation with continuous spectrum

$$\left[1 + \frac{e_a}{\varepsilon_0 k^2 m_a} \int_{-\infty}^{\infty} \frac{P}{w - \boldsymbol{k} \cdot \boldsymbol{v}} \boldsymbol{k} \cdot \frac{\partial f_{a0}}{\partial \boldsymbol{v}} \mathrm{d}\boldsymbol{v} \right] + \frac{e_a}{\varepsilon_0 k^2} \lambda = 0$$

5.4 Landau damping: Irreversible Phenomenon Caused by Reversible Equation

1. Fourier-Laplace transformed Vlasov-Poisson equation

$$(\omega - \boldsymbol{k} \cdot \boldsymbol{v}) f_{e1\boldsymbol{k}\omega}(\boldsymbol{v}) = \mathrm{i} f_{e1\boldsymbol{k}}(\boldsymbol{v}, t = 0) + \frac{e}{m_e} \varphi_{\boldsymbol{k}\omega} \boldsymbol{k} \cdot \frac{\partial f_{e0}}{\partial \boldsymbol{v}}$$

$$\mathrm{i} \varepsilon_0 k^2 \varphi_{\boldsymbol{k}\omega} = -e \int_{-\infty}^{\infty} f_{e1\boldsymbol{k}}(\boldsymbol{v}, \omega) \mathrm{d}\boldsymbol{v}$$

2. Solution of electrostatic potential

$$\varphi_{\boldsymbol{k}\omega} = -\frac{ie}{\varepsilon_0 k^2 K(\omega, \boldsymbol{k})} \int_{-\infty}^{\infty} \frac{f_{e1k}(\boldsymbol{v}, t = 0)}{\omega - \boldsymbol{k} \cdot \boldsymbol{v}} \mathrm{d}\boldsymbol{v} ,$$

$$K(\boldsymbol{k}, \omega) = 1 + \frac{\omega_{pe}^2}{n_e k^2} \int \frac{\boldsymbol{k} \cdot \partial f_{e0} / \partial \boldsymbol{v}}{\omega - \boldsymbol{k} \cdot \boldsymbol{v}} \mathrm{d}\boldsymbol{v}$$

3. Landau damping rate

$$\omega_i = -\frac{K_i(\boldsymbol{k}, \omega_r)}{\partial K_r(\boldsymbol{k}, \omega_r) / \partial \omega_r} ,$$

$$K_i(\boldsymbol{k}, \omega_r) = -\pi \frac{\omega_{pe}^2}{k^2} \frac{\partial f_{e0}}{\partial v} \Big|_{u=\omega_r/k} ;$$

$$K_r(\boldsymbol{k}, \omega) = 1 + \frac{\omega_{pe}^2}{n_e k^2} P \int \frac{\boldsymbol{k} \cdot \partial f_{e0}/\partial \boldsymbol{v}}{\omega_r - \boldsymbol{k} \cdot \boldsymbol{v}} \mathrm{d}\boldsymbol{v}$$

4. Electric field damped by phase mixing

$$\boldsymbol{E}_{\boldsymbol{k}}(t) = -\frac{e\boldsymbol{k}}{2\pi\varepsilon_0 k^2} \int_{-\infty}^{\infty} \mathrm{d}\boldsymbol{v}\, f_{e1\boldsymbol{k}}(\boldsymbol{v}, t=0) \int_{-\infty+i\omega_{I0}}^{\infty+i\omega_{I0}} \frac{\exp(-\mathrm{i}\omega t) d\omega}{K(\omega, \boldsymbol{k})(\omega - \boldsymbol{k} \cdot \boldsymbol{v})}$$

5.5 Coulomb Logarithm: Collective Behavior in the Coulomb Field

1. Shielded Coulomb potential

$$\phi = \frac{e}{4\pi\varepsilon_0 r} e^{-r/\lambda_D}$$

2. Debye length

$$\lambda_D^{-2} = \lambda_{De}^{-2} + \sum \lambda_{Di}^{-2} ,$$

$$\lambda_{De}^2 = (\varepsilon_0 kT/e^2 n_e)^{0.5} \quad (= 7.43 \times 10^3 [T_e(\mathrm{eV})/n_e(\mathrm{m}^{-3})]^{0.5}\ [\mathrm{m}]) ,$$

$$\lambda_{Di}^2 = (\varepsilon_0 kT/e^2 Z_i^2 n_i)^{0.5}$$

3. Impact parameter and scattering angle

$$b = b_0 \cot\left(\frac{\theta}{2}\right)$$

4. Landau parameter

$$b_0 = e_a e_b/(4\pi\varepsilon_0 m_{ab} u^2) = 7.2 \times 10^{-10} Z_a Z_b / E_r(\mathrm{eV})(\mathrm{m})$$

5. Reduced mass

$$m_{ab} = m_a m_b/(m_a + m_b)$$

6. Differential cross section

$$\sigma(\theta) = b(\mathrm{d}b/\mathrm{d}\theta)/\sin\theta$$

7. Rutherford cross section

$$\sigma(\theta) = \frac{b_0^2}{\sin^4(\theta/2)}$$

8. Velocity of species a

$$\boldsymbol{v}_a = \boldsymbol{V} + m_b \boldsymbol{u}_{ab}/(m_a + m_b)$$

9. Change of relative velocity

$$\Delta \boldsymbol{u}_{ab} = u_{ab} \sin\theta \boldsymbol{n} - 2\sin^2(\theta/2)\boldsymbol{u}_{ab}$$
$$(\sin^2(\theta/2) = b_0^2/(b_0^2 + b^2))$$

10. Change of velocity of species a

$$\Delta \boldsymbol{v}_a = -4\pi b_0^2 \Delta\phi_b \boldsymbol{u}_{ab} \frac{m_{ab}}{m_a} \int\limits_0^{\lambda_D} \frac{b}{b^2 + b_0^2} \mathrm{d}b$$

11. Coulomb logarithm

$$\ln \Lambda \equiv \ln(\lambda_D/b)$$

5.6 Fokker–Planck Equation: Statistics of Soft Coulomb Collision

1. Integral equation of velocity distribution function for Markov process

$$f_a(\boldsymbol{v}, t) = \int \mathrm{d}\Delta\boldsymbol{v}\, f_a(\boldsymbol{v} - \Delta\boldsymbol{v}, t - \Delta t) P(\boldsymbol{v} - \Delta\boldsymbol{v}; \Delta\boldsymbol{v}, \Delta t)$$

2. Collision term for Markov process

$$C(f_a) = -\frac{\partial}{\partial \boldsymbol{v}} \cdot \left(\frac{\langle \Delta\boldsymbol{v} \rangle}{\Delta t} f_a \right) + \frac{\partial^2}{\partial \boldsymbol{v} \partial \boldsymbol{v}} : \left(\frac{\langle \Delta\boldsymbol{v} \Delta\boldsymbol{v} \rangle}{2\Delta t} f_a \right)$$

3. Expression for changing rate of velocity

$$\left\langle \frac{\Delta \boldsymbol{v}_a}{\Delta t} \right\rangle = -\sum_b \frac{e_a^2 e_b^2 \ln \Lambda}{4\pi m_a^2 \varepsilon_0^2} \left(1 + \frac{m_a}{m_b}\right) \int \frac{\boldsymbol{u}_{ab}}{u_{ab}^3} f_b(\boldsymbol{v}_b) \mathrm{d}\boldsymbol{v}_b$$

4. Expression for changing rate of perpendicular velocity

$$\left\langle \frac{\Delta v_\perp^2}{\Delta t} \right\rangle = \sum_b \frac{e_a^2 e_b^2}{4\pi m_a^2 \varepsilon_0^2} \left(\frac{1}{2} + \ln \Lambda\right) \int \frac{1}{u_{ab}} f_b(\boldsymbol{v}_b) \mathrm{d}\boldsymbol{v}_b$$

5. Expression for changing rate of parallel velocity

$$\left\langle \frac{\Delta v_\parallel^2}{\Delta t} \right\rangle = \sum_b \frac{e_a^2 e_b^2}{16\pi m_a^2 \varepsilon_0^2} \int \frac{1}{u_{ab}} f_b(\boldsymbol{v}_b) \mathrm{d}\boldsymbol{v}_b$$

6. Slowing down rate

$$\left\langle \frac{\Delta \boldsymbol{v}_a}{\Delta t} \right\rangle = \sum_b \frac{e_a^2 e_b^2 \ln \Lambda}{4\pi m_a^2 \varepsilon_0^2} \frac{\partial h_{ab}(\boldsymbol{v}_a)}{\partial \boldsymbol{v}_a} ,$$

$$h_{ab}(\boldsymbol{v}_a) = \left(1 + \frac{m_a}{m_b}\right) \int \frac{f_b(\boldsymbol{v}_b)}{u_{ab}} \mathrm{d}\boldsymbol{v}_b$$

7. Velocity space diffusion tensor

$$\left\langle \frac{\Delta \boldsymbol{v}_a \Delta \boldsymbol{v}_a}{2\Delta t} \right\rangle = \sum_b \frac{e_a^2 e_b^2 \ln \Lambda}{8\pi m_a^2 \varepsilon_0^2} \int \frac{u_{ab}^2 \boldsymbol{I} - \boldsymbol{u}_{ab}\boldsymbol{u}_{ab}}{u_{ab}^3} f_b(\boldsymbol{v}_b)\, \mathrm{d}\boldsymbol{v}_b ,$$

$$= \sum_b \frac{e_a^2 e_b^2 \ln \Lambda}{8\pi m_a^2 \varepsilon_0^2} \frac{\partial^2 g_{ab}(\boldsymbol{v}_a)}{\partial vec v_a \partial \boldsymbol{v}_a}$$

$$g_{ab}(\boldsymbol{v}_a) = \int u_{ab} f_b(\boldsymbol{v}_b)\, \mathrm{d}\boldsymbol{v}_b$$

8. Collision term using Rosenbluth potentials

$$C(f_a) = \sum_b \frac{e_a^2 e_b^2 \ln \Lambda}{4\pi m_a^2 \varepsilon_0^2} \left[-\frac{\partial}{\partial \boldsymbol{v}_a} \cdot \left(\frac{\partial h_{ab}}{\partial \boldsymbol{v}_a} f_a \right) + \frac{1}{2} \frac{\partial^2}{\partial \boldsymbol{v}_a \partial \boldsymbol{v}_a} : \left(\frac{\partial^2 g_{ab}}{\partial \boldsymbol{v}_a \partial \boldsymbol{v}_a} f_a \right) \right]$$

9. Landau form of collision integral

$$C(f_a) = \sum_b \frac{e_a^2 e_b^2 \ln \Lambda}{8\pi \varepsilon_0^2 m_a} \frac{\partial}{\partial \boldsymbol{v}} \cdot \int \mathrm{d}\boldsymbol{v}_b \boldsymbol{U} \cdot \left[\frac{f_b(\boldsymbol{v}_b)}{m_a} \frac{\partial f_a(\boldsymbol{v})}{\partial \boldsymbol{v}} - \frac{f_a(\boldsymbol{v})}{m_b} \frac{\partial f_b(\boldsymbol{v}_b)}{\partial \boldsymbol{v}_b} \right]$$

10. Balescu–Lenard collision term

$$C(f_a) = -\sum_b \frac{e_a^2 e_b^2}{8\pi \varepsilon_0^2 m_a} \frac{\partial}{\partial \boldsymbol{v}_a} \cdot \int \mathrm{d}\boldsymbol{v}_b \boldsymbol{K}_{ab}(\boldsymbol{v}_a, \boldsymbol{v}_b)$$

$$\cdot \left[f_a \frac{1}{m_b} \frac{\partial f_b}{\partial \boldsymbol{v}_b} - f_b \frac{1}{m_a} \frac{\partial f_a}{\partial \boldsymbol{v}_a} \right] ,$$

$$K_{ab}(\boldsymbol{v}_a, \boldsymbol{v}_b) = \int \mathrm{d}\boldsymbol{k} \delta(\boldsymbol{k} \cdot (\boldsymbol{v}_a - \boldsymbol{v}_b)) \frac{\boldsymbol{k}\boldsymbol{k}}{k^4 \left|\kappa(\boldsymbol{k}, \boldsymbol{k} \cdot \boldsymbol{v}_a)\right|^2} ,$$

$$\kappa(\boldsymbol{k}, \omega) = 1 + \frac{e_b^2}{\varepsilon_0 m_b k^2} \int \mathrm{d}\boldsymbol{v} \frac{\boldsymbol{k} \cdot \partial f_b / \partial \boldsymbol{v}}{\omega - \boldsymbol{k} \cdot \boldsymbol{v}}$$

5.7 Gyro-center Kinetic Theory: Drift and Gyro Kinetic Theory

1. Kinetic equation

$$\frac{\partial f}{\partial t} + \boldsymbol{v} \cdot \frac{\partial f}{\partial \boldsymbol{x}} + (\boldsymbol{E} + \boldsymbol{v} \times \boldsymbol{B}) \cdot \frac{\partial f}{\partial \boldsymbol{v}} = C(f)$$

2. Guiding center Poisson bracket

$$\{X, Y\} = \frac{e_a}{m_a}\left(\frac{\partial X}{\partial \theta}\frac{\partial Y}{\partial \mu} - \frac{\partial X}{\partial \mu}\frac{\partial Y}{\partial \theta}\right) - \frac{\boldsymbol{b}}{e_a B_\parallel^*}\cdot \nabla X \times \nabla Y$$
$$+ \frac{\boldsymbol{B}^*}{m_a B_\parallel^*}\left(\nabla X \frac{\partial Y}{\partial v_\parallel} - \frac{\partial X}{\partial v_\parallel}\nabla Y\right)$$

3. Orbit equation in guiding center coordinates $\boldsymbol{z} = (\boldsymbol{r}, v_\parallel, \mu, \theta)$

$$\frac{\mathrm{d}\mu}{\mathrm{d}t} = \{\mu, H\} = 0\,, \quad \frac{\mathrm{d}\theta}{\mathrm{d}t} = \{\theta, H\}\ ,$$
$$\frac{\mathrm{d}v_\parallel}{\mathrm{d}t} = \{v_\parallel, H\} = -\frac{\boldsymbol{B}^*}{m_a B_\parallel^*}\nabla H\ ,$$
$$\frac{\mathrm{d}\boldsymbol{r}}{\mathrm{d}t} = \{\boldsymbol{r}, H\} = \frac{\boldsymbol{b}}{e_a B_\parallel^*}\times \nabla H + \frac{\boldsymbol{B}^*}{m_a B_\parallel^*}\frac{\partial H}{\partial v_\parallel}$$

4. Drift kinetic equation

$$\frac{\partial F}{\partial t} + \{F, H\} = C(F)$$

or

$$\frac{\partial F}{\partial t} + \dot{\boldsymbol{r}}\cdot\frac{\partial F}{\partial \boldsymbol{r}} + \dot{v}_\parallel \frac{\partial F}{\partial v_\parallel} = C(F)$$

5. Perturbed electrostatic and vector potentials

$$(\delta\varphi, \delta\boldsymbol{A})$$

6. Perturbed Lagrangian

$$\delta L\,\mathrm{d}t = e_a\delta_*\boldsymbol{A}\cdot(\mathrm{d}\boldsymbol{r} + \mathrm{d}\boldsymbol{\rho}) - e_a\delta_*\varphi\,\mathrm{d}t = -\delta H\,\mathrm{d}t\ ,$$
$$\delta_*\boldsymbol{A} = \delta\boldsymbol{A}(\boldsymbol{r} + \boldsymbol{\rho}), \delta_*\varphi = \delta\varphi(\boldsymbol{r} + \boldsymbol{\rho})$$

7. Perturbed Hamiltonian

$$\delta H = e_a\delta_*\varphi - e_a\delta_*\boldsymbol{A}\cdot\boldsymbol{v}$$

8. Coordinate transformation

$$\boldsymbol{z} = (\boldsymbol{r}, v_\parallel, \mu, \theta) \Rightarrow \bar{\boldsymbol{z}} = (\bar{\boldsymbol{r}}, \bar{v}_\parallel, \bar{\mu}, \bar{\theta})$$

9. Perturbation expansion of Hamiltonian

$$\bar{H} = \bar{H}_0 + \bar{H}_1 + \bar{H}_2 + \,\ldots$$
$$\bar{H}_0 = \frac{1}{2}m_a\bar{v}_\parallel^2 + \bar{\mu}B(\bar{\boldsymbol{r}}) + e_a\bar{\Phi}(\bar{\boldsymbol{r}}, t)\ ,$$

$$\bar{H}_1 = \delta H - \frac{\mathrm{d}S_1}{\mathrm{d}t} \,,$$

$$\bar{H}_2 = \frac{e_a^2}{2m_a} |\delta_* \boldsymbol{A}|^2 - \frac{1}{2}\{S_1, \delta H\} - \frac{\mathrm{d}S_2}{\mathrm{d}t}$$

10. Solution of $\bar{H} = \bar{H}_0 + \bar{H}_1 + \bar{H}_2 + \ldots$

$$\bar{H}_1 = e_a \langle \delta_* \varphi \rangle - e_a \langle \delta_* \boldsymbol{A} \rangle \cdot \boldsymbol{b} v_\parallel - e_a \langle \delta_* \boldsymbol{A} \cdot \boldsymbol{v}_\perp \rangle \,,$$

$$\bar{H}_2 = \frac{e_a^2}{2m_a} \left\langle |\delta_* \boldsymbol{A}|^2 \right\rangle - \frac{1}{2} \langle \{S_1, \delta H\} \rangle$$

11. Solution of coordinate transformation

$$\bar{z}_a = z_a + \{S_1, z_a\} + e_a \delta_* \boldsymbol{A} \cdot \{\boldsymbol{r} + \boldsymbol{\rho}, z_a\} + \ldots$$

12. Gyro kinetic equation

$$\frac{\partial \bar{F}}{\partial t} + \{\bar{F}, \bar{H}\} = \bar{C}(\bar{F})$$

or

$$\frac{\partial \bar{F}}{\partial t} + \dot{\bar{\boldsymbol{r}}} \cdot \frac{\partial \bar{F}}{\partial \bar{\boldsymbol{r}}} + \dot{\bar{v}}_\parallel \frac{\partial \bar{F}}{\partial \bar{v}_\parallel} = \bar{C}(\bar{F})$$

13. Gyro kinetic orbit equations

$$\frac{\mathrm{d}\bar{v}_\parallel}{\mathrm{d}t} = \{\bar{v}_\parallel, \bar{H}\} = -\frac{\boldsymbol{B}^*}{m_a B_\parallel^*} \bar{\nabla} \bar{H} \,,$$

$$\frac{\mathrm{d}\bar{\boldsymbol{r}}}{\mathrm{d}t} = \{\bar{\boldsymbol{r}}, \bar{H}\} = \frac{\boldsymbol{b}}{e_a B_\parallel^*} \times \bar{\nabla} \bar{H} + \frac{\boldsymbol{B}^*}{m_a B_\parallel^*} \frac{\partial \bar{H}}{\partial \bar{v}_\parallel}$$

Chapter 6
Magnetohydrodynamic Stability: Energy Principle, Flow, and Dissipation

6.1 Stability: Introduction

1. Evolution equation in dynamical system

$$\frac{\mathrm{d}\boldsymbol{X}}{\mathrm{d}t} = N(\boldsymbol{X})$$

2. Linearized evolution equation

$$L\boldsymbol{\xi} = \lambda \boldsymbol{\xi}$$
$$(L = N'(\boldsymbol{X}_0) \,, \; \boldsymbol{\xi} = \boldsymbol{X} - \boldsymbol{X}_0)$$

3. Regular matrix

$$\boldsymbol{L}\boldsymbol{L}^* = \boldsymbol{L}^*\boldsymbol{L}$$

4. Unitary transformation

$$\boldsymbol{U}^{-1}\boldsymbol{L}\boldsymbol{U}$$

5. Self-adjoint matrix

$$\boldsymbol{L}^* = \boldsymbol{L}$$

6. Position operator (with continuous spectrum)

$$Au(x) = xu(x)$$

7. Eigenvalue equation

$$Au = \lambda u$$

8. Inverse operator of eigenvalue equation

$$(\lambda I - A)^{-1}$$

6.2 Ideal Magnetohydrodynamics: Action Principles and the Hermitian Operator

1. Action integral of ideal MHD equation

$$S = \int_{t_1}^{t_2} \mathrm{d}t \int L \,\mathrm{d}V$$

2. Lagrangian density of ideal MHD

$$L = \frac{1}{2}\rho \boldsymbol{v}^2 - \frac{P}{\gamma - 1} - \frac{\boldsymbol{B}^2}{2\mu_0}$$

3. Perturbed MHD relations

$$\begin{aligned}
\delta \boldsymbol{v} &= \boldsymbol{v} \cdot \nabla \boldsymbol{\xi} - \boldsymbol{\xi} \cdot \nabla \boldsymbol{v} + \partial \boldsymbol{\xi}/\partial t \ , \\
\delta \rho &= -\nabla \cdot (\rho \boldsymbol{\xi}) \ , \\
\delta P &= -\gamma P \nabla \cdot \boldsymbol{\xi} - \boldsymbol{\xi} \cdot \nabla P \ , \\
\delta \boldsymbol{B} &= \nabla \times (\boldsymbol{\xi} \times \boldsymbol{B})
\end{aligned}$$

4. Variation of action integral

$$\delta S = -\int_{t_1}^{t_2} \mathrm{d}t \int \mathrm{d}V \delta \boldsymbol{\xi} \cdot \left[\frac{\partial(\rho \boldsymbol{v})}{\partial t} + \nabla \cdot (\rho \boldsymbol{v}\boldsymbol{v}) + \nabla P - \boldsymbol{J} \times \boldsymbol{B} \right]$$

5. Action integral of toroidal equilibrium

$$S = \int L \, \mathrm{d}V = \int \left[\frac{B^2}{2\mu_0} + \frac{P}{\gamma - 1} \right] \mathrm{d}V$$

6. Variation of action integral for equilibrium

$$\delta S = -\int \boldsymbol{\xi} \cdot \left[\mu_0^{-1} (\nabla \times \boldsymbol{B}) \times \boldsymbol{B} - \nabla P \right] \mathrm{d}V$$

7. Linearized ideal MHD equation

$$\begin{aligned} \rho \partial^2 \boldsymbol{\xi} / \partial t^2 &= \delta \boldsymbol{J} \times \boldsymbol{B} + \boldsymbol{J} \times \delta \boldsymbol{B} - \nabla \delta P \\ &= \mu_0^{-1} \{ \nabla \times [\nabla \times (\boldsymbol{\xi} \times \boldsymbol{B})] \} \times \boldsymbol{B} + \mu_0^{-1} (\nabla \times \boldsymbol{B}) \times [\nabla \times (\boldsymbol{\xi} \times \boldsymbol{B})] \\ &\quad + \nabla [\gamma P \nabla \cdot \boldsymbol{\xi} + \boldsymbol{\xi} \cdot \nabla P] \end{aligned}$$

8. Hermitian property of MHD operator

$$\int \boldsymbol{\eta} \cdot F(\boldsymbol{\xi}) \, \mathrm{d}V = \int \boldsymbol{\xi} \cdot F(\boldsymbol{\eta}) \, \mathrm{d}V$$

9. Hermitian structure of MHD operator

$$\begin{aligned} \int \boldsymbol{\eta} \cdot \boldsymbol{F}(\boldsymbol{\xi}) \, \mathrm{d}V = -\int \mathrm{d}V \Bigg[& \frac{1}{\mu_0} (\boldsymbol{B} \cdot \nabla \boldsymbol{\xi}_\perp) \cdot (\boldsymbol{B} \cdot \nabla \boldsymbol{\eta}_\perp) + \gamma P (\nabla \cdot \boldsymbol{\xi})((\nabla \cdot \boldsymbol{\eta}) \\ & + \frac{B^2}{\mu_0} (\nabla \cdot \boldsymbol{\xi}_\perp + 2 \boldsymbol{\xi}_\perp \cdot \boldsymbol{\kappa})(\nabla \cdot \boldsymbol{\eta}_\perp + 2 \boldsymbol{\eta}_\perp \cdot \boldsymbol{\kappa}) \\ & - \frac{4B^2}{\mu_0} (\boldsymbol{\xi}_\perp \cdot \boldsymbol{\kappa})(\boldsymbol{\eta}_\perp \cdot \boldsymbol{\kappa}) + (\boldsymbol{\eta}_\perp \boldsymbol{\xi}_\perp : \nabla \nabla) \left(P + \frac{B^2}{2\mu_0} \right) \Bigg] \end{aligned}$$

6.3 Energy Principle: Potential Energy and Spectrum

1. Energy conservation law

$$\frac{1}{2} \int \rho \left(\frac{\partial \boldsymbol{\xi}}{\partial t} \right)^2 \mathrm{d}V = \frac{1}{2} \int \boldsymbol{\xi} \cdot F(\boldsymbol{\xi}) \, \mathrm{d}V$$

2. Kinetic energy

$$\delta K = (1/2) \int \rho (\partial \boldsymbol{\xi} / \partial t)^2 \mathrm{d}V$$

3. Potential energy

$$\delta W = -(1/2) \int \boldsymbol{\xi} \cdot F(\boldsymbol{\xi}) \, \mathrm{d}V$$

4. Energy integral of Furth

$$\delta W(\xi) = \int \mathrm{d}V[\delta W_{\mathrm{SA}} + \delta W_{\mathrm{MS}} + \delta W_{\mathrm{SW}} + \delta W_{\mathrm{IC}} + \delta W_{\mathrm{KI}}]$$

Shear Alfven wave energy	$\delta W_{\mathrm{SA}} = \boldsymbol{B}_1^2/2\mu_0$,
Magneto sonic wave energy	$\delta W_{\mathrm{MS}} = \boldsymbol{B}^2(\nabla\cdot\boldsymbol{\xi}_\perp + 2\boldsymbol{\xi}_\perp\cdot\boldsymbol{\kappa})^2/2\mu_0$,
Sound wave energy	$\delta W_{\mathrm{SW}} = \gamma P(\nabla\cdot\boldsymbol{\xi})^2/2$,
Interchange energy	$\delta W_{\mathrm{EX}} = (\boldsymbol{\xi}_\perp\cdot\nabla P)(\boldsymbol{\xi}_\perp\cdot\boldsymbol{\kappa})/2$,
Kink free energy	$\delta W_{\mathrm{KI}} = -J_\parallel \boldsymbol{b}\cdot(\boldsymbol{B}_{1\perp}\times\boldsymbol{\xi}_\perp)/2$,
	$\boldsymbol{B}_1 = \nabla\times(\boldsymbol{\xi}\times\boldsymbol{B})$

5. Eigenvalue

$$\omega^2 = -\int \boldsymbol{\xi}^*\cdot\boldsymbol{F}(\boldsymbol{\xi})\,\mathrm{d}V\Big/\int\rho\,|\boldsymbol{\xi}|^2\,\mathrm{d}V$$

6. Orthogonality of eigen function

$$\left(\omega_m^2 - \omega_n^2\right)\int \rho\boldsymbol{\xi}_m\cdot\boldsymbol{\xi}_n\,\mathrm{d}V = 0$$

7. Laplace transformed linear MHD equation

$$[\lambda - \boldsymbol{F}/\rho]\boldsymbol{\xi} = \boldsymbol{a}$$

8. Eigen mode equation of linear MHD equation

$$\boldsymbol{\xi} = [\lambda - \boldsymbol{F}/\rho]^{-1}\boldsymbol{a}$$

6.4 Newcomb equation: Euler–Lagrange Equation of Ideal MHD

Newcomb equation for cylindrical plasma

1. Energy integral of cylindrical plasma

$$W = \frac{\pi}{2\mu_0}\int_0^a\left[f\left|\frac{\mathrm{d}\xi}{\mathrm{d}r}\right|^2 + g\,|\xi|^2\right]\mathrm{d}r + W_a + W_v\ ,$$

$$f = \frac{r(kB_z + (m/r)B_\theta)^2}{k^2 + (m/r)^2}\ ,$$

$$g = \frac{1}{r}\frac{r(kB_z - (m/r)B_\theta)^2}{k^2 + (m/r)^2} + r(kB_z + (m/r)B_\theta)^2 - \frac{2B_\theta}{r}\frac{\mathrm{d}(rB_\theta)}{\mathrm{d}r} - \frac{\mathrm{d}}{\mathrm{d}r}\left(\frac{k^2B_z^2 - (m/r)^2B_\theta^2}{k^2 + (m/r)^2}\right)$$

2. Newcomb equation

$$\frac{\mathrm{d}}{\mathrm{d}r}\left(f\frac{\mathrm{d}\xi}{\mathrm{d}r}\right) - g\xi = 0$$

3. Suydam criterion

$$q'(r)/q(r)^2 + 8\mu_0 P'(r)/rB_z^2 > 0$$

4. Mercier criterion

$$r(\mathrm{d}\ln q/\mathrm{d}r)^2/4 + 2\mu_0(\mathrm{d}P/\mathrm{d}r)(1-q^2)/B_z^2 > 0$$

5. Magnetic shear

$$s = r(\mathrm{d}q/\mathrm{d}r)$$

Newcomb equation for axisymmetric plasma

6. Grad–Shafranov equation for $r = [2R_0 \int_0^{\psi} (q/F)\,\mathrm{d}\psi]^{1/2}$

$$\frac{\partial}{\partial r}\left[r\frac{\mathrm{d}\psi}{\mathrm{d}r}|\nabla r|^2\right] + \frac{\partial(\nabla r\cdot\nabla\theta)}{\partial\theta}\frac{\mathrm{d}\psi}{\mathrm{d}r} = -\mu_0 R^2\frac{\mathrm{d}P}{\mathrm{d}\psi} - F\frac{\mathrm{d}F}{\mathrm{d}\psi}$$

7. Energy integral

$$W_p = \frac{\pi}{2\mu_0}\int_0^a \mathrm{d}r \int_0^{2\pi} \mathrm{d}\theta\, L\left(X, \frac{\partial X}{\partial\theta}, \frac{\partial X}{\partial r}, V, \frac{\partial V}{\partial\theta}\right)$$

8. Euler–Lagrange equation for V

$$\frac{\partial}{\partial\theta}\left[\frac{\partial L}{\partial(\partial V/\partial\theta)}\right] - \frac{\partial L}{\partial V} = 0$$

9. Solvable condition

$$\int_0^{2\pi} \frac{\partial L}{\partial V}\mathrm{d}\theta = 0$$

10. Reduced 1-dimensional energy integral

$$W_p = \frac{\pi^2}{\mu_0}\int_0^a L\left(\boldsymbol{X}, \frac{\mathrm{d}\boldsymbol{X}}{\mathrm{d}r}\right)\mathrm{d}r$$

11. Euler–Lagrange equation for $\boldsymbol{X}$

$$\frac{\mathrm{d}}{\mathrm{d}r}\frac{\partial L}{\partial(\mathrm{d}\boldsymbol{X}/\mathrm{d}r)} - \frac{\partial L}{\partial \boldsymbol{X}} = 0$$

12. 2D Newcomb equation

$$\frac{\mathrm{d}}{\mathrm{d}r} \boldsymbol{f} \frac{\mathrm{d}\boldsymbol{X}}{\mathrm{d}r} + \boldsymbol{g} \frac{\mathrm{d}\boldsymbol{X}}{\mathrm{d}r} + \boldsymbol{h}\boldsymbol{X} = 0$$

6.5 Tension of Magnetic Field: Kink and Tearing

1. Energy integral in Tokamak ordering

$$W_p = \frac{\pi^2 B_\zeta^2}{\mu_0 R_0} \left\{ \int_0^a \left[\left(r \frac{\mathrm{d}\xi}{\mathrm{d}r} \right)^2 + (m^2 - 1)\xi^2 \right] \left(\frac{n}{m} - \frac{1}{q} \right)^2 r\,\mathrm{d}r \right\} ,$$

$$W_v = \frac{\pi^2 B_\zeta^2}{\mu_0 R_0} \left[\frac{2}{q_a} \left(\frac{n}{m} - \frac{1}{q_a} \right) + (1 + m\lambda) \left(\frac{n}{m} - \frac{1}{q_a} \right)^2 \right] a^2 \xi_a^2 ,$$

$$\lambda = \left(1 + (a/b)^{2m}\right) / \left(1 - (a/b)^{2m}\right)$$

2. Ohm's law

$$\gamma B_r - \frac{B_\theta}{r}(m - nq)\mathrm{i}v_r = \frac{\eta}{\mu_0} \frac{\mathrm{d}^2 B_r}{\mathrm{d}r^2}$$

3. Diffusion equation for helical flux

$$\frac{\partial \psi}{\partial t} = \frac{\eta}{\mu_0} \frac{\partial^2 \psi}{\partial r^2}$$

4. Rutherford equation

$$\frac{\mathrm{d}w}{\mathrm{d}t} = 1.66 \frac{\eta}{\mu_0} (\Delta'(w) - \alpha w)$$

5. Newcomb equation for ψ in Tokamak ordering

$$\frac{1}{r} \frac{\mathrm{d}}{\mathrm{d}r} \left(r \frac{\mathrm{d}\psi}{\mathrm{d}r} \right) - \frac{m^2}{r^2} \psi - \frac{\mu_0 \mathrm{d}J/\mathrm{d}r}{B_\theta (1 - nq/m)} \psi = 0$$

6.6 Curvature of Magnetic Field: Ballooning and Quasi-mode Expansion

1. Expression of displacement by stream function

$$\boldsymbol{\xi}_\perp = \frac{\mathrm{i}\boldsymbol{B} \times \nabla \Phi}{B^2}$$

2. Eikonal form of stream function

$$\Phi = F(\psi, \theta)\mathrm{e}^{-\mathrm{i}n\alpha} , \quad \alpha = \zeta - q\theta$$

Magnetic field in Crebsch coordinates

$$\boldsymbol{B} = \nabla\alpha \times \nabla\psi$$

3. Shear Alfven term for ballooning modes

$$\delta W_{\mathrm{SA}} = \frac{\boldsymbol{B}_1^2}{2\mu_0} \sim \frac{(\nabla\alpha)^2}{2\mu_0 B^2}|\boldsymbol{B}\cdot\nabla F|^2$$

4. Ballooning term for ballooning modes

$$\delta W_{\mathrm{EX}} = (\boldsymbol{\xi}_\perp\cdot\nabla P)(\boldsymbol{\xi}_\perp\cdot\boldsymbol{\kappa})/2 \sim P'(\psi)[(\boldsymbol{B}\times\nabla\alpha)\cdot\boldsymbol{\kappa}/B^2]|F|^2$$

5. Energy integral for ballooning modes

$$W_p = \frac{1}{2\mu_0}\int\left[\frac{|\nabla\alpha|^2}{B^2}(\boldsymbol{B}\cdot\nabla F)^2 - 2\mu_0 P'(\psi)\kappa_w F^2\right]\mathrm{d}V$$

6. Euler–Lagrange equation for ballooning modes

$$J^{-1}\frac{\partial}{\partial\theta}\left[\frac{|\nabla\alpha|^2}{JB^2}\frac{\partial F}{\partial\theta}\right] + \mu_0 P'(\psi)\kappa_w F = 0$$

7. Quasi-mode expansion of F

$$F(\psi,\theta) = \sum_{N=-\infty}^{\infty} F_1(\theta - 2\pi N)$$

8. Euler–Lagrange equation for quasi-mode

$$J^{-1}\frac{\partial}{\partial y}\left[\frac{|\nabla\alpha|^2}{JB^2}\frac{\partial F_1(y)}{\partial y}\right] + \mu_0 P'(\psi)\kappa_w F_1(y) = 0$$

9. Double periodic Fourier expansion of Φ

$$\Phi = \sum_{k=-\infty}^{\infty}\Phi_k(q)\mathrm{e}^{\mathrm{i}(m+k)\theta}\mathrm{e}^{-\mathrm{i}n\zeta}$$

10. Ballooning mode expression using radial Fourier integral

$$\Phi = \frac{1}{2\pi}\sum_{k=-\infty}^{\infty}\mathrm{e}^{\mathrm{i}k\theta}\mathrm{e}^{-\mathrm{i}n\alpha}\int_{-\infty}^{\infty}F_0(s)\mathrm{e}^{\mathrm{i}s(nq-k)}\,\mathrm{d}s$$

6.7 Flow: Non-Hermitian Frieman–Rotenberg Equation

1. MHD equilibrium equations with flow

$$\rho(\boldsymbol{u}\cdot\nabla)\boldsymbol{u}+\nabla P-\boldsymbol{J}\times\boldsymbol{B}=0\,,$$
$$\nabla\times(\boldsymbol{u}\times\boldsymbol{B})=0\,,$$
$$\boldsymbol{B}=\nabla\zeta\times\nabla\psi+F\nabla\zeta$$

2. Flow relation

$$\boldsymbol{u}\times\boldsymbol{B}=-\nabla\Phi\,,$$
$$\nabla\Phi=\Omega(\psi)\nabla\psi\,,$$
$$\boldsymbol{u}=\frac{\Phi_M}{\rho}\boldsymbol{B}+R^2\Omega\nabla\zeta$$

3. Surface quantities

$$F=F(\psi),\ P(\psi,R)=P_0(\psi)\exp\left[\frac{M}{2T}R^2\Omega^2\right]$$

where $\boldsymbol{B}=\nabla\zeta\times\nabla\psi+F\nabla\zeta$

4. Grad–Shafranov equation with toroidal flow

$$\Delta^*\psi=-\mu_0R^2\partial P(\psi,R)/\partial\psi-FF'(\psi)$$

5. Action integral for ideal MHD with flow

$$S=\int L\,\mathrm{d}V\,\mathrm{d}t\,,$$
$$L=\frac{1}{4}\rho\dot{\boldsymbol{\xi}}^2-\rho\boldsymbol{\xi}\cdot(\boldsymbol{u}\cdot\nabla)\dot{\boldsymbol{\xi}}+\frac{1}{2}\rho\boldsymbol{\xi}\cdot\boldsymbol{F}(\boldsymbol{\xi})$$

6. Generalized momentum

$$\boldsymbol{p}\equiv\partial L/\partial\boldsymbol{\xi}=\rho(\partial\boldsymbol{\xi}/\partial t)+\rho\boldsymbol{u}\cdot\nabla\boldsymbol{\xi}$$

7. Hamiltonian for flow MHD

$$H=\frac{1}{2\rho}[\boldsymbol{p}-\rho\boldsymbol{u}\cdot\nabla\boldsymbol{\xi}]^2-\frac{1}{2}\rho\boldsymbol{\xi}\cdot\boldsymbol{F}(\boldsymbol{\xi})$$

8. Hamilton equation

$$\mathrm{d}\boldsymbol{p}/\mathrm{d}t=-\partial H/\partial\boldsymbol{\xi}\,,$$
$$\mathrm{d}\boldsymbol{p}/\mathrm{d}t=\boldsymbol{F}(\boldsymbol{\xi})-\rho\boldsymbol{u}\cdot\nabla[(\boldsymbol{p}/\rho)-\boldsymbol{u}\cdot\nabla\boldsymbol{\xi}]$$

9. Frieman–Rotenberg equation

$$\rho\frac{\partial^2\boldsymbol{\xi}}{\partial t^2}+2\rho(\boldsymbol{u}\cdot\nabla)\frac{\partial\boldsymbol{\xi}}{\partial t}=\boldsymbol{F}(\boldsymbol{\xi}),$$
$$\boldsymbol{F}(\boldsymbol{\xi})=\boldsymbol{F}_{\mathrm{s}}(\boldsymbol{\xi})+\boldsymbol{F}_{\mathrm{d}}(\boldsymbol{\xi}),$$
$$\boldsymbol{F}_{\mathrm{s}}(\boldsymbol{\xi})=\nabla[\boldsymbol{\xi}\cdot\nabla P+\gamma P\nabla\cdot\boldsymbol{\xi}]+(\nabla\times\boldsymbol{B}_1)\times\boldsymbol{B}+\boldsymbol{J}\times\boldsymbol{B}_1,$$
$$\boldsymbol{F}_{\mathrm{d}}(\boldsymbol{\xi})=\nabla\cdot[\rho\boldsymbol{\xi}(\boldsymbol{u}\cdot\nabla)\boldsymbol{u}-\rho\boldsymbol{u}(\boldsymbol{u}\cdot\nabla)\boldsymbol{\xi}],$$
$$\boldsymbol{B}_1=\nabla\times(\boldsymbol{\xi}\times\boldsymbol{B})$$

Chapter 7 Wave Dynamics: Propagation and Resonance in Inhomogeneous Plasma

7.1 Eikonal Equation: Dynamics of Wave Propagation

1. Eikonal equation

$$\omega=-\frac{\partial\zeta}{\partial t},$$
$$\boldsymbol{k}=\frac{\partial\zeta}{\partial\boldsymbol{x}}$$

2. Eikonal relation

$$\frac{\partial\boldsymbol{k}}{\partial t}=-\frac{\partial\omega}{\partial\boldsymbol{x}}$$

3. Dispersion relation and group velocity

$$\omega=\Omega(\boldsymbol{k},\boldsymbol{x},t),$$
$$\frac{\partial\omega}{\partial\boldsymbol{x}}=\boldsymbol{v}_{\mathrm{g}}\cdot\frac{\partial\boldsymbol{k}}{\partial\boldsymbol{x}}+\left.\frac{\partial\Omega}{\partial\boldsymbol{x}}\right|_{\boldsymbol{k}}$$

4. Hamilton equation for wave propagation

$$\frac{\mathrm{d}\boldsymbol{x}}{\mathrm{d}t}=\left(\frac{\partial\Omega}{\partial\boldsymbol{k}}\right)_{\boldsymbol{x}},$$
$$\frac{\mathrm{d}\boldsymbol{k}}{\mathrm{d}t}=-\left(\frac{\partial\Omega}{\partial\boldsymbol{x}}\right)_{\boldsymbol{k}}$$

5. Lagrangian for wave propagation

$$\mathcal{L}=\boldsymbol{k}\cdot\boldsymbol{x}-\Omega$$

6. Ohm's law

$$\hat{\boldsymbol{J}} = \boldsymbol{\sigma} \hat{\boldsymbol{E}}$$

7. Dielectric tensor

$$\boldsymbol{K}(w, \boldsymbol{k}) = \boldsymbol{I} + \frac{\mathrm{i}\boldsymbol{\sigma}}{\varepsilon_0 \omega}$$

8. Electric field equation

$$\begin{aligned} &\boldsymbol{M} \cdot \hat{\boldsymbol{E}} = 0\,, \\ &\boldsymbol{M} = (\boldsymbol{k}\boldsymbol{k} - k^2 \boldsymbol{I})/k_0^2 + \boldsymbol{K}\,, \\ &\boldsymbol{K} = \boldsymbol{I} + \frac{\mathrm{i}\boldsymbol{\sigma}}{\varepsilon_0 \omega}\,, \quad k_0 = \omega / c \end{aligned}$$

9. Poynting theorem ($\boldsymbol{S} = \boldsymbol{E} \times \boldsymbol{B}/\mu_0$)

$$\frac{\partial}{\partial t}\left(\frac{\boldsymbol{B}^2}{2\mu_0} + \frac{\varepsilon_0 \boldsymbol{E}^2}{2}\right) = -\boldsymbol{J} \cdot \boldsymbol{E} - \nabla \cdot \boldsymbol{S}$$

10. Wave energy equation

$$\frac{\partial \varepsilon}{\partial t} + \nabla \cdot \hat{\boldsymbol{S}} = Q$$

11. Wave energy

$$\begin{aligned} &\varepsilon = \frac{1}{2}\left(\varepsilon_0 \hat{\boldsymbol{E}}^* \cdot \frac{\partial(\omega \boldsymbol{e}_h)}{\partial \omega} \cdot \hat{\boldsymbol{E}} + \frac{1}{\mu_0} \boldsymbol{B}^* \cdot \boldsymbol{B}\right) \quad \text{or} \quad \varepsilon = \frac{\varepsilon_0}{2} \hat{\boldsymbol{E}}^* \cdot \frac{\partial(\omega \boldsymbol{M}_h)}{\partial \omega} \cdot \hat{\boldsymbol{E}}\,, \\ &\boldsymbol{\varepsilon}_h = (\boldsymbol{\varepsilon} + \boldsymbol{\varepsilon}^+)/2\,, \quad \boldsymbol{M}_h = (\boldsymbol{M} + \boldsymbol{M}^+)/2\,, \quad \boldsymbol{\sigma}_h = (\boldsymbol{\sigma} + \boldsymbol{\sigma}^+)/2\,, \\ &Q = \hat{\boldsymbol{E}}^* \cdot \boldsymbol{\sigma}_h \cdot \hat{\boldsymbol{E}}, \quad \hat{\boldsymbol{S}} = \mathrm{Re}(\hat{\boldsymbol{E}}^* \times \hat{\boldsymbol{B}})/\mu_0 \end{aligned}$$

12. Wave energy and action

$$\varepsilon = \omega \mathcal{J}$$

13. Wave action

$$\mathcal{J} = \frac{\varepsilon_0}{2} \hat{\boldsymbol{E}}^* \cdot \frac{\partial \boldsymbol{M}_h}{\partial \omega} \cdot \hat{\boldsymbol{E}}$$

14. Wave kinetic equation

$$\frac{\partial \mathcal{J}}{\partial t} + [\mathcal{J}, \Omega] = 0$$

15. Poisson bracket

$$[\mathcal{J}, \Omega] = \frac{\partial \Omega}{\partial \boldsymbol{k}} \cdot \frac{\partial \mathcal{J}}{\partial \boldsymbol{x}} - \frac{\partial \Omega}{\partial \boldsymbol{x}} \cdot \frac{\partial \mathcal{J}}{\partial \boldsymbol{k}}$$

7.2 Lagrange Wave Dynamics: Ideal and Dissipative Systems

1. Action principle for wave packet

$$S = \int \mathrm{d}t \int L \,\mathrm{d}\boldsymbol{x}$$

$$L = L_M(\boldsymbol{A}, \Phi) + \sum_a L_a(\boldsymbol{\xi}_a, \boldsymbol{A}, \Phi)$$

$$L_M(\boldsymbol{A}, \Phi) = \varepsilon_0 \left[\frac{\partial \boldsymbol{A}}{\partial t} + \nabla \Phi \right]^2 - \frac{1}{\mu_0} (\nabla \times \boldsymbol{A})^2$$

$$L_a(\boldsymbol{\xi}_a, \boldsymbol{A}, \Phi) = n_a \left[\frac{m_a}{2} \dot{\boldsymbol{\xi}}_a^2 + e_a \left(\dot{\boldsymbol{\xi}}_a \cdot \boldsymbol{A}(\boldsymbol{x} + \boldsymbol{\xi}_a, t) - \Phi(\boldsymbol{x} + \boldsymbol{\xi}_a, t) \right) \right]$$

2. Lagrangian for linear wave

$$[L]_{\mathrm{lin}} = \varepsilon_0 \hat{\boldsymbol{E}}^* \cdot \boldsymbol{M} \cdot \hat{\boldsymbol{E}}$$

3. Euler–Lagrange equation

$$\frac{\partial}{\partial t} \left[\frac{\partial L}{\partial \omega} \right] + \frac{\partial}{\partial \boldsymbol{x}} \cdot \left[\frac{\partial L}{\partial \boldsymbol{k}} \right] = 0$$

4. Momentum conjugate to eikonal

$$\mathcal{J} = \frac{\partial L}{\partial \omega} = \frac{\partial L}{\partial \dot{\zeta}} = \varepsilon_0 \hat{\boldsymbol{E}}^* \cdot \frac{\partial \boldsymbol{M}}{\partial \omega} \cdot \hat{\boldsymbol{E}}$$

5. Number conservation in a wave packet

$$\frac{\partial \mathcal{J}}{\partial t} + \frac{\partial}{\partial \boldsymbol{x}} \cdot (\boldsymbol{v}_{\mathrm{g}} \mathcal{J}) = 0$$

6. Lagrangian for MHD wave

$$L = \frac{1}{2} \rho \omega^2 |A|^2 - \frac{1}{2} \rho \Omega(\boldsymbol{k}, \boldsymbol{x}, t)^2 |A|^2$$

7. Euler–Lagrange equation

$$\frac{\partial}{\partial t} \left[\frac{\partial L}{\partial \omega} \right] + \frac{\partial}{\partial \boldsymbol{x}} \cdot \left[\frac{\partial L}{\partial \boldsymbol{k}} \right] = 0$$

8. Conservation of number of wave ($N = \frac{\partial L}{\partial \omega}$)

$$\frac{\partial N}{\partial t} + \frac{\partial}{\partial \boldsymbol{x}} \cdot (\boldsymbol{v}_{\mathrm{g}} N) = 0$$

Hamilton Mechanics Using Conjugate Variables

9. Ordinary differential equation

$$\frac{\mathrm{d}\boldsymbol{x}}{\mathrm{d}t} = \boldsymbol{f}(\boldsymbol{x}, t)$$

10. Lagrangian from ordinary differential equation

$$L = \boldsymbol{p} \cdot \left(\frac{\mathrm{d}\boldsymbol{x}}{\mathrm{d}t} - \boldsymbol{f} \right) = \boldsymbol{p} \cdot \frac{\mathrm{d}\boldsymbol{x}}{\mathrm{d}t} - H$$

11. Hamiltonian

$$H = \boldsymbol{p} \cdot \boldsymbol{f}(\boldsymbol{x}, t)$$

12. Hamilton equation

$$\frac{\mathrm{d}\boldsymbol{x}}{\mathrm{d}t} = \frac{\partial H}{\partial \boldsymbol{p}} = \boldsymbol{f}(\boldsymbol{x}, t) ,$$
$$\frac{\mathrm{d}\boldsymbol{p}}{\mathrm{d}t} = -\frac{\partial H}{\partial \boldsymbol{x}} = \frac{\partial \boldsymbol{f}(\boldsymbol{x}, t)}{\partial \boldsymbol{x}} \cdot \boldsymbol{p}$$

7.3 Plasma as a Dielectric Medium: Cold and Hot Plasmas

1. Dielectric tensor

$$\boldsymbol{K}(w, \boldsymbol{k}) = \boldsymbol{I} + \frac{\mathrm{i}\sigma}{\varepsilon_0 \omega}$$

2. Dispersion relation

$$M = \det(\boldsymbol{M}) = 0 ,$$
$$\boldsymbol{M} \cdot \boldsymbol{E} = 0 , \quad \boldsymbol{M} = (\boldsymbol{k}\boldsymbol{k} - \boldsymbol{I}) \left(\frac{kc}{\omega} \right)^2 + \boldsymbol{K}$$

Cold Plasma

3. Dielectric tensor for cold wave

$$\boldsymbol{K} = \begin{bmatrix} S & -\mathrm{i}D & 0 \\ \mathrm{i}D & S & 0 \\ 0 & 0 & P \end{bmatrix} ,$$

$$S = 1 - \sum_a \frac{\omega_{pa}^2}{\omega^2 - \Omega_a^2}, \quad D = \sum_a \frac{\Omega_a}{\omega} \frac{\omega_{pa}^2}{\omega^2 - \Omega_a^2},$$

$$P = 1 - \sum_a \frac{\omega_{pa}^2}{\omega^2}, \quad \omega_{pa}^2 = \frac{n_a e_a^2}{\varepsilon_0 m_a}, \quad \Omega_a = \frac{e_a B}{m_a}$$

4. Refractivity

$$n = kc/\omega$$

5. Dispersion relation for cold wave

$$[S \sin^2\theta + P\cos^2\theta]n^4 - [RL\sin^2\theta + PS(1+\cos^2\theta)]n^2 + PRL = 0,$$

$$R = \frac{S+D}{2}, \quad L = \frac{S-D}{2}$$

6. Cut-off ($n = 0$) condition

$$PRL = 0$$

7. Resonance ($n = \infty$) condition

$$\tan^2\theta = -\frac{P}{S}$$

8. Time symmetry of cold wave dielectric tensor

$$\boldsymbol{K}(-\omega) = \boldsymbol{K}^*(\omega)$$

9. Onsager symmetry of cold wave dielectric tensor

$$\boldsymbol{K}(-\boldsymbol{B}) = \boldsymbol{K}^t(\boldsymbol{B})$$

10. Hermite symmetry of cold wave dielectric tensor

$$\boldsymbol{K} = \boldsymbol{K}^+$$

(+: complex conjugate)

Hot Plasma

11. Integral form of perturbed velocity distribution function

$$f_{a1k}(\boldsymbol{x}, \boldsymbol{v}, t) =$$

$$-\frac{e_a}{m_a}\int_{-\infty}^{t}\left(\boldsymbol{E}_1(\boldsymbol{x}(t'), t') + \frac{1}{\omega}\boldsymbol{v}(t') \times (\boldsymbol{k} \times \boldsymbol{E}_1(\boldsymbol{x}(t'), t'))\right) \cdot \frac{\partial f_0(\boldsymbol{v}(t'))}{\partial \boldsymbol{v}} \mathrm{d}t'$$

12. Unperturbed particle orbit

$$x(t') = x(t) + \frac{v_\perp}{\Omega}\left(\sin(\theta + \Omega(t'-t)) - \sin\theta\right) ,$$
$$y(t') = y(t) - \frac{v_\perp}{\Omega}\left(\cos(\theta + \Omega(t'-t)) - \cos\theta\right) ,$$
$$z(t') = z(t) + v_z(t'-t)$$

13. Perturbed velocity distribution function

$$f_{a1k}(\boldsymbol{x},\boldsymbol{v},t) =$$
$$-\frac{e_a}{m_a}\mathrm{e}^{\mathrm{i}(k_x x + k_z z - \omega t)} \int_{-\infty}^{t} \left((1 - \frac{\boldsymbol{k}\cdot\boldsymbol{v}(t')}{\omega})\boldsymbol{E} + (\boldsymbol{v}(t')\cdot\boldsymbol{E})\frac{\boldsymbol{k}}{\omega} \right) \cdot \frac{\partial f_0(\boldsymbol{v}(t'))}{\partial \boldsymbol{v}}$$
$$\times \exp\left(\mathrm{i}\frac{k_x v_\perp}{\Omega}\left(\sin(\theta + \Omega(t'-t)) - \sin\theta\right) + \mathrm{i}(k_z v_z - \omega)(t'-t) \right) \mathrm{d}t'$$

14. Dielectric tensor for Maxwell distribution

$$\boldsymbol{K} = \boldsymbol{I} + \sum_{a=i,e} \frac{\omega_{pa}^2}{\omega^2}\Bigg[2\eta_{0a}^2 \lambda_{Ta} \boldsymbol{L}$$
$$+ \sum_n \left(\zeta_{0a} Z(\zeta_{na}) - \left(1 - \frac{1}{\lambda_{Ta}}\right)(1 + \zeta_{na} Z(\zeta_{na})) \right) \exp(-b_a) \boldsymbol{X}_{na} \Bigg] ,$$

$$\boldsymbol{X}_{na} = \begin{bmatrix} (n^2/b_a) I_n & \mathrm{i}n(I_n' - I_n) & -2nI_n\lambda_{Ta}^{1/2}\eta_{na}/\alpha_a \\ -\mathrm{i}n(I_n' - I_n) & (n^2/b_a + 2b_a)I_n - 2b_a I_n' & \mathrm{i}\lambda_{Ta}^{1/2}\eta_{na}\alpha_a(I_n' - I_n) \\ -2nI_n\lambda_{Ta}^{1/2}\eta_{na}/\alpha_a & -\mathrm{i}\lambda_{Ta}^{1/2}\eta_{na}\alpha_a(I_n' - I_n) & 2\lambda_{Ta}\eta_{na}^2 I_n \end{bmatrix} ,$$

$$\eta_{na} = \frac{\omega + n\Omega_a}{k_z v_{T\parallel a}} , \quad \lambda_{Ta} = \frac{T_{\parallel a}}{T_{\perp a}} , \quad b_a = \frac{1}{2}\left(\frac{k_\perp v_{T\perp a}}{\Omega_a}\right)^2 , \quad v_{Ta} = \left(\frac{2T_a}{m_a}\right)^{1/2} ,$$

$$\zeta_{na} = \frac{\omega - k_\parallel u_{\parallel a} + n\Omega_a}{k_z v_{T\parallel a}} , \quad \alpha_a = \frac{k_x v_{T\perp a}}{\Omega_a} , \quad Z(\zeta) = \frac{1}{\sqrt{\pi}} \int_{-\infty}^{\infty} \frac{\exp(-\beta^2)}{\beta - \zeta}$$

Causality and translational symmetry in plasma waves

15. Principle of superposition, translational invariance

$$\boldsymbol{J}(\boldsymbol{x},t) = \frac{1}{2\pi} \iint \mathrm{d}\boldsymbol{x}'\mathrm{d}t' \boldsymbol{\sigma}(\boldsymbol{x} - \boldsymbol{x}', t - t')\boldsymbol{E}(\boldsymbol{x}', t') ,$$
$$\boldsymbol{J}(\omega,\boldsymbol{k}) = \boldsymbol{\sigma}(\omega,\boldsymbol{k})\boldsymbol{E}(\omega,\boldsymbol{k})$$

7.4 Non-uniform plasma: Alfven Wave Resonance and Continuous Spectrum

1. Perpendicular refractivity

$$n_\perp^2 = \frac{(R - n_\parallel^2)(L - n_\parallel^2)}{S - n_\parallel^2}$$

2. Alfven wave resonance condition

$$\omega = k_\parallel V_A$$

3. Cold plasma wave equation

$$\frac{c^2}{\omega^2}\frac{\mathrm{d}^2 E}{\mathrm{d}x^2} + \frac{(R - n_\parallel^2)(L - n_\parallel^2)}{S - n_\parallel^2} E = 0$$

4. Wave equation near resonance

$$\frac{\mathrm{d}^2 E}{\mathrm{d}y^2} + \frac{\lambda^2(y^2 - 1)}{y + \mathrm{i}\varepsilon} E = 0$$

$$\text{where } \lambda^2 = \left| \frac{D^3\omega^2}{c^2(\mathrm{d}S(0)/\mathrm{d}x)^2} \right|$$

5. Dispersion relation of kinetic Alfven wave

$$\omega^2 = k_\parallel^2 V_A^2 \left[1 + k_\perp^2 \rho_i^2 \left(\frac{3}{4} + \frac{T_e}{T_i} \right) \right] \qquad \left((k_\perp \rho_i)^2 \ll 1 \right)$$

7.5 Drift Waves: Universal Waves in Confined Plasma

1. Dispersion relation of simple drift wave

$$\omega(\omega - \omega_*) = k_\parallel^2 C_s^2$$

2. Sound velocity

$$C_s = [(Z_i T_e + \gamma_i T_i)/m_i]^{1/2}$$

3. Diamagnetic drift frequency

$$\omega_* = -\frac{k_\perp T_e}{eB}\frac{\mathrm{d}\ln n_e}{\mathrm{d}r}$$

4. Dispersion relation of ITG drift wave

$$\omega(\omega - \omega_*) = k_\parallel^2 C_s^2 \left[1 + \frac{\omega_*}{\omega} \frac{Z_i}{\gamma_i \tau + Z_i} \left(\eta_i - \frac{\gamma_i - Z_i}{Z_i}\right)\right]$$

5. Critical temperature for ITG (example)

$$\omega^2 \sim -\frac{k_\parallel^2 C_s^2}{2.5\tau + 1} \left(\eta_i - \frac{2}{3}\right)$$

$$\text{where } \eta_i = \frac{\mathrm{d} \ln T_i}{\mathrm{d} \ln n_e}$$

6. Polarization drift

$$\boldsymbol{v}_{pa} = -\frac{m_a}{eB^2} \frac{\mathrm{d}\boldsymbol{E}}{\mathrm{d}t}$$

7. Boltzmann relation for electron

$$\frac{\tilde{n}_e}{n_e} = \frac{e\tilde{\Phi}}{T_e}$$

8. Drift wave dispersion with polarization drift

$$\omega^2(1 + \tau k_\perp^2 \rho_i^2/2) - \omega\omega_* = k_\parallel^2 C_s^2$$

9. Hasegawa–Mima equation

$$\frac{\partial \Phi_{\boldsymbol{k}}(t)}{\partial t} + \mathrm{i}\omega_{\boldsymbol{k}^*} \Phi_{\boldsymbol{k}}(t) = \sum_{\boldsymbol{k}=\boldsymbol{k}_1+\boldsymbol{k}_2} V_{\boldsymbol{k}_1,\boldsymbol{k}_2} \Phi_{\boldsymbol{k}1}(t) \Phi_{\boldsymbol{k}2}(t) ,$$

$$\omega_{\boldsymbol{k}^*} = \frac{\omega_*}{1 + \tau k_\perp^2 \rho_i^2/2} ,$$

$$V_{\boldsymbol{k}_1,\boldsymbol{k}_2} = \frac{\rho_s^2}{(1 + \tau k^2 \rho_s^2) B} (\boldsymbol{k}_1 x \boldsymbol{k}_2) \cdot \boldsymbol{e}_z [k_2^2 - k_1^2]$$

$$\rho_s = (T_e/m_i)^{1/2}/\Omega_i$$

Chapter 8 Collisional Transport: Neoclassical Transport in a Closed Magnetic Configuration

8.1 Collisionless Plasma: Moment Equation and Neoclassical Viscosity

1. Momentum balance equation

$$m_a n_a \frac{\mathrm{d}\boldsymbol{u}_a}{\mathrm{d}t} = e_a n_a(\boldsymbol{E} + \boldsymbol{u}_a \times \boldsymbol{B}) - \nabla P_a - \nabla \cdot \boldsymbol{\Pi}_a + \boldsymbol{F}_{a1} + \boldsymbol{M}_a$$

2. Heat flux balance equation

$$m_a \frac{\partial}{\partial t}\left(\frac{\boldsymbol{q}_a}{T_a}\right) = \frac{e_a}{T_a}\boldsymbol{q}_a \times \boldsymbol{B} - \frac{5}{2} n_a \nabla T_a - \nabla \cdot \boldsymbol{\Theta}_a + \boldsymbol{F}_{a2} + \boldsymbol{Q}_a$$

3. CGL expression of viscous tensor

$$\boldsymbol{\Pi}_a = (P_{\parallel a} - P_{\perp a})\left(\boldsymbol{bb} - \frac{1}{3}\boldsymbol{I}\right) + O(\delta^2)$$

4. CGL expression of heat viscous tensor

$$\boldsymbol{\Theta}_a = (\Theta_{\parallel a} - \Theta_{\perp a})\left(\boldsymbol{bb} - \frac{1}{3}\boldsymbol{I}\right) + O(\delta^2)$$

5. First order perpendicular flow

$$\boldsymbol{u}_{\perp a}^{(1)} = \frac{\boldsymbol{E} \times \boldsymbol{B}}{B^2} + \frac{\boldsymbol{b} \times \nabla P_a}{m_a n_a \Omega_a}$$

6. First order perpendicular heat flow

$$\boldsymbol{q}_{\perp a}^{(1)} = \frac{5}{2} P_a \frac{\boldsymbol{b} \times \nabla T_a}{m_a \Omega_a}$$

7. Averaged parallel momentum balance

$$\langle \boldsymbol{B} \cdot \nabla \cdot \boldsymbol{\Pi}_a \rangle = \langle \boldsymbol{B} \cdot \boldsymbol{F}_{a1} \rangle + e_a n_a \langle \boldsymbol{B} \cdot \boldsymbol{E} \rangle + \langle \boldsymbol{B} \cdot \boldsymbol{M}_a \rangle$$

8. Averaged parallel heat flux balance

$$\langle \boldsymbol{B} \cdot \nabla \cdot \boldsymbol{\Theta}_a \rangle = \langle \boldsymbol{B} \cdot \boldsymbol{F}_{a2} \rangle + \langle \boldsymbol{B} \cdot \boldsymbol{Q}_a \rangle$$

9. Particle density

$$n_a \equiv \int f_a(\boldsymbol{x}, \boldsymbol{v}, t)\mathrm{d}\boldsymbol{v}$$

10. Flow

$$\boldsymbol{u}_a \equiv \frac{1}{n_a} \int \boldsymbol{v} f_a(\boldsymbol{x}, \boldsymbol{v}, t)\mathrm{d}\boldsymbol{v}$$

11. Heat flow

$$\boldsymbol{q}_a \equiv \int (\boldsymbol{v}-\boldsymbol{u}_a)\frac{m_a}{2}\left|\boldsymbol{v}-\boldsymbol{u}_a\right|^2 f_a(\boldsymbol{x},\boldsymbol{v},t)\mathrm{d}\boldsymbol{v}$$

12. Pressure

$$P_a \equiv \int \frac{m_a}{3}\left|\boldsymbol{v}-\boldsymbol{u}_a\right|^2 f_a(\boldsymbol{x},\boldsymbol{v},t)\mathrm{d}\boldsymbol{v}$$

13. Viscous stress tensor

$$\boldsymbol{\Pi}_a \equiv \int m_a\left(\boldsymbol{v}\boldsymbol{v}-\frac{1}{3}v^2\boldsymbol{I}\right) f_a(\boldsymbol{x},\boldsymbol{v},t)\mathrm{d}\boldsymbol{v}$$

14. Heat viscous tensor

$$\boldsymbol{\Theta}_a \equiv \int m_a\left(\boldsymbol{v}\boldsymbol{v}-\frac{1}{3}v^2\boldsymbol{I}\right)\left(\frac{m_a v^2}{2T_a}-\frac{5}{2}\right) f_a(\boldsymbol{x},\boldsymbol{v},t)\mathrm{d}\boldsymbol{v}$$

15. Friction force

$$\boldsymbol{F}_{a1} \equiv \int m_a\boldsymbol{v}C(f_a)\mathrm{d}\boldsymbol{v}$$

16. Heat friction

$$\boldsymbol{F}_{a2} \equiv \int m_a\boldsymbol{v}\left(\frac{m_a v^2}{2T_a}-\frac{5}{2}\right)C(f_a)\mathrm{d}\boldsymbol{v}$$

17. Parallel flow

$$v_\parallel \equiv \boldsymbol{b}\cdot\boldsymbol{v}\ , \quad u_{\parallel a} \equiv \boldsymbol{b}\cdot\boldsymbol{u}_a$$

18. Parallel pressure

$$P_{\parallel a} \equiv \int m_a\left(v_\parallel - u_{\parallel a}\right)^2 f_a(\boldsymbol{x},\boldsymbol{v},t)\mathrm{d}\boldsymbol{v}$$

19. Perpendicular pressure

$$P_{\perp a} \equiv \int \frac{m_a}{2}(\boldsymbol{v}_\perp - \boldsymbol{u}_{\perp a})^2 f_a(\boldsymbol{x},\boldsymbol{v},t)\mathrm{d}\boldsymbol{v}$$

20. Parallel heat viscosity

$$\Theta_{\parallel a} \equiv \int m_a(v_\parallel - u_{\parallel a})^2\left(\frac{m_a v^2}{2T_a}-\frac{5}{2}\right) f_a(\boldsymbol{x},\boldsymbol{v},t)\mathrm{d}\boldsymbol{v}$$

21. Perpendicular heat viscosity

$$\Theta_{\perp a} \equiv \int \frac{m_a}{2}(\boldsymbol{v}_\perp - \boldsymbol{u}_{\perp a})^2\left(\frac{m_a v^2}{2T_a}-\frac{5}{2}\right) f_a(\boldsymbol{x},\boldsymbol{v},t)\mathrm{d}\boldsymbol{v}$$

22. Laguerre expansion of distribution function

$$f_a(\boldsymbol{v}) = f_{aM}(\boldsymbol{v}) + f_{a1}(\boldsymbol{v}) + f_{a2}(\boldsymbol{v})\ ,$$

$$f_{aM}(\boldsymbol{v}) = \frac{n_a(\psi)}{\pi^{3/2}v_{Ta}^3}\exp\left(-v^2/v_{Ta}^2\right)\ ,$$

$$f_{a2}(\boldsymbol{v}) = 2\frac{\boldsymbol{v}\boldsymbol{v} - \frac{v^2}{3}\boldsymbol{I}}{m_a n_a v_{Ta}^4} : \left[\boldsymbol{\Pi}_a + (\boldsymbol{\Theta}_a + \boldsymbol{\Pi}_a)\left(1 - \frac{2x_a^2}{7}\right)\right] f_{aM}(\boldsymbol{v})$$

8.2 Incompressible Flow: First Order Flow on a Magnetic Surface

1. First order perpendicular flow

$$\begin{aligned}
\boldsymbol{u}_{\perp a}^{(1)} &= \frac{1}{B}\left[\frac{\mathrm{d}\Phi}{\mathrm{d}\psi} + \frac{1}{e_a n_a}\frac{\mathrm{d}P_a}{\mathrm{d}\psi}\right]\boldsymbol{b} \times \nabla\psi \,, \\
\boldsymbol{q}_{\perp a}^{(1)} &= \frac{5P_a}{2e_a B}\frac{\mathrm{d}T_a}{\mathrm{d}\psi}\boldsymbol{b} \times \nabla\psi
\end{aligned}$$

2. First order perpendicular flow is on the flux surface

$$\begin{aligned}
\boldsymbol{u}_{\perp a}^{(1)} \cdot \nabla\psi &= 0 \,, \\
\boldsymbol{q}_{\perp a}^{(1)} \cdot \nabla\psi &= 0
\end{aligned}$$

3. First order flow

$$\begin{aligned}
\boldsymbol{u}_a^{(1)} &= \boldsymbol{u}_{\perp a}^{(1)} + u_{\parallel a}\boldsymbol{b} \,, \\
\boldsymbol{q}_a^{(1)} &= \boldsymbol{q}_{\perp a}^{(1)} + q_{\parallel a}\boldsymbol{b}
\end{aligned}$$

4. Poloidal flow

$$\begin{aligned}
\frac{\boldsymbol{u}_a^{(1)} \cdot \nabla\theta}{\boldsymbol{B} \cdot \nabla\theta} &= u_{a\theta}^*(\psi) \,, \\
\frac{\boldsymbol{q}_a^{(1)} \cdot \nabla\theta}{\boldsymbol{B} \cdot \nabla\theta} &= q_{a\theta}^*(\psi)
\end{aligned}$$

5. Flow relation

$$\begin{aligned}
B u_{a\theta}^*(\psi) &= u_{\parallel a} - V_{1a} \,, \\
B q_{a\theta}^*(\psi) &= q_{\parallel a} - \frac{5}{2}P_a V_{2a}
\end{aligned}$$

6. Thermodynamic forces

$$\begin{aligned}
V_{1a} &= -\frac{\boldsymbol{u}_{\perp a}^{(1)} \cdot \nabla\theta}{\boldsymbol{b} \cdot \nabla\theta} = -\frac{F(\psi)}{B}\left(\frac{\mathrm{d}\Phi}{\mathrm{d}\psi} + \frac{1}{e_a n_a}\frac{\mathrm{d}P_a}{\mathrm{d}\psi}\right) \,, \\
V_{2a} &= -\frac{2}{5P_a}\frac{\boldsymbol{q}_{\perp a}^{(1)} \cdot \nabla\theta}{\boldsymbol{b} \cdot \nabla\theta} = -\frac{F(\psi)}{e_a B}\frac{\mathrm{d}T_a}{\mathrm{d}\psi}
\end{aligned}$$

7. Expression for first order flows

$$\boldsymbol{u}_a^{(1)} = u_{\|a}\boldsymbol{b} + \boldsymbol{u}_{\perp a}^{(1)} = u_{\|a}\boldsymbol{b} + \frac{BV_{1a}}{F(\psi)}\frac{\nabla\psi\times\boldsymbol{b}}{B} ,$$
$$\boldsymbol{q}_a^{(1)} = q_{\|a}\boldsymbol{b} + \boldsymbol{q}_{\perp a}^{(1)} = q_{\|a}\boldsymbol{b} + \frac{5P_a}{2}\frac{BV_{2a}}{F(\psi)}\frac{\nabla\psi\times\boldsymbol{b}}{B}$$

7.1 Corollary

$$\boldsymbol{u}_a^{(1)} = u_{a\theta}^*(\psi)\boldsymbol{B} + \frac{BV_{1a}}{F(\psi)}R^2\nabla\zeta ,$$
$$\boldsymbol{q}_a^{(1)} = q_{a\theta}^*(\psi)\boldsymbol{B} + \frac{5P_a}{2}\frac{BV_{2a}}{F(\psi)}R^2\nabla\zeta$$

8. Axisymmetric relation

$$\boldsymbol{b}\times\nabla\psi = F(\psi)\boldsymbol{b} - R^2 B\nabla\zeta$$

9. Toroidal flows

$$u_{a\zeta}^{(1)} = u_{a\theta}^*(\psi)B_\zeta + \frac{BV_{1a}}{F(\psi)}R ,$$
$$q_{a\zeta}^{(1)} = q_{a\theta}^*(\psi)B_\zeta + \frac{5P_a}{2}\frac{BV_{2a}}{F(\psi)}R$$

10. Flux surface averaged flow relation

$$\langle B^2\rangle u_{a\theta}^*(\psi) = \langle Bu_{\|a}\rangle - \langle BV_{1a}\rangle ,$$
$$\langle B^2\rangle q_{a\theta}^*(\psi) = \langle Bq_{\|a}\rangle - \frac{5P_a}{2}\langle BV_{2a}\rangle$$

11. Expression for local toroidal flows

$$u_{a\zeta}^{(1)} = \frac{B_\zeta}{\langle B^2\rangle}\langle Bu_{\|a}\rangle + \left[1 - \frac{B_\zeta^2}{\langle B^2\rangle}\right]\frac{BV_{1a}}{B_\zeta} ,$$
$$q_{a\zeta}^{(1)} = \frac{B_\zeta}{\langle B^2\rangle}\langle Bq_{\|a}\rangle + \frac{5P_a}{2}\left[1 - \frac{B_\zeta^2}{\langle B^2\rangle}\right]\frac{BV_{2a}}{B_\zeta}$$

8.3 Friction and Viscous Forces: Momentum and Heat Flow Balance

1. Friction–flow relations

$$\begin{bmatrix}\boldsymbol{F}_{a1}\\ \boldsymbol{F}_{a2}\end{bmatrix} = \sum_b \begin{pmatrix} l_{11}^{ab} & -l_{12}^{ab}\\ -l_{21}^{ab} & l_{22}^{ab}\end{pmatrix}\begin{bmatrix}\boldsymbol{u}_b^{(1)}\\ 2\boldsymbol{q}_b^{(1)}/5P_b\end{bmatrix}$$

2. Friction coefficient

$$l_{ij}^{ab} = \frac{m_a n_a}{\tau_{aa}} \left[\left(\sum_k \frac{\tau_{aa}}{\tau_{ak}} M_{ak}^{i-1,j-1} \right) \delta_{ab} + \frac{\tau_{aa}}{\tau_{ab}} N_{ab}^{i-1,j-1} \right] ,$$

$$M_{ab}^{00} = -\left(1 + \frac{m_a}{m_b}\right) \left(1 + x_{ab}^2\right)^{-3/2} ,$$

$$M_{ab}^{01} = M_{ab}^{10} = -\frac{3}{2} \left(1 + \frac{m_a}{m_b}\right) \left(1 + x_{ab}^2\right)^{-5/2} ,$$

$$M_{ab}^{11} = -\left(\frac{13}{4} + 4x_{ab}^2 + \frac{15}{2} x_{ab}^4\right) \left(1 + x_{ab}^2\right)^{-5/2} ,$$

$$N_{ab}^{00} = \left(1 + \frac{m_a}{m_b}\right) \left(1 + x_{ab}^2\right)^{-3/2} ,$$

$$N_{ab}^{10} = \frac{3}{2} \left(1 + \frac{m_a}{m_b}\right) \left(1 + x_{ab}^2\right)^{-5/2} ,$$

$$N_{ab}^{01} = \frac{3}{2} \frac{T_a}{T_b} x_{ab}^{-1} \left(1 + \frac{m_b}{m_a}\right) \left(1 + x_{ba}^2\right)^{-5/2} ,$$

$$N_{ab}^{11} = \frac{27}{4} \frac{T_a}{T_b} x_{ab}^2 \left(1 + x_{ab}^2\right)^{-5/2} ,$$

$$\tau_{ab} = \frac{3\pi^{3/2} \varepsilon_0^2 m_a^2 v_{Ta}^3}{n_b e_a^2 e_b^2 \ln \Lambda} ,$$

$$x_{ab}^2 = \frac{m_a T_b}{m_b T_a} , \quad v_{Ta} = \sqrt{\frac{2T_a}{m_a}}$$

3. Symmetry relations of friction coefficient

$$l_{ij}^{ab} = l_{ji}^{ba} ,$$

$$M_{ab}^{ij} = M_{ab}^{ji} ,$$

$$N_{ab}^{j0} = -M_{ab}^{j0} ,$$

$$N_{ab}^{ij} = \frac{T_a v_{Ta}}{T_b v_{Tb}} N_{ba}^{ji}$$

4. Viscous force-flow relation

$$\begin{bmatrix} \langle \boldsymbol{B} \cdot \nabla \cdot \boldsymbol{\Pi}_a \rangle \\ \langle \boldsymbol{B} \cdot \nabla \cdot \boldsymbol{\Theta}_a \rangle \end{bmatrix} = \langle B^2 \rangle \begin{bmatrix} \mu_{a1} & \mu_{a2} \\ \mu_{a2} & \mu_{a3} \end{bmatrix} \begin{bmatrix} u_{a\theta}^*(\psi) \\ 2q_{a\theta}^*(\psi)/5P_a \end{bmatrix}$$

5. Viscosity coefficients

$$\mu_{a1} = K_{11}^a ,$$
$$\mu_{a2} = K_{12}^a - \frac{5}{2} K_{11}^a ,$$
$$\mu_{a3} = K_{22}^a - 5K_{12}^a + \frac{25}{4} K_{11}^a$$

6. Velocity partitioned viscosity coefficient

$$K_{ij}^a = \frac{m_a n_a}{\tau_{aa}} \frac{f_t}{f_c} \left\{ x_a^{2(i+j-2)} \nu_{\rm tot}^a(v) \tau_{aa} \right\} ,$$
$$\{A(v)\} = \frac{8}{3\pi^{1/2}} \int_0^\infty \exp(-x_a^2) x_a^4 A(x_a v_a) \mathrm{d}x_a ,$$
$$\nu_{\rm tot}^a(v) = \frac{\nu_D^a(v)}{\left[1 + 2.48 \nu_a^* \nu_D^a(v) \tau_{aa}/x_a\right]\left[1 + 1.96 \nu_T^a(v)/x_a \omega_{Ta}\right]} ,$$
$$\nu_D^a(v) = \frac{3\sqrt{\pi}}{4} \frac{1}{\tau_{aa}} \sum_b \frac{n_b Z_b^2}{n_a Z_a^2} \frac{\Phi(x_b) - G(x_b)}{x_a^3} ,$$
$$\nu_E^a(v) = \frac{3\sqrt{\pi}}{4} \frac{1}{\tau_{aa}} \sum_b \frac{n_b Z_b^2}{n_a Z_a^2} \left[4 \left(\frac{T_a}{T_b} + x_{ab}^{-2} \right) \frac{G(x_b)}{x_a} - \frac{2\Phi(x_b)}{x_a^3} \right] ,$$
$$\Phi(x) = \frac{2}{\sqrt{\pi}} \int_0^x \exp(-u^2) \mathrm{d}u , \quad G(x) = \left[\Phi(x) - x\Phi'(x)\right]/2x^2 ,$$
$$\nu_T^a(v) = 3\nu_D^a(v) + \nu_E^a(v) , \quad x_a = v/v_{Ta} , \quad \omega_{Ta} = v_{Ta}/L_c$$

6. Collisionality

$$\nu_a^* \equiv \frac{1}{\varepsilon^{1.5} \omega_{Ta} \tau_{aa}} \sim \left(\frac{R}{r}\right)^{1.5} \frac{Rq}{v_{Ta} \tau_{aa}} ,$$
$$\nu_a^* = 4.9 \times 10^{-18} \frac{R(m) q n_a(m^{-3}) Z_a^4 \ln \Lambda}{\varepsilon^{1.5} T_a^2(\mathrm{eV})}$$

7. Circulating particle fraction

$$f_c = \frac{3\langle B^2 \rangle}{4} \int_0^{1/B_{\max}} \frac{\lambda \mathrm{d}\lambda}{\left\langle \sqrt{1 - \lambda B} \right\rangle}$$

8. Parallel momentum and heat flow balance equation

$$\begin{bmatrix} \mu_{a1} & \mu_{a2} \\ \mu_{a2} & \mu_{a3} \end{bmatrix} \begin{bmatrix} \langle Bu_{\parallel a}\rangle - BV_{1a} \\ \langle 2Bq_{\parallel a}/5P_a\rangle - BV_{2a} \end{bmatrix} =$$

$$\sum_b \begin{bmatrix} l_{11}^{ab} & -l_{12}^{ab} \\ -l_{21}^{ab} & l_{22}^{ab} \end{bmatrix} \begin{bmatrix} \langle Bu_{\parallel b}\rangle \\ \langle 2Bq_{\parallel b}/5P_b\rangle \end{bmatrix} + \begin{bmatrix} e_a n_a \langle BE_\parallel\rangle \\ 0 \end{bmatrix} + \begin{bmatrix} \langle BM_{a\parallel}\rangle \\ \langle BQ_{a\parallel}\rangle \end{bmatrix}$$

8.4 Parallel Current: Generalized Ohm's Law

1. Parallel momentum and heat flow balance equation (Matrix form)

$$\boldsymbol{M}(\boldsymbol{U}_\parallel - \boldsymbol{V}_\perp) = \boldsymbol{L}\boldsymbol{U}_\parallel + \boldsymbol{E}^* + \boldsymbol{S}_\parallel ,$$

$$\boldsymbol{L} = \begin{bmatrix} l_{11}^{ee} & l_{11}^{ei} & l_{11}^{eI} & -l_{12}^{ee} & -l_{12}^{ei} & -l_{12}^{eI} \\ l_{11}^{ie} & l_{11}^{ii} & l_{11}^{iI} & -l_{12}^{ie} & -l_{12}^{ii} & -l_{12}^{iI} \\ l_{11}^{Ie} & l_{11}^{Ii} & l_{11}^{II} & -l_{12}^{Ie} & -l_{12}^{Ii} & -l_{12}^{II} \\ -l_{21}^{ee} & -l_{21}^{ei} & -l_{21}^{eI} & l_{22}^{ee} & l_{22}^{ei} & l_{22}^{eI} \\ -l_{21}^{ie} & -l_{21}^{ii} & -l_{21}^{iI} & l_{22}^{ie} & l_{22}^{ii} & l_{22}^{iI} \\ -l_{21}^{Ie} & -l_{21}^{Ii} & -l_{21}^{II} & l_{22}^{Ie} & l_{22}^{Ii} & l_{22}^{II} \end{bmatrix}, \quad \boldsymbol{M} = \begin{bmatrix} \mu_{e1} & 0 & 0 & \mu_{e2} & 0 & 0 \\ 0 & \mu_{i1} & 0 & 0 & \mu_{i2} & 0 \\ 0 & 0 & \mu_{I1} & 0 & 0 & \mu_{I2} \\ \mu_{e2} & 0 & 0 & \mu_{e3} & 0 & 0 \\ 0 & \mu_{i2} & 0 & 0 & \mu_{i3} & 0 \\ 0 & 0 & \mu_{I2} & 0 & 0 & \mu_{I3} \end{bmatrix}$$

$$\boldsymbol{U}_\parallel = \begin{bmatrix} \langle Bu_{\parallel e}\rangle \\ \langle Bu_{\parallel i}\rangle \\ \langle Bu_{\parallel I}\rangle \\ 2\langle Bq_{\parallel e}\rangle/5P_e \\ 2\langle Bq_{\parallel i}\rangle/5P_i \\ 2\langle Bq_{\parallel I}\rangle/5P_I \end{bmatrix}, \quad \boldsymbol{V}_\perp = \begin{bmatrix} BV_{1e} \\ BV_{1i} \\ BV_{1I} \\ BV_{2e} \\ BV_{2i} \\ BV_{2I} \end{bmatrix},$$

$$\boldsymbol{E}^* = \langle BE_\parallel\rangle \begin{bmatrix} -en_e \\ eZ_in_i \\ eZ_In_I \\ 0 \\ 0 \\ 0 \end{bmatrix}, \quad \boldsymbol{S}_\parallel = \begin{bmatrix} \langle BM_e\rangle \\ \langle BM_i\rangle \\ \langle BM_I\rangle \\ \langle BQ_e\rangle \\ \langle BQ_i\rangle \\ \langle BQ_I\rangle \end{bmatrix}$$

2. Parallel flow–thermodynamical force relation

$$\begin{aligned}
\boldsymbol{U}_{\|} &= (\boldsymbol{M}-\boldsymbol{L})^{-1}\boldsymbol{M}\boldsymbol{V}_{\perp} + (\boldsymbol{M}-\boldsymbol{L})^{-1}\boldsymbol{E}^{*} + (\boldsymbol{M}-\boldsymbol{L})^{-1}\boldsymbol{S}_{\|}\,,\\
U_{\|a} &= \sum_{b}\left(\alpha_{ab}V_{\perp b} + c_{ab}(E_b^{*} + S_{\|b})\right),\\
\boldsymbol{\alpha} &= (\boldsymbol{M}-\boldsymbol{L})^{-1}\boldsymbol{M} \text{ and } \boldsymbol{c} = (\boldsymbol{M}-\boldsymbol{L})^{-1}
\end{aligned}$$

3. Generalized Ohm's law

$$\begin{aligned}
\langle \boldsymbol{B}\cdot\boldsymbol{J}\rangle &= \sum_{a=e,i,I} e_a n_a \langle \boldsymbol{B}\cdot\boldsymbol{u}_a\rangle\\
&= \sum_{a=e,i,I} e_a n_a \left\{\sum_{b=1}^{6}\left[(\boldsymbol{M}-\boldsymbol{L})^{-1}\boldsymbol{M}\right]_{ab} V_{\perp b}\right.\\
&\qquad + \sum_{b=1}^{3}\left[(\boldsymbol{M}-\boldsymbol{L})^{-1}\right]_{ab} e_b n_b \langle BE_{\|}\rangle\\
&\qquad \left. + \sum_{b=1}^{6}\left[(\boldsymbol{M}-\boldsymbol{L})^{-1}\right]_{ab} S_{\|b}\right\}\\
&= \langle \boldsymbol{B}\cdot\boldsymbol{J}\rangle_{\text{bs}} + \sigma_{\|}^{\text{NC}}\langle \boldsymbol{B}\cdot\boldsymbol{E}\rangle + \langle \boldsymbol{B}\cdot\boldsymbol{J}\rangle_{bd} + \langle \boldsymbol{B}\cdot\boldsymbol{J}\rangle_{\text{RFCD}}
\end{aligned}$$

8.5 Trapped Particle Effect: Electrical Conductivity

1. Spitzer electrical conductivity

$$\begin{aligned}
\sigma_{\|}^{\text{Spitzer}} &= \frac{n_e e^2 \tau_{ee}}{m_e}\frac{3.4(1.13 + Z_{\text{eff}})}{Z_{\text{eff}}(2.67 + Z_{\text{eff}})}\,,\\
Z_{\text{eff}} &= \frac{1}{n_e}\sum_{b=i,I} n_b Z_b^2\,,\\
\tau_{ee} &= \frac{6\sqrt{2}\pi^{3/2}\varepsilon_0^2 m_e^{1/2} T_e^{3/2}}{n_e e^4 \ln\Lambda} = 2.74\times 10^{-4}\frac{T_e[\text{keV}]^{3/2}}{n_e[\text{m}^{-3}]\ln\Lambda}[\text{sec}]
\end{aligned}$$

1′. Spitzer electrical conductivity (moment equation)

$$\sigma_{\|}^{\text{Spitzer}} = -\sum_{a=e,i,I}\sum_{b=e,i,I} e_a n_a e_b n_b \boldsymbol{L}_{ab}^{-1}$$

2. Neoclassical electrical conductivity

Hirshman–Hawryluk–Birge form

$$\sigma_{\parallel}^{\mathrm{NC}} = \sigma_{\parallel}^{\mathrm{Spitzer}} \left[1 - \frac{f_t}{1+\xi\nu_e^*}\right]\left[1 - \frac{C_R f_t}{1+\xi\nu_e^*}\right],$$

$$C_R(Z_{\mathrm{eff}}) = \frac{0.56}{Z_{\mathrm{eff}}}\frac{3-Z_{\mathrm{eff}}}{3+Z_{\mathrm{eff}}}, \quad \xi(Z_{\mathrm{eff}}) = 0.58 + 0.2Z_{\mathrm{eff}},$$

$$f_t = 1 - \frac{(1-\varepsilon)^2}{\left(1+1.46\varepsilon^{1/2}\right)\sqrt{1-\varepsilon^2}}$$

Moment equation

$$\sigma_{\parallel}^{\mathrm{NC}} = \sum_{a=e,i,I}\sum_{b=e,i,I} e_a n_a e_b n_b \left[(\boldsymbol{M}-\boldsymbol{L})^{-1}\right]_{ab}$$

8.6 Thermodynamic Force: Bootstrap Current

1. Bootstrap current expression

$$\langle \boldsymbol{B}\cdot\boldsymbol{J}\rangle_{\mathrm{bs}} = -F(\psi)n_e(\psi)\sum_{a=e,i,I}\frac{1}{|Z_a|}\left[L_{31}^a\frac{1}{n_a(\psi)}\frac{\mathrm{d}P_a(\psi)}{\mathrm{d}\psi} + L_{32}^a\frac{\mathrm{d}T_a(\psi)}{\mathrm{d}\psi}\right],$$

$$L_{31}^a = \sum_{a=e,i,I}\frac{|Z_a|}{Z_a}\frac{Z_b}{n_e}\alpha_{ab},$$

$$L_{32}^a = \sum_{a=e,i,I}\frac{|Z_a|}{Z_a}\frac{Z_b n_b}{n_e}\alpha_{a,b+3}$$

3. Galeev-Sagdeev simple expression

$$J_{\mathrm{bootstrap}} \sim -\varepsilon^{1/2}\frac{T_e}{B_p}\frac{\mathrm{d}n_e}{\mathrm{d}r} \sim -\varepsilon^{1/2}\frac{1}{B_p}\frac{\mathrm{d}P_e}{\mathrm{d}r}$$

4. Simplified electron generalized Ohm's law

$$J_{\parallel e} = \frac{e^2 n_e}{m_e\nu_{ei}}\frac{1}{\hat{\mu}_{e1}+1}E_{\parallel} - \frac{\hat{\mu}_{e1}}{\hat{\mu}_{e1}+1}\frac{1}{B_p}\frac{\mathrm{d}P_e}{\mathrm{d}r}$$

8.7 Momentum Input: Beam-driven Currents

1. Expression for beam-driven current

$$\langle \boldsymbol{B}\cdot\boldsymbol{J}\rangle_{\text{bd}} = \sum_{a=e,i,I,f} e_a n_a \left[(\boldsymbol{M}-\boldsymbol{L})^{-1}\right]_{af} S_{\parallel f} \; ,$$

$$\langle \boldsymbol{B}\cdot\boldsymbol{J}\rangle_{\text{bd}} = \langle \boldsymbol{B}\cdot\boldsymbol{J}\rangle_{\text{fast}} + \langle \boldsymbol{B}\cdot\boldsymbol{J}\rangle_{\text{shield}} \; ,$$

$$\langle \boldsymbol{B}\cdot\boldsymbol{J}\rangle_{\text{fast}} = e_f n_f \left[(\boldsymbol{M}-\boldsymbol{L})^{-1}\right]_{ff} S_{\parallel f} \; ,$$

$$\langle \boldsymbol{B}\cdot\boldsymbol{J}\rangle_{\text{shield}} = \sum_{a=e,i,I} e_a n_a \left[(\boldsymbol{M}-\boldsymbol{L})^{-1}\right]_{af} S_{\parallel f}$$

2. Stacking factor

$$F = \frac{\langle \boldsymbol{B}\cdot\boldsymbol{J}\rangle_{\text{bd}}}{\langle \boldsymbol{B}\cdot\boldsymbol{J}\rangle_{\text{fast}}} = \frac{\sum_{a=e,i,I,f} e_a n_a \left[(\boldsymbol{M}-\boldsymbol{L})^{-1}\right]_{af}}{e_f n_f \left[(\boldsymbol{M}-\boldsymbol{L})^{-1}\right]_{ff}}$$

3. Fast ion Fokker–Planck equation

$$\tau_{se}\frac{\partial f_b}{\partial t} = v^{-2}\frac{\partial}{\partial v}\left[\left(v_c^3+v^3\right) f_b\right] + \frac{\beta v_c^3}{v^3\eta\langle v/v_\parallel\rangle}\frac{\partial}{\partial\eta}\left[\frac{(1-\eta^2)}{\eta}\left\langle\frac{v_\parallel}{v}\right\rangle\frac{\partial f_b}{\partial\eta}\right]$$
$$+ \tau_{se} S(v,\eta) \; ,$$

$$\tau_{se} = \frac{3(2\pi)^{3/2}\varepsilon_0^2 M_b T_e^{3/2}}{e^4 Z_b^2 n_e m_e^{1/2}\ln\Lambda} = \frac{0.200 A_b T_e[\text{keV}]^{3/2}}{Z_b^2 n_e[10^{20}\text{m}^{-3}]\ln\Lambda}[\text{s}] \; ,$$

$$E_c = \left(\frac{9\pi m_p}{16 m_e}\right)^{1/3}\left[\sum_j \frac{n_j Z_j^2}{n_e A_j}\right]^{2/3} A_b T_e = 14.8\bar{Z}^{2/3} A_b^{1/3} T_e[\text{keV}][\text{keV}]$$

$$\langle v/v_\parallel\rangle = (2/\pi)K[(\eta_t/\eta)^2] \, , \quad \langle v_\parallel/v\rangle = (2/\pi)E[(\eta_t/\eta)^2] \, , \quad v_c = (2E_c/m_b)^{1/2}$$

4. Solution of Fokker–Planck equation (Gaffey)

$$f_b(v,\eta) = S_0\tau_{se}\sum_n a_n(v)c_n(\eta) \; ,$$

$$a_n = \begin{cases} \dfrac{k_n}{v^3+v_c^3}\left[\dfrac{v^3(v_b^3+v_c^3)}{v_b^3(v^3+v_c^3)}\right]^{\beta\lambda n/3} & v < v_b \\[2ex] \dfrac{k_n}{v_b^3+v_c^3}\exp\left[-\dfrac{2E_b(1+(v_c/v_b))^3}{T_e+(v_c/v_b)^3 T_i}\dfrac{v-v_b}{v_b}\right] & v > v_b \end{cases}$$

5. Equation for c_n

$$\frac{1}{\eta\langle v/v_\parallel\rangle}\frac{\mathrm{d}}{\mathrm{d}\eta}\left[(1-\eta^2)\left\langle\frac{v_\parallel}{v}\right\rangle\frac{\mathrm{d}c_n}{\mathrm{d}\eta}\right]+\lambda_n c_n=0$$

6. Fast ion current

$$\langle \boldsymbol{B}\cdot\boldsymbol{J}\rangle_{\text{fast}} = eZ_f\int_0^\infty v^3\mathrm{d}v\int_{-1}^{1}\mathrm{d}\eta f_b(v,\eta,\rho)H(v,\eta,\rho)\ ,$$

$$H(v,\eta,\rho)=\int\frac{B}{v}\mathrm{d}s\int\frac{\mathrm{d}s}{v_\parallel}$$

7. Friction and viscous matrix elements with fast ion components

$$l_{11}^{fe}=l_{11}^{ef}=(n_e m_e/\tau_{ee})\left(Z_f^2 n_f/n_e\right)$$

$$l_{11}^{fk}=l_{11}^{kf}=(3\sqrt{\pi}n_k m_k/4\tau_{kk})\left(Z_f^2 n_f/Z_k^2 n_k\right)(1+m_k/m_f)$$
$$\left(\tau_s v_{Tk}^3/3\tau_{th}v_c^3\right)\ (k=i,I)$$

$$l_{11}^{fe}=l_{11}^{ef}=(n_e m_e/\tau_{ee})\left(Z_f^2 n_f/n_e\right)$$

$$l_{11}^{ff}=-\left(l_{11}^{fe}+l_{11}^{fi}+l_{11}^{fI}\right),\ l_{21}^{ef}=-1.5l_{11}^{ef}\ ,$$
$$l_{21}^{if}=l_{21}^{If}=l_{12}^{fe}=l_{12}^{fi}=l_{12}^{fI}=0$$

$$l_{11}^{kk}=l_{11}^{kk}(th)-l_{11}^{kf}\quad (k=e,i,I),\quad l_{21}^{ee}=l_{21}^{ee}(th)-l_{21}^{ef}\ ,$$

$\tau_{se}=3(2\pi)^{3/2}\varepsilon_0^2 M_b T_e^{3/2}/e^4 Z_b^2 n_e m_e^{1/2}\ \ln\Lambda$: slowing down time

$\tau_{th}=(\tau_{se}/3)\ln((E_{b0}/E_c)^{1.5}+1)$: thermalization time

$\mu_{f1}=\dfrac{f_t}{f_c}\dfrac{n_e m_e}{\tau_{ee}}\dfrac{Z_f^2 n_f}{n_e}\dfrac{\hat{Z}v_c^3}{\hat{v}_c^3}$: fast ion viscosity coefficient

$p_f=\int\dfrac{m_f v^2}{3}f_f(\boldsymbol{v})\,\mathrm{d}\boldsymbol{v}$: fast ion pressure

$$\hat{Z}=\frac{\sum Z_i^2 n_i}{\sum Z_i^2 n_i m_f/m_i},\quad \hat{v}_c^3=\frac{\tau_{se}}{3\tau_{th}}\frac{n_f m_f}{2p_f}\int_0^{v_b}\frac{v\,\mathrm{d}v}{v^3+v_c^3}\ .$$

$$\boldsymbol{L} = \begin{bmatrix} l_{11}^{ee} & l_{11}^{ei} & l_{11}^{eI} & l_{11}^{ef} & -l_{12}^{ee} & -l_{12}^{ee} & -l_{12}^{ee} \\ l_{11}^{ie} & l_{11}^{ii} & l_{11}^{iI} & l_{11}^{if} & -l_{12}^{ie} & -l_{12}^{ii} & -l_{12}^{iI} \\ l_{11}^{Ie} & l_{11}^{Ii} & l_{11}^{II} & l_{11}^{If} & -l_{12}^{Ie} & -l_{12}^{Ii} & -l_{12}^{II} \\ -l_{11}^{Ie} & -l_{11}^{Ii} & -l_{11}^{II} & -l_{11}^{ff} & 0 & 0 & 0 \\ -l_{21}^{ee} & -l_{21}^{ei} & -l_{21}^{eI} & -l_{21}^{ef} & l_{22}^{ee} & l_{22}^{ei} & l_{22}^{eI} \\ -l_{21}^{ie} & -l_{21}^{ii} & -l_{21}^{iI} & 0 & l_{22}^{ie} & l_{22}^{ii} & l_{22}^{iI} \\ -l_{21}^{Ie} & -l_{21}^{Ii} & -l_{21}^{II} & 0 & l_{22}^{Ie} & l_{22}^{Ii} & l_{22}^{II} \end{bmatrix},$$

$$\boldsymbol{M} = \begin{bmatrix} \mu_{e1} & 0 & 0 & 0 & \mu_{e2} & 0 & 0 \\ 0 & \mu_{i1} & 0 & 0 & 0 & \mu_{i2} & 0 \\ 0 & 0 & \mu_{I1} & 0 & 0 & 0 & \mu_{I2} \\ 0 & 0 & 0 & \mu_{f1} & 0 & 0 & 0 \\ \mu_{e2} & 0 & 0 & 0 & \mu_{e3} & 0 & 0 \\ 0 & \mu_{i2} & 0 & 0 & 0 & \mu_{i3} & 0 \\ 0 & 0 & \mu_{I2} & 0 & 0 & 0 & \mu_{I3} \end{bmatrix}$$

8.8 "Rotation": Toroidal Rotation and Flow Relations among Ions, Impurities, and Electrons

1. Fluid equation of motion

$$\sum_a m_a n_a \frac{\mathrm{d}\boldsymbol{u}_a}{\mathrm{d}t} = \boldsymbol{J} \times \boldsymbol{B} - \nabla P - \sum_a \nabla \cdot \boldsymbol{\Pi}_a + \sum_a \boldsymbol{M}_a$$

2. Momentum conservation and charge neutrality

$$\sum_a \boldsymbol{F}_{a1} = 0\,,$$

$$\sum_a e_a n_a = 0$$

3. Toroidal angular momentum balance equation

$$\sum_a m_a \left\langle n_a R \frac{\mathrm{d}u_{a\zeta}}{\mathrm{d}t} \right\rangle = \left\langle \sum_a R^2 \nabla\zeta \cdot \nabla \cdot \boldsymbol{\Pi}_a \right\rangle + \left\langle \sum_a R^2 \nabla\zeta \cdot \boldsymbol{M}_a \right\rangle$$

4. Conservation law

$$\langle R^2 \nabla\zeta \cdot \nabla \cdot \boldsymbol{\Pi}_a \rangle = 0$$

5. Local impurity toroidal flow

$$u_{I\zeta}^{(1)} = \frac{B_\zeta}{\langle B^2 \rangle} \langle \boldsymbol{B} \cdot \boldsymbol{u}_I \rangle + \left[1 - \frac{B_\zeta^2}{\langle B^2 \rangle} \right] R \left(\frac{\mathrm{d}\Phi}{\mathrm{d}\psi} + \frac{1}{e_I n_I} \frac{\mathrm{d}P_I}{\mathrm{d}\psi} \right) ,$$

$$\langle \boldsymbol{B} \cdot \boldsymbol{u}_I \rangle = -F(\psi) \sum_{b=e,i,I} \left[\alpha_{Ib} \left(\frac{\mathrm{d}\Phi}{\mathrm{d}\psi} + \frac{1}{e_b n_b} \frac{\mathrm{d}P_b}{\mathrm{d}\psi} \right) + \alpha_{I,b+3} \frac{1}{e_b} \frac{\mathrm{d}T_b}{\mathrm{d}\psi} \right] + \sum_{b=e,i,I} e_b n_b c_{Ib} \langle \boldsymbol{B} \cdot \boldsymbol{E} \rangle$$

6. Radial electric field from impurity toroidal flow

$$\frac{\mathrm{d}\Phi}{\mathrm{d}\psi} = \frac{1}{\sum\limits_{b=e,i,I} \alpha_{Ib}^*} \left[-\frac{\langle B^2 \rangle}{R B_\zeta^2} u_{\zeta I} - \sum_{b=e,i,I} \left\{ \frac{\alpha_{Ib}^*}{e_b n_b} \frac{\mathrm{d}P_b}{\mathrm{d}\psi} + \alpha_{I,b+3} \frac{1}{e_b} \frac{\mathrm{d}T_b}{\mathrm{d}\psi} \right\} + \sum_{b=e,i,I} e_b n_b c_{Ib} \langle \mathbf{B} \cdot \boldsymbol{E} \rangle \right]$$

7. Flux surface averaged electron flow

$$\langle \boldsymbol{B} \cdot \boldsymbol{u}_e \rangle = -F(\psi) \sum_{b=e,i,I} \left[\alpha_{eb} \left(\frac{\mathrm{d}\Phi}{\mathrm{d}\psi} + \frac{1}{e_b n_b} \frac{\mathrm{d}P_b}{\mathrm{d}\psi} \right) + \alpha_{e,b+3} \frac{1}{e_b} \frac{\mathrm{d}T_b}{\mathrm{d}\psi} \right] + \sum_{b=e,i,I} e_b n_b c_{eb} \langle \boldsymbol{B} \cdot \boldsymbol{E} \rangle$$

8. Flux surface averaged ion flow

$$\langle \boldsymbol{B} \cdot \boldsymbol{u}_i \rangle = -F(\psi) \sum_{b=e,i,I} \left[\alpha_{ib} \left(\frac{\mathrm{d}\Phi}{\mathrm{d}\psi} + \frac{1}{e_b n_b} \frac{\mathrm{d}P_b}{\mathrm{d}\psi} \right) + \alpha_{i,b+3} \frac{1}{e_b} \frac{\mathrm{d}T_b}{\mathrm{d}\psi} \right] + \sum_{b=e,i,I} e_b n_b c_{ib} \langle \boldsymbol{B} \cdot \boldsymbol{E} \rangle$$

9. Momentum and heat flow balance equation for impurity and ion

$$\hat{\boldsymbol{\mu}}^I \cdot \boldsymbol{u}_\theta^I = \hat{\boldsymbol{L}}_{II} \cdot \boldsymbol{u}_\parallel^I + \hat{\boldsymbol{L}}_{Ii} \cdot \boldsymbol{u}_\parallel^i ,$$

$$\hat{\boldsymbol{\mu}}^i \cdot \boldsymbol{u}_\theta^i = \hat{\boldsymbol{L}}_{iI} \cdot \boldsymbol{u}_\parallel^I + \hat{\boldsymbol{L}}_{ii} \cdot \boldsymbol{u}_\parallel^i ,$$

$$\hat{\boldsymbol{\mu}}^a = \begin{bmatrix} \hat{\mu}_{a1} & \hat{\mu}_{a2} \\ \hat{\mu}_{a2} & \hat{\mu}_{a3} \end{bmatrix} , \quad \hat{\boldsymbol{L}}_{ab} = \begin{bmatrix} \hat{l}_{11}^{ab} & -\hat{l}_{12}^{ab} \\ -\hat{l}_{21}^{ab} & \hat{l}_{22}^{ab} \end{bmatrix} ,$$

$$\boldsymbol{u}_\theta^a = \begin{bmatrix} \langle B^2 \rangle u_{a\theta}^*(\psi) \\ 2 \langle B^2 \rangle q_{a\theta}^*(\psi) / 5 P_a \end{bmatrix} , \quad \boldsymbol{u}_\parallel^a = \begin{bmatrix} \langle B u_{\parallel b} \rangle \\ \langle 2 B q_{\parallel b} / 5 P_b \rangle \end{bmatrix}$$

10. Normalized friction coefficient

$$\hat{l}_{ij}^{ab} = \frac{\tau_{aa}}{m_a n_a} l_{ij}^{ab}$$

11. Normalized viscosity coefficient

$$\hat{\mu}_{ai} = \frac{\tau_{aa}}{m_a n_a} \mu_{ai}$$

12. Flow relation

$$\boldsymbol{u}_\theta^a = \boldsymbol{u}_\parallel^a - \boldsymbol{V}^a$$

13. Thermodynamic force

$$\boldsymbol{V}^a = \begin{bmatrix} BV_{1a} \\ BV_{2a} \end{bmatrix}$$

14. Ion and impurity flows

$$\begin{aligned}
\boldsymbol{u}_\parallel^i &= \begin{bmatrix} 1 & K_1 \\ 0 & K_2 \end{bmatrix} \boldsymbol{V}^i \,, \\
\boldsymbol{u}_\parallel^I &= \begin{bmatrix} 1 & K_1 + 1.5K_2 \\ 0 & 0 \end{bmatrix} \boldsymbol{V}^i \,, \\
K_1 &= \frac{\gamma \hat{\mu}_{i2}}{D} \,, \quad K_2 = \frac{\hat{\mu}_{i1}\hat{\mu}_{i3} - \hat{\mu}_{i2}^2}{D} \,, \\
D &= \hat{\mu}_{i1}(\hat{\mu}_{i3} + \gamma) - \hat{\mu}_{i2}^2 \,, \\
\gamma &= 2^{0.5} + \alpha \,, \quad \alpha = n|Z|^2/n_i Zi^2
\end{aligned}$$

15. Simplified parallel and poloidal rotations of ion and impurity

$$\begin{aligned}
u_{\parallel i} &\approx -\frac{1}{B_\theta}\left[\frac{\mathrm{d}\Phi}{\mathrm{d}r} + \frac{1}{eZ_i n_i}\frac{\mathrm{d}P_i}{\mathrm{d}r} + \frac{K_1}{eZ_i}\frac{\mathrm{d}T_i}{\mathrm{d}r}\right] \,, \\
u_{\parallel I} &\approx -\frac{1}{B_\theta}\left[\frac{\mathrm{d}\Phi}{\mathrm{d}r} + \frac{1}{eZ_i n_i}\frac{\mathrm{d}P_i}{\mathrm{d}r} + \frac{K_1 + 1.5K_2}{eZ_i}\frac{\mathrm{d}T_i}{\mathrm{d}r}\right] \,, \\
u_{i\theta} &\approx -\frac{B_\zeta}{e\langle B^2\rangle}\frac{K_1}{Z_i}\frac{\mathrm{d}T_i}{\mathrm{d}r} \,, \\
u_{I\theta} &\approx -\frac{B_\zeta}{e\langle B^2\rangle}\left(\frac{1}{Z_I n_I}\frac{\mathrm{d}P_I}{\mathrm{d}r} - \frac{1}{Z_i n_i}\frac{\mathrm{d}P_i}{\mathrm{d}r} + \frac{K_1 + 1.5K_2}{Z_i}\frac{\mathrm{d}T_i}{\mathrm{d}r}\right)
\end{aligned}$$

8.9 Neoclassical Transport: Transport Across the Magnetic Field

1. Cross field transport flux

$$\begin{bmatrix} \langle n_a \boldsymbol{u}_{a\perp} \cdot \nabla\psi \rangle \\ \left\langle \frac{\boldsymbol{q}_{a\perp}}{T_a} \cdot \nabla\psi \right\rangle \end{bmatrix} = \begin{bmatrix} \Gamma_a^{cl} \\ \frac{q_a^{cl}}{T_a} \end{bmatrix} + \begin{bmatrix} \Gamma_a^{\mathrm{NC}} \\ \frac{q_a^{\mathrm{NC}}}{T_a} \end{bmatrix}$$

2. Classical flux

$$\begin{bmatrix} \Gamma_a^{cl} \\ \frac{q_a^{cl}}{T_a} \end{bmatrix} = \left\langle \frac{\boldsymbol{B} \times \nabla\psi}{e_a B^2} \cdot \begin{bmatrix} \boldsymbol{F}_{a1} \\ \boldsymbol{F}_{a2} \end{bmatrix} \right\rangle$$

3. Neoclassical flux

$$\begin{bmatrix} \Gamma_a^{\mathrm{NC}} \\ \frac{q_a^{\mathrm{NC}}}{T_a} \end{bmatrix} = \left\langle \frac{\boldsymbol{B} \times \nabla\psi}{e_a B^2} \cdot \begin{bmatrix} \nabla P_a + \nabla \cdot \boldsymbol{\Pi}_a - e_a n_a \boldsymbol{E} \\ \frac{5}{2} n_a \nabla T_a + \nabla \cdot \boldsymbol{\Theta}_a \end{bmatrix} \right\rangle$$

4. NC particle flux

$$\Gamma_a^{\mathrm{NC}} = -\frac{F(\psi)}{e_a} \left\langle \left(\frac{1}{B^2} - \frac{1}{\langle B^2 \rangle} \right) \boldsymbol{B} \cdot \boldsymbol{F}_{a1} \right\rangle - \frac{F(\psi)}{e_a \langle B^2 \rangle} \langle \boldsymbol{B} \cdot \nabla \cdot \boldsymbol{\Pi}_a \rangle + \Gamma_a^E$$

5. Electrical flux

$$\Gamma_a^E = \frac{F(\psi) \langle n_a \boldsymbol{B} \cdot \boldsymbol{E} \rangle}{\langle B^2 \rangle} - \left\langle R^2 \nabla\zeta \cdot n_a \boldsymbol{E}_A \right\rangle, \quad \boldsymbol{E}_A = -\frac{\partial \boldsymbol{A}}{\partial t}$$

6. NC heat flux

$$\frac{q_a^{\mathrm{NC}}}{T_a} = -\frac{F(\psi)}{e_a} \left\langle \left(\frac{1}{B^2} - \frac{1}{\langle B^2 \rangle} \right) \boldsymbol{B} \cdot \boldsymbol{F}_{a2} \right\rangle - \frac{F(\psi)}{e_a \langle B^2 \rangle} \langle \boldsymbol{B} \cdot \nabla \cdot \boldsymbol{\Theta}_a \rangle$$

7. Classical flux by thermodynamic force

$$\begin{bmatrix} \Gamma_a^{cl} \\ \frac{q_a^{cl}}{T_a} \end{bmatrix} = \left\langle \frac{|\nabla\psi|^2}{B^2} \right\rangle \sum_b \frac{1}{e_a e_b} \begin{bmatrix} l_{11}^{ab} & -l_{12}^{ab} \\ -l_{21}^{ab} & l_{22}^{ab} \end{bmatrix} \begin{bmatrix} P_b'(\psi)/n_b \\ T_b'(\psi) \end{bmatrix}$$

8. Pfirsch–Schlüter flux

$$\begin{bmatrix} \Gamma_a^{ps} \\ \frac{q_a^{ps}}{T_a} \end{bmatrix} = -\frac{F(\psi)}{e_a} \left\langle \left(\frac{1}{B^2} - \frac{1}{\langle B^2 \rangle} \right) \begin{bmatrix} \boldsymbol{B} \cdot \boldsymbol{F}_{a1} \\ \boldsymbol{B} \cdot \boldsymbol{F}_{a2} \end{bmatrix} \right\rangle$$

9. Pfirsch–Schlüter flux by thermodynamic force

$$\begin{bmatrix} \Gamma_a^{ps} \\ \frac{q_a^{ps}}{T_a} \end{bmatrix} = \frac{F(\psi)^2}{e_a} \left(\left\langle \frac{1}{B^2} \right\rangle - \frac{1}{\langle B^2 \rangle} \right) \sum_b \frac{1}{e_b} \begin{bmatrix} l_{11}^{ab} & -l_{12}^{ab} \\ -l_{21}^{ab} & l_{22}^{ab} \end{bmatrix} \begin{bmatrix} P_b'(\psi)/n_b \\ T_b'(\psi) \end{bmatrix}$$

10. Banana-plateau flux

$$\begin{bmatrix} \Gamma_a^{\mathrm{bp}} \\ \frac{q_a^{\mathrm{bp}}}{T_a} \end{bmatrix} = -\frac{F(\psi)}{e_a \langle B^2 \rangle} \begin{pmatrix} \langle \boldsymbol{B} \cdot \nabla \cdot \boldsymbol{\Pi}_a \rangle \\ \langle \boldsymbol{B} \cdot \nabla \cdot \boldsymbol{\Theta}_a \rangle \end{pmatrix}$$

11. Banana-plateau flux by thermodynamic force

$$\begin{bmatrix} \Gamma_a^{\rm bp} \\ \frac{q_a^{\rm bp}}{T_a} \end{bmatrix} = -\sum_{b=e,i,I} \begin{bmatrix} K_{11}^{ab} & K_{12}^{ab} \\ K_{21}^{ab} & K_{22}^{ab} \end{bmatrix} \begin{bmatrix} P_a'(\psi)/n_a \\ T_a'(\psi) \end{bmatrix} + \begin{bmatrix} g_{1a} \\ g_{2a} \end{bmatrix} \langle BE_\parallel \rangle$$

8.10 Neoclassical Ion Thermal Diffusivity: Ion Thermal Diffusivity by Coulomb Collisions

1. Classical and Pfirsch–Schlüter flux

$$\begin{bmatrix} \Gamma_a^{c+ps} \\ \frac{q_a^{c+ps}}{T_a} \end{bmatrix} = \left(\langle R^2 \rangle - \frac{F(\psi)^2}{\langle B^2 \rangle} \right) \frac{m_a n_a}{e^2 \psi'(\rho) \tau_{aa}}$$

$$\times \sum_b \frac{1}{Z_a Z_b} \begin{bmatrix} \hat{l}_{11}^{ab} & -\hat{l}_{12}^{ab} \\ -\hat{l}_{21}^{ab} & \hat{l}_{22}^{ab} \end{bmatrix} \begin{bmatrix} (\mathrm{d}P_b/\mathrm{d}\rho)/n_b \\ \mathrm{d}T_b/\mathrm{d}\rho \end{bmatrix}$$

2. Minor radius defined using toroidal flux

$$\rho = (\phi/\phi_a)^{1/2} a$$

3. Ion thermal diffusivity for classical and Pfirsch–Schlüter

$$\chi_a^{\rm cps} = -\frac{\langle \boldsymbol{q}_{a\perp} \cdot \nabla\rho \rangle}{\left\langle |\nabla\rho|^2 \right\rangle n_a \mathrm{d}T_a/\mathrm{d}\rho} = \sqrt{\varepsilon} \rho_{pa}^2 \nu_{aa} \hat{K}_{2a}^{\rm cps} ,$$

$$\rho_{pa} = \frac{m_a v_{Ta}}{e_a B_{p1}} , \quad B_{p1}^2 = \frac{B_0^2}{F(\psi)^2} |\nabla\psi|^2 ,$$

$$\hat{K}_{2a}^{\rm cps} = \frac{B_0^2 \left(\hat{l}_{21}^{aa} - \hat{l}_{22}^{aa} \right)}{2\sqrt{\varepsilon} \langle B^2 \rangle} \left(\frac{\langle R^2 \rangle \langle B^2 \rangle}{F(\psi)^2} - 1 \right)$$

4. Ion thermal diffusivity for classical and Pfirsch–Schlüter including impurity collisions

$$\hat{K}_{2a}^{\rm cps} = \frac{B_0^2}{2\sqrt{\varepsilon} \langle B^2 \rangle} \left(\frac{\langle R^2 \rangle \langle B^2 \rangle}{F(\psi)^2} - 1 \right) \sum_{b=i,I} \frac{Z_a}{Z_b} \left(\hat{l}_{21}^{ab} - \hat{l}_{22}^{ab} \right)$$

5. Ion thermal diffusivity for classical and Pfirsch–Schlüter including impurity channel loss

$$\chi_{i\,(\rm tot)}^{\rm cps} = \frac{\langle (\boldsymbol{q}_{i\perp} + \mathbf{q}_{I\perp}) \cdot \nabla\rho \rangle}{\left\langle |\nabla\rho|^2 \right\rangle (n_i + n_I) \mathrm{d}T_a/\mathrm{d}\rho} = \sqrt{\varepsilon} \rho_{pi}^2 \nu_{ii} \left[f_i \hat{K}_{2i}^{\rm cps} + (1 - f_i) \alpha \hat{K}_{2I}^{\rm cps} \right] ,$$

$$f_i = n_i/(n_i + n_I) , \quad \alpha = Z_I^2 n_I / Z_i^2 n_i$$

6. Ion thermal diffusivity for banana-plateau

$$\chi_a^{\mathrm{bp}} = -\frac{\left\langle \boldsymbol{q}_{a\perp}^{\mathrm{bp}} \cdot \nabla\rho \right\rangle}{\left\langle |\nabla\rho|^2 \right\rangle n_a \mathrm{d}T_a/\mathrm{d}\rho} = \sqrt{\varepsilon}\rho_{pa}^2 \nu_{aa} \hat{K}_{2a}^{\mathrm{bp}} ,$$

$$\hat{K}_{2a}^{\mathrm{bp}} = \frac{B_0^2\left(\hat{\mu}_{3a}(1-\alpha_{a+3,a}+\alpha_{a+3,a+3}) + \hat{\mu}_{2a}(1+\alpha_{a,a}-\alpha_{a,a+3})\right)}{2\sqrt{\varepsilon}\langle B^2\rangle} ,$$

$$\hat{\mu}_{3a} = \frac{\tau_{aa}}{m_a n_a}\mu_{3a} , \quad \hat{\mu}_{2a} = \frac{\tau_{aa}}{m_a n_a}\mu_{2a}, \quad \alpha_{ij} = \left[(\boldsymbol{M}-\boldsymbol{L})^{-1}\boldsymbol{M}\right]_{ij}$$

7. Neoclassical ion thermal diffusivity

$$\chi_a^{\mathrm{NC}} = \sqrt{\varepsilon}\rho_{pa}^2 \nu_{aa} \hat{K}_{2a}^{\mathrm{NC}} , \quad \hat{K}_{2a}^{\mathrm{NC}} = \hat{K}_{2a}^{\mathrm{cps}} + \hat{K}_{2a}^{\mathrm{bp}}$$

8. Neoclassical thermal diffusivity including impurity channel loss

$$\chi_{i\,(\mathrm{tot})}^{\mathrm{NC}} = \frac{\langle(\boldsymbol{q}_{i\perp}+\mathbf{q}_{I\perp})\cdot\nabla\rho\rangle}{\left\langle|\nabla\rho|^2\right\rangle(n_i+n_I)\mathrm{d}T_i/\mathrm{d}\rho} = \sqrt{\varepsilon}\rho_{pi}^2\nu_{ii}\left[f_i\hat{K}_{2i}^{\mathrm{NC}} + (1-f_i)\alpha\hat{K}_{2I}^{\mathrm{NC}}\right]$$

Chapter 9
Turbulence in the Plasma: Self-organized Criticality and its Local Breakdown

9.1 Concepts of Nonlinear Dynamics: Dynamical System and Attractor

1. Second order linear dynamics

$$\frac{\mathrm{d}^2X}{\mathrm{d}t^2} + b\frac{\mathrm{d}X}{\mathrm{d}t} + cX = 0$$

2. Addition of nonlinear term produces limit cycle

$$\frac{\mathrm{d}^2X}{\mathrm{d}t^2} - \mu\frac{\mathrm{d}X}{\mathrm{d}t} + X + \left(\frac{\mathrm{d}X}{\mathrm{d}t}\right)^3 = 0$$

9.2 Self-organized Criticality: Turbulent Transport and Critical Temperature Gradient

1. Hydrodynamic equation to produce Bernard cell

$$\frac{\partial u}{\partial t} + \frac{1}{P_r}(u\cdot\nabla u) = -\frac{1}{P_r}\nabla p + \nabla^2 u + RkT ,$$

$$P_r\frac{\partial T}{\partial t} + (u\cdot\nabla)T = \nabla^2 , \quad T\nabla\cdot u = 0$$

$R = (g\alpha d^4/\kappa\nu)\,|\Delta T/\Delta z|$: Rayleigh number, $P_r = \kappa/\nu$: Prandtl number, g: gravity constant, α: thermal expansion coefficient, κ: thermal conductivity, ν: kinematic viscosity, d: vertical height, $\Delta T/\Delta z = (T_1 - T_2)/d$

9.3 Chaotic Attractor: Three-Wave Interaction in Drift Wave Turbulence

1. Wave number relation in three-wave interaction

$$\boldsymbol{k} = \boldsymbol{k}_1 + \boldsymbol{k}_2$$

2. Hasegawa–Mima equation

$$(1 - \rho_s^2\nabla^2)\frac{\partial\tilde{\Phi}}{\partial t} + v_{\mathrm{de}}\frac{\partial\tilde{\Phi}}{\partial y} - \left[\tilde{\Phi}, \rho_s^2\nabla^2\tilde{\Phi}\right] = 0\,,$$

$$\rho_s^2 = \frac{m_i T_e}{e^2 B^2}\,, \quad \left[\tilde{\Phi}, \tilde{\Psi}\right] = \boldsymbol{e}_z \cdot \nabla\tilde{\Phi} \times \nabla\tilde{\Psi}$$

3. Conservation of mass, energy, and enstrophy in HM equation

$$\mathcal{M} = \int [\tilde{\Phi} - \rho_s^2\nabla_\perp^2\tilde{\Phi}]\mathrm{d}x\,\mathrm{d}y\,,$$

$$\mathcal{W} = \frac{1}{2}\int \left[\tilde{\Phi}^2 + \rho_s^2\left(\nabla_\perp\tilde{\Phi}\right)^2\right]\mathrm{d}x\,\mathrm{d}y\,,$$

$$\mathcal{U} = \frac{1}{2}\int \left[\left(\nabla_\perp\tilde{\Phi}\right)^2 - \rho_s^2\left(\nabla_\perp^2\tilde{\Phi}\right)^2\right]\mathrm{d}x\,\mathrm{d}y$$

4. Nonlinear drift wave equation with dissipation

$$(1 + \mathcal{L})\frac{\partial\tilde{\Phi}}{\partial t} + v_{\mathrm{de}}\frac{\partial\tilde{\Phi}}{\partial y} + \hat{\gamma}_i\tilde{\Phi} + \left[\tilde{\Phi}, \mathcal{L}\tilde{\Phi}\right] = 0$$

$$\mathcal{L} = \mathcal{L}_h + \mathcal{L}_{ah} = -\rho_s^2\nabla^2 + \delta_0\left(c_0 + \rho_s^2\nabla^2\right)\frac{\partial}{\partial y}$$

5. Amplitude and phase equations for three-wave interaction

$$\frac{\mathrm{d}a_j(t)}{\mathrm{d}t} = \gamma_j a_j - A a_k a_l (F_j\cos\zeta + G_j\sin\zeta)$$

$$\frac{\mathrm{d}\zeta}{\mathrm{d}t} = -\Delta\omega + A\sum_{jkl}\frac{a_k a_l}{a_j}(F_j\sin\zeta - G_j\cos\zeta)$$

6. Kolmogolov spectrum for 3-dimensional uniform turbulence

$$F(k) = Ck^{-5/3}$$

9.4 Structure Formation: Turbulence Suppression by Shear Flow and Zonal Flow

1. Pump wave

$$\tilde{\Phi}_0(\boldsymbol{x},t) = \exp(in\zeta - \mathrm{i}\omega_0 t)\sum_m \tilde{\varphi}_0(m-nq)\exp(im\theta) + \text{c.c.}$$

2. Zonal flow

$$\tilde{\Phi}_{\mathrm{ZF}}(\boldsymbol{x},t) = \exp(\mathrm{i}q_r r - \mathrm{i}\Omega t)\tilde{\varphi}_{\mathrm{ZF}} + \text{c.c.}$$

3. Side band waves

$$\tilde{\Phi}_+(\boldsymbol{x},t) = \exp(-in\zeta - \mathrm{i}\omega_0 t + \mathrm{i}q_r r - \mathrm{i}\Omega t)\sum_m \tilde{\varphi}_+(m-nq)\exp(im\theta) + \text{c.c.}$$

$$\tilde{\Phi}_-(\boldsymbol{x},t) = \exp(-in\zeta + \mathrm{i}\omega_0 t + \mathrm{i}q_r r - \mathrm{i}\Omega t)\sum_m \tilde{\varphi}_-(m-nq)\exp(im\theta) + \text{c.c.}$$

4. Zonal flow evolution equation

$$\frac{\partial V_{\theta,\mathrm{ZF}}}{\partial t} = \frac{\partial}{\partial r}\langle \tilde{v}_\theta \tilde{v}_r\rangle - \gamma_{\mathrm{damp}} V_{\theta,\mathrm{ZF}}$$

$V_{\theta,\mathrm{ZF}}$: velocity of zonal flow

$\tilde{v}_\theta$, $\tilde{v}_r$: poloidal and radial fluctuating velocity

γ_{damp}: damping rate of zonal flow

Vector formulae

1. $\nabla \times \nabla\phi = 0$
2. $\nabla \cdot (\nabla \times \boldsymbol{a}) = 0$
3. $\nabla \cdot (\nabla\phi \times \nabla\psi) = 0$
4. $\boldsymbol{a} \cdot (\boldsymbol{b} \times \boldsymbol{c}) = \boldsymbol{b} \cdot (\boldsymbol{c} \times \boldsymbol{a}) = \boldsymbol{c} \cdot (\boldsymbol{a} \times \boldsymbol{b})$
5. $\boldsymbol{a} \times (\boldsymbol{b} \times \boldsymbol{c}) = \boldsymbol{b}(\boldsymbol{a} \cdot \boldsymbol{c}) - \boldsymbol{c}(\boldsymbol{a} \cdot \boldsymbol{b})$
6. $(\boldsymbol{a} \times \boldsymbol{b}) \cdot (\boldsymbol{c} \times \boldsymbol{d}) = (\boldsymbol{a} \cdot \boldsymbol{c})(\boldsymbol{b} \cdot \boldsymbol{d}) - (\boldsymbol{a} \cdot \boldsymbol{d})(\boldsymbol{b} \cdot \boldsymbol{c})$
7. $\nabla \cdot (\phi\boldsymbol{a}) = \phi\nabla \cdot \boldsymbol{a} + \boldsymbol{a} \cdot \nabla\phi$
8. $\nabla \times (\phi\boldsymbol{a}) = \nabla\phi \times \boldsymbol{a} + \phi\nabla \times \boldsymbol{a}$
9. $\nabla(\boldsymbol{a} \cdot \boldsymbol{c}) = \boldsymbol{c} \cdot \nabla\boldsymbol{a} + \boldsymbol{c} \times (\nabla \times \boldsymbol{a}) + \boldsymbol{a} \cdot \nabla\boldsymbol{c} + \boldsymbol{a} \times (\nabla \times \boldsymbol{c})$
10. $\nabla(\boldsymbol{a} \cdot \boldsymbol{c}) = \boldsymbol{c} \cdot \nabla\boldsymbol{a} + \boldsymbol{c} \times \nabla \times \boldsymbol{a}$ (If $\boldsymbol{c}$ is constant)

11. $\nabla \cdot (\boldsymbol{a} \times \boldsymbol{b}) = \boldsymbol{b} \cdot (\nabla \times \boldsymbol{a}) - \boldsymbol{a} \cdot (\nabla \times \boldsymbol{b})$

12. $\nabla \times (\boldsymbol{a} \times \boldsymbol{b}) = \boldsymbol{a}(\nabla \cdot \boldsymbol{b}) - \boldsymbol{b}(\nabla \cdot \boldsymbol{a}) + (\boldsymbol{b} \cdot \nabla)\boldsymbol{a} - (\boldsymbol{a} \cdot \nabla)\boldsymbol{b}$

13. $\nabla \times (\nabla \times \boldsymbol{a}) = \nabla(\nabla \cdot \boldsymbol{a}) - \nabla^2 \boldsymbol{a}$

Tensor Algebra

Let $\boldsymbol{a}$ and $\boldsymbol{b}$ are vectors, we define "dyad" $\boldsymbol{ab}$ by using arbitrary vector $\boldsymbol{x}$ as follows,

$$\boldsymbol{x} \cdot \boldsymbol{ab} \equiv (\boldsymbol{x} \cdot \boldsymbol{a})\boldsymbol{b}, \quad \text{or} \quad \boldsymbol{ab} \cdot \boldsymbol{x} \equiv \boldsymbol{a}(\boldsymbol{b} \cdot \boldsymbol{x}), \tag{1}$$

where $(\cdot)$ is dot product of two vectors. Consider decomposition of dyad $\boldsymbol{ab}$ by gradient vector $\boldsymbol{e}^i = \nabla u^i$, and tangent vector $\boldsymbol{e}_i = \partial \boldsymbol{x}/\partial u^i$, where it should be noted that $\boldsymbol{e}^i$ and $\boldsymbol{e}_i$ are not unit vector and sub- and superscript i correspond to the location of u^i in the definition. A vector $\boldsymbol{a}$ can be expressed as $\boldsymbol{a} = a_i \boldsymbol{e}^i$ or $\boldsymbol{a} = a^i \boldsymbol{e}_i$. Therefore, dyad $\boldsymbol{ab}$ is decomposed as

$$\begin{aligned} \boldsymbol{ab} = a_i b_j \boldsymbol{e}^i \boldsymbol{e}^j, \quad \text{or} \quad \boldsymbol{ab} = a_i b^j \boldsymbol{e}_i \boldsymbol{e}^j, \quad \text{or} \quad \boldsymbol{ab} = a^i b_j \boldsymbol{e}^i \boldsymbol{e}_j, \\ \text{or} \quad \boldsymbol{ab} = a^i b^j \boldsymbol{e}_i \boldsymbol{e}_j. \end{aligned} \tag{2}$$

General 2nd order tensor $\boldsymbol{\Theta}$ is defined as,

$$\boldsymbol{\Theta} = \theta^{ij} \boldsymbol{e}_i \boldsymbol{e}_j = \theta^i_j \boldsymbol{e}_i \boldsymbol{e}^j = \theta^j_i \boldsymbol{e}^i \boldsymbol{e}_j = \theta_{ij} \boldsymbol{e}^i \boldsymbol{e}^j. \tag{3}$$

This general 2nd order tensor can't be written as a single "dyad" since it has 9 components but can be written as a sum of three dyads. Because of this, general 2nd order tensor is called "dyadic". It is now trivial that dyad is dyadic but dyadic is not necessarily dyad. Dot product of a vector $\boldsymbol{a}$ and a tensor $\boldsymbol{\Theta}$ is defined using orthogonal relation $\boldsymbol{e}^k \cdot \boldsymbol{e}_i = \delta_{ki}$ as,

$$\boldsymbol{a} \cdot \boldsymbol{\Theta} = a_i \theta^{ij} \boldsymbol{e}_j = a^i \theta^j_i \boldsymbol{e}_j = a_i \theta^i_j \boldsymbol{e}^j = a^i \theta_{ij} \boldsymbol{e}^j \tag{4}$$

$$\boldsymbol{\Theta} \cdot \boldsymbol{a} = \theta^{ij} a_j \boldsymbol{e}_i = \theta^i_j a^j \boldsymbol{e}_i = \theta^j_i a_j \boldsymbol{e}^i = \theta_{ij} a^j \boldsymbol{e}^i. \tag{5}$$

Dot product of a tensor with two vectors $\boldsymbol{a}$, $\boldsymbol{b}$ from both side $\boldsymbol{a} \cdot \boldsymbol{\Theta} \cdot \boldsymbol{b}$ is a scalar and is given by various forms as expected from (4) and (5). An example is shown as follows,

$$\boldsymbol{a} \cdot \boldsymbol{\Theta} \cdot \boldsymbol{b} = a_i \theta^{ij} \boldsymbol{e}_j \cdot b_k \boldsymbol{e}^k = a_i \theta^{ij} b_j. \tag{6}$$

"Double dot product" of two tensors $\boldsymbol{F}$ and $\boldsymbol{G}$ as a scalar is defined as follows,

$$\boldsymbol{F} : \boldsymbol{G} = f^{ij} g_{ji} = f_{ij} g^{ji} = f^j_i g^i_j = f^i_j g^j_i. \tag{7}$$

Here it should be noted that scripts for g are reversed in comparison with those for f. The gradient operator ∇ is defined as follows,

$$\nabla = \nabla u^i (\partial/\partial u^i) = \boldsymbol{e}^i \partial/\partial u^i \, . \tag{8}$$

The divergence of tensor $\boldsymbol{\Theta}$ is defined as,

$$\nabla \cdot \boldsymbol{\Theta} = \left[\boldsymbol{e}^k \partial/\partial u^k\right] \cdot \left[\theta^{ij} \boldsymbol{e}_i \boldsymbol{e}_j\right] . \tag{9}$$

This includes $\partial \boldsymbol{e}_j/\partial u^k$ or $\boldsymbol{e}^k \cdot \partial \boldsymbol{e}_i/\partial u^k$, which has to be expressed in terms of Christoffel symbols of first kind $((\partial \boldsymbol{e}_i/\partial u^k)_j)$ and second kind $((\partial \boldsymbol{e}_i/\partial u^k)^j)$, where $\partial \boldsymbol{e}_i/\partial u^k$ is expanded as $\partial \boldsymbol{e}_i/\partial u^k = (\partial \boldsymbol{e}_i/\partial u^k)_j \boldsymbol{e}^j = (\partial \boldsymbol{e}_i/\partial u^k)^j \boldsymbol{e}_j$. Taking u^k derivative of orthogonal relation $\boldsymbol{e}^i \cdot \boldsymbol{e}_j = \delta_{ij}$, we obtain $(\partial \boldsymbol{e}^i/\partial u^k \cdot \boldsymbol{e}_j) = -(\boldsymbol{e}^i \cdot \partial \boldsymbol{e}_j/\partial u^k)$ or

$$(\partial \boldsymbol{e}^i/\partial u^k)_j = -(\partial \boldsymbol{e}_j/\partial u^k)^i \, . \tag{10}$$

Now, we can derive a vector formula $\nabla \cdot (\boldsymbol{a} \cdot \boldsymbol{F}) = \boldsymbol{a} \cdot \nabla \cdot \boldsymbol{F} + \boldsymbol{F} : \nabla \boldsymbol{a}$ where $\boldsymbol{a}$ is an arbitrary vector and $\boldsymbol{F}$ is an arbitrary symmetric tensor. Proof in Cartesian (Descartes) coordinates is simple. Since $\boldsymbol{a} \cdot \boldsymbol{F} = a_i F_{ij}$, we obtain,

$$\begin{aligned}\nabla \cdot (\boldsymbol{a} \cdot \boldsymbol{F}) &= \partial a_i F_{ij}/\partial x^j = a_i \partial F_{ij}/\partial x^j + F_{ij} \partial a_i/\partial x^j \\ &= a_i \partial F_{ji}/\partial x^j + F_{ij} \partial a_i/\partial x^j = \boldsymbol{a} \cdot \nabla \cdot \boldsymbol{F} + \boldsymbol{F} : \nabla \boldsymbol{a} \, .\end{aligned}$$

Here symmetry $F_{ij} = F_{ji}$ is used. Proof of vector formula $\nabla \cdot (\boldsymbol{a} \cdot \boldsymbol{F}) = \boldsymbol{a} \cdot \nabla \cdot \boldsymbol{F} + \boldsymbol{F} : \nabla \boldsymbol{a}$ in general curvilinear coordinates u^i is more lengthy but is useful to understand how to use tensor algebra. Let $\boldsymbol{e}^i = \nabla u^i$ and $\boldsymbol{e}_i = \partial \boldsymbol{x}/\partial u^i$, ∇ operator in general curvilinear coordinates is given as $\nabla = \boldsymbol{e}^i \partial/\partial u^i$. Dot product $\boldsymbol{a} \cdot \boldsymbol{F}$ is given as $\boldsymbol{a} \cdot \boldsymbol{F} = a_k \boldsymbol{e}^k \cdot F^{ij} \boldsymbol{e}_i \boldsymbol{e}_j = a_i F^{ij} \boldsymbol{e}_j$. Therefore, $\nabla \cdot (\boldsymbol{a} \cdot \boldsymbol{F})$ is given by,

$$\begin{aligned}\nabla \cdot (\boldsymbol{a} \cdot \boldsymbol{F}) &= \boldsymbol{e}^k \partial/\partial u^k \cdot \left[a_i F^{ij} \boldsymbol{e}_j\right] \\ &= \boldsymbol{e}^k \cdot \left[\left(\partial a_i F^{ij}/\partial u^k\right) \boldsymbol{e}_j + a_i F^{ij} \partial \boldsymbol{e}_j/\partial u^k\right] .\end{aligned}$$

Since $\partial \boldsymbol{e}_j/\partial u^k = (\partial \boldsymbol{e}_j/\partial u^k)^n \boldsymbol{e}_n$, this equation is expressed as follows,

$$\begin{aligned}\nabla \cdot (\boldsymbol{a} \cdot \boldsymbol{F}) &= \boldsymbol{e}^k \cdot \left[\left(\partial a_i F^{ij}/\partial u^k\right) \boldsymbol{e}_j + a_i F^{ij} \left(\partial \boldsymbol{e}_j/\partial u^k\right)^n \boldsymbol{e}_n\right] \\ &= \left(\partial a_i F^{ij}/\partial u^j\right) + a_i F^{ij} \left(\partial \boldsymbol{e}_j/\partial u^k\right)^k \\ &= F^{ij} \partial a_i/\partial u^j + a_i \partial F^{ij}/\partial u^j + a_i F^{ij} \left(\partial \boldsymbol{e}_j/\partial u^k\right)^k .\end{aligned} \tag{11}$$

The divergence $\nabla \cdot \boldsymbol{F}$ is given by,

$$\begin{aligned}\nabla \cdot \boldsymbol{F} &= \boldsymbol{e}^k \partial/\partial u^k \cdot \left[F^{ij} \boldsymbol{e}_i \boldsymbol{e}_j\right] \\ &= \boldsymbol{e}^k \cdot \left[\partial F^{ij}/\partial u^k \boldsymbol{e}_i \boldsymbol{e} j + F^{ij} \left\{\left(\partial \boldsymbol{e}_i/\partial u^k\right)^n \boldsymbol{e}_n \boldsymbol{e}_j + \left(\partial \boldsymbol{e}_j/\partial u^k\right)^n \boldsymbol{e}_i \boldsymbol{e}_n\right\}\right] .\end{aligned}$$

Here, we used a relation $\partial \boldsymbol{e}_i / \partial u^k = (\partial \boldsymbol{e}_i / \partial u^k)^n \boldsymbol{e}_n$. Using orthogonal relation, we have

$$\nabla \cdot \boldsymbol{F} = \partial F^{ij} / \partial u^i \boldsymbol{e}_j + F^{ij} \left\{ \left(\partial \boldsymbol{e}_i / \partial u^k \right)^k \boldsymbol{e}_j + \left(\partial \boldsymbol{e}_j / \partial u^i \right)^n \boldsymbol{e}_n \right\} .$$

Therefore, $\boldsymbol{a} \cdot \nabla \cdot \boldsymbol{F}$ is given by

$$\begin{aligned} \boldsymbol{a} \cdot \nabla \cdot \boldsymbol{F} &= a_m \boldsymbol{e}^m \cdot \left[\partial F^{ij} / \partial u^i \boldsymbol{e}_j + F^{ij} \left\{ \left(\partial \boldsymbol{e}_i / \partial u^k \right)^k \boldsymbol{e}_j + \left(\partial \boldsymbol{e}_j / \partial u^i \right)^n \boldsymbol{e}_n \right\} \right] \\ &= a_j \partial F^{ij} / \partial u^i + F^{ij} \left\{ a_j \left(\partial \boldsymbol{e}_i / \partial u^k \right)^k + a_n \left(\partial \boldsymbol{e}_j / \partial u^i \right)^n \right\} . \end{aligned} \tag{12}$$

The double dot product $\boldsymbol{F} : \nabla \boldsymbol{a}$ is given by,

$$\boldsymbol{F} : \nabla \boldsymbol{a} = F^{ij} \left\{ \left(\partial a_i / \partial u^j \right) - a_k \left(\partial \boldsymbol{e}_i / \partial u^j \right)^k \right\} . \tag{13}$$

Where we used (note (10) for last line)

$$\begin{aligned} \nabla \boldsymbol{a} &= (\boldsymbol{e}^j \partial / \partial u^j) \left(a_i \boldsymbol{e}^i \right) = \left(\partial a_i / \partial u^j \right) \boldsymbol{e}^j \boldsymbol{e}^i + a_i \left(\partial \boldsymbol{e}^i / \partial u^j \right)_k \boldsymbol{e}^j \boldsymbol{e}^k \\ &= \left(\partial a_i / \partial u^j \right) \boldsymbol{e}^j \boldsymbol{e}^i + a_k \left(\partial \boldsymbol{e}^k / \partial u^j \right)_i \boldsymbol{e}^j \boldsymbol{e}^i \\ &= \left(\partial a_i / \partial u^j \right) \boldsymbol{e}^j \boldsymbol{e}^i - a_k \left(\partial \boldsymbol{e}_i / \partial u^j \right)^k \boldsymbol{e}^j \boldsymbol{e}^i . \end{aligned}$$

The 1st term of (11) is same with first term of (13). The 2nd term of (11) is same with 1st term of (12) due to symmetry of $\boldsymbol{F}$. The 3rd term of (11) is same with 2nd term of (21) due to symmetry of $\boldsymbol{F}$. The 3rd term of (12) cancels the 2nd term of (13). Therefore, proof of $\nabla \cdot (\boldsymbol{a} \cdot \boldsymbol{F}) = \boldsymbol{a} \cdot \nabla \cdot \boldsymbol{F} + \boldsymbol{F} : \nabla \boldsymbol{a}$ is concluded.

Name Index

Subject Index

J

K

L

M

N

O

P

Q

R

MIX
Papier aus verantwortungsvollen Quellen
Paper from responsible sources
FSC® C105338

If you have any concerns about our products,
you can contact us on
ProductSafety@springernature.com

In case Publisher is established outside the EU,
the EU authorized representative is:
Springer Nature Customer Service Center GmbH
Europaplatz 3, 69115 Heidelberg, Germany

Printed by Libri Plureos GmbH
in Hamburg, Germany